ACKNOWLEDGMENTS

We gratefully acknowledge the support for this conference
by the Energy Research and Development Administration, the
Environmental Protection Agency and the University of Rochester.
Special appreciation is extended to Judy Havalack for taking
care of the conference details and to Jane Leadbeter for the
conference correspondence and assistance in the preparation of
much of the material for the book.

The Conference Committee
Taft Y. Toribara, Chairman
James R. Coleman
Barton E. Dahneke
Isaac Feldman
Morton Miller

PREFACE

The principal emphasis of the Department of Radiation Biology and Biophysics is on biological problems. Techniques for measuring are considered very necessary but the development of them is usually left to someone else. Therefore it is a little unusual for the department to sponsor a conference which is devoted mostly to methodology.

Environmental Pollution is a very popular topic now, and one notices that there are a number of scientific conferences devoted to the topic. Furthermore, part of every conference is devoted to measurements of pollutants. So the question becomes one of what should be different about our conference. To start with there are two unique features here: The first is the limited attendance which should provide more meaningful discussion; the second is the availability to the world of all the information in book form after the conference. We gave considerable thought to the contents of the conference which would take advantage of the unique features.

Therefore, we decided to look to the future and present material here that is not in routine use. The search for pollutants has just begun, and their presence cannot be established without some means of detection. Many substances are not known to be toxic because no one has studied them. The necessary information can only be obtained if techniques for detection and measurement are available.

In order to provide the proper setting for the conference, you will note that our opening session concerns a presentation of the problem - who decides what is a pollutant? Who decides what to do about it? There are many other questions, many with political implications, which will be discussed. A quantitative basis on which to make many judgments is what measurements provide.

Actually no biological study can be carried out without a measurement of some kind. It is encouraging that due recognition is being given to methodology by a department of a medical center.

I am certain that there will be many techniques presented here which would not otherwise have been known to many. The availability of this information in book form after the conference should provide a useful starting place for those in search of tools to use in their research.

A final word concerning the organization of the sessions. Since the only unifying theme is "techniques" it was somewhat difficult to group the papers, but we managed to make the session titles general enough to encompass all subjects. It almost appears as if the techniques are being presented in the order of increasing cost of the equipment. As is obvious the real measure of the value of a technique is its usefulness and not the cost. We have here instruments which are very simple and inexpensive to those of such size and cost that they are available in only a limited number of areas in this country. We are certain that all techniques will be utilized.

Speaking in advance for the organizing committee, I will say with no show of modesty that we are ourselves very impressed with the program we have organized and with the group of speakers we have selected. Now we want all of you participants to do your parts to make us look good.

Taft Y. Toribara
Conference Chairman

CONTENTS

SESSION I: SPECIFICATION OF
ANALYTICAL PROBLEMS
James P. Lodge, Chairman

SESSION II: MORE FAMILIAR PRINCIPLES
Isaac Friedman, Chairman

WELCOMING REMARKS

     It is my pleasure to welcome you to the Medical Center of the
University of Rochester for the Tenth Rochester International
Conference on Environmental Toxicity sponsored by the Department
of Radiation Biology and Biophysics. Everyone is deeply concerned
with the condition of the environment today and for the future,
but that was not the case when the conferences were started ten
years ago. We should credit the foresight of the then co-chairmen
of the department for starting this series of conferences.

     The Department of Radiation Biology and Biophysics incorpora-
ting the Energy Research and Development Agency Project has a long
record of productive research into the mechanisms of biological
damage by radioactive and toxic materials. The department began
as part of the Manhattan Project, devoted to the study of health
related problems associated with nuclear weapons. Subsequently,
under Atomic Energy Commission sponsorship, the department was
active in the study of toxic materials associated with nuclear
power, and since then, with the support of the Energy Research and
Development Agency, the department has studied effects of toxic
materials associated with non-nuclear power generation. It is
important to note that during this long history, the guiding
principle of the department was that the key to comprehending and
alleviating the effects of toxic pollutants was to fully understand
the fundamental biological processes involved. Consequently, the
department has become an internationally recognized resource for
the study of toxic pollutants.

     The topic selected for this year, "Modern Techniques for the
Detection and Measurement of Environmental Pollutants" is of basic
importance. This addresses a basic premise not usually given the
prominence it should command which is that every studv must start
with methodology and wherever possible with a quantitative measurement. Since the interest in an environmental pollutant is its
toxicity, the measurement of its level is of prime importance. If
a method for the measurement does not exist, then one must be
developed. The interest of the department in pursuing the develop-

ment of good methodology provides a most appropriate setting for
this year's conference.

It is my understanding that you have been asked to look to
the possible future uses of the various techniques which will be
presented.   This approach is certainly a sensible one since it
seems probable that many toxic pollutants may be as yet unspecified
because no one has looked for them, or the tools for their identification are not familiar to the interested investigators.   It is
also a certainty that new pollutants will be the subjects of future
study, and the techniques discussed here may be applicable to the
pursuit of such studies.

The number of participants in this conference has been kept
small in order that all can participate fully in the discussions.
Past conferences have shown that a fruitful exchange of ideas has
occurred with this format and I recommend it for your use.

I hope that all of you will have an enjoyable, stimulating
and enlightening time at this conference.

J. Lowell Orbison
Dean, School of Medicine
and Dentistry
The University of Rochester
Rochester, New York

WELCOME FROM THE DEPARTMENT OF RADIATION BIOLOGY AND BIOPHYSICS

I'd like to welcome you on behalf of the department, faculty, and staff people. It's a real pleasure to have you in Rochester and carry on the tradition of ten years.

Two obligations I have. One is to thank Taft Toribara and his committee for the hard work that's gone into the conference, and to assure you that if you have any needs during the conference, that you may turn to Taft and his people or our office.

Secondly, it's a pleasure to acknowledge the support of E.R.D.A. and the Environmental Protection Agency in past conferences as well as the present one.

As you're all aware, our department is a very interdiscipline kind of department in approach. There are many different kinds of activities going on, and they've been brought to focus on a number of problems, including the one that you're hearing about in this present conference. If you come across students who ask you questions, that will, I think, represent the highest ideal in our department: namely people whose interest may not be in a given field, but who have curiosity to reach over to new ones.

We hope the conference will be very stimulating to you.

Thank you.

Paul L. LaCelle, M.D.

SESSION I:

SPECIFICATION

OF ANALYTICAL PROBLEMS

AIR POLLUTANT MONITORING: OR HOW TO IMPROVE ON AN OUIJA BOARD

James P. Lodge, Jr.

Consultant in Atmospheric Chemistry

385 Broadway, Boulder, Colorado 80303

ABSTRACT

Recent developments in air pollutant monitoring equipment
have provided, really for the first time, instruments of reasonable stability, precision, and for the most part specificity.
However, this comfortable situation did not arise spontaneously,
or without many false starts and evolutionary changes. Based on
previous historical studies and personal reminiscence, it is
instructive to trace the evolution of knowledge of air pollutants
from the situation of the first several millennia of air pollution,
during which the nature of air pollutants were obvious, through
a period of utter confusion caused by the proliferation of new
pollutants, to the present situation where, once again, the
nature of air pollutants is becoming better understood.

In parallel with this change, there have been interesting
changes in pollution regulations, in control technology, and in
research and monitoring. The present confused situation, in
which "environmentalists" are simultaneously heroes and villains,
in which nearly everyone is angry with everyone else, and research
support is being cut when a clearer view of the entire problem is
almost in sight, derives in part from demography, in part from
perceived opposition of energy and environmental needs, but
largely from the large scale collapse of philosophical/theological
systems. A detailed examination of these is beyond the scope of
the present paper.

If I am to speak to the direction of this conference, it must
be in the context of the parallel developments of air pollution,
control legislation, air pollution control technology and air
monitoring; for these have grown together, by no means always in
step, but certainly always changing. At an earlier and simpler
time, the coupling among these areas was fairly loose. Today
the cost of each advance (if it is an advance) is large enough
that even the optimum path to a cleaner environment is expensive
indeed, and the incremental cost of deviation from the optimum
can rapidly become astronomical.

In this context, my title may be taken in two senses: first,
that any real data on airborne pollutants, if taken seriously,
can help in charting that optimum or near-optimum path; and
second, that we are only now reaching the point where our air
quality data may conceivably represent a significant improvement
over a combination of intuition and guesswork.

I have written elsewhere (2) about Ug, the caveman who first
brought fire into his cave, thereby both founding civilization
and inventing air pollution. For pollution must be considered
as the unfavorable environmental accumulation of the metabolic
products of our culture (3). As I have pointed out, we are now
in the United States producing wastes which, in a subsistence
culture, would correspond to a population of something around
one hundred billion.

Up until early in this century, air pollution was a simple
matter. It had changed little from the mid-seventeenth century,
when John Evelyn wrote his classic tract, "Fumifugium" (1). At
that time, entirely on the basis of his unaided senses, he
correctly diagnosed the problems of London as soot and "sulfurous
vapors" resulting from coal combustion. (It might also be noted
that acts of Parliament were already (and are still, so far as I
know) in force forbidding the burning of "sea-coal" in London,
dating from as early as the thirteenth century.) It was this
problem that the first modern pollution legislation addressed,
starting in the middle of the nineteenth century. But people
stubbornly insisted on keeping warm, even if the results were a
dirty and smelly environment. I myself remember when the first
harbinger of winter was enough environmental sulfur dioxide to
taste in the crisp morning air of central Illinois. In a small
agricultural town, that sulfur dioxide (and the accompanying soot)
came entirely from home heating. In many manufacturing towns,
similar concentrations were commonplace the year around, and were
viewed as a sign of prosperity.

Nevertheless, the beginnings of air pollution legislation
were already the best part of a century in the past. New York
had electrified its railroads. St. Louis and Chicago had made

black smoke illegal, even though this illegality carried no
penalty, and had begun some halting steps toward tall chimneys.
Nearly a decade earlier the U.S. Bureau of Mines had first issued
that very first source monitoring device known as the Ringelmann
chart; I would guess that, without it, few local air pollution
programs could function even today. Ironically, it was not issued
to prevent air pollution but to aid in teaching firemen to
achieve a decent yield of heat from the coal they stoked. (Conceivably there is a lesson here for those who strive to place
environmental concerns in an antagonist position to the energy
problems of the day.) It was at about the same time that, in
the course of investigating vegetation destruction around the
enormous smelter in Trail, B.C., Moyer D. Thomas developed the
first true air quality monitor, the Autometer. The legendary
"Chicago Study" was 20 years in the past, but no one had quite
gotten around to doing anything about it. The cleanup of
Pittsburgh was underway. The cleanup of St. Louis was in the
earliest formative stages. (Lest it be thought I am singling
out Chicago as uniquely retarded, it should be noted that, in
"Fumifugium," Londoners had a report on their pollution virtually
as complete in its way as the "Chicago Report" in its; the
cleanup of Chicago required just about 50 years, and that of
London almost precisely 300 years.

     The 1940's showed a fascinating mixed trend in those factors
affecting air quality. The vast pace of industrial construction
in wartime led to major new pollution sources being located with
total disregard for environmental impact. The "Rubbertown"
complex was built rather precisely upwind of the main part of
Louisville, Kentucky. Many people have dated the beginnings of
the Los Angeles smog from the building of a butadiene plant in
the middle of Los Angeles. A singularly inefficient sulfuric
acid plant was built immediately adjacent to downtown Denver.
All of these were constructed in great haste and with little
regard to leakage, actual or potential. These scenes undoubtedly
occurred in many other places all across the U.S.; given the
atmosphere of the day, it is entirely possible that we should
have lost the War if these things had not been done. Furthermore,
it was unpatriotic to complain.

     On the other hand, the immediate postwar period saw the
first real proliferation of Continent-spanning natural gas lines,
the consequent decline of coal burning for home heating, and a
precipitous drop in emissions of soot and sulfur compounds. This
was just in time, because this was also the real beginning of the
period of urban sprawl on a vast scale. The bulk of the housing
was insubstantial and poorly insulated against either heat or
cold. The problem was compounded by a series of fads in housing
style, with the construction of sprawling ranch homes in New
England and Cape Cod cottages in the Southwest. Without cheap gas
and power, these would have been uninhabitable.

As a final ingredient, the spending power generated by the wartime economy but bottled up by the lack of consumer goods met the industrial capacity that resulted from conversion of the war industries to consumer goods, and the result was a spree of consumption with few parallels in history. I suspect that, from the standpoint of a future historian of a couple of centuries hence, that spree is still going on. The truly unique ingredient was, of course, the automobile. Following the wartime suspension of auto construction, new cars proliferated so rapidly that the end of wartime demand for butadiene gave Los Angeles no respite whatever; on the contrary, the situation there continually worsened. A strictly engineering approach of assuming that the problem in Los Angeles was identical to that in London, New York and Chicago led to a program to control sulfur emissions, totally without effect. At the end of the decade, A. J. Haagen-Smit, with a few qualitative experiments, first elucidated the chemical nature of the Los Angeles "smog," thereby both drawing the ire of the auto and petroleum industries and laying the foundations of modern atmospheric chemistry.

Perhaps the next benchmark in air pollution history may be taken as 1955. The fatal air pollution episodes of 1948 in Donora, Pennsylvania and of 1952 in London had supervened, massive research was still underway to attempt to refute Haagen-Smit's theories, and the U.S. Congress passed the first Federal Air Pollution Act. There was substantial public awareness that too much air pollution could be injurious to your health. Haagen-Smit had developed an automatic impinger that exposed 24 portions of phenolphthalin solution to the atmosphere daily, yielding hourly average values of something called oxidant. The first continuous instruments for that same chemical entity using potassium iodide had been developed. The nitrogen dioxide method of Saltzman had just been published. The Thomas Autometer was in continuing use, and had been joined by a coulometric instrument of marginal sensitivity called the Titrilog. The earliest non-dispersive infrared analyzers for carbon monoxide were available, and were (with considerable inaccuracy) measuring ambient carbon monoxide levels by pressurization to levels that were probably mildly hazardous for those who had to work near them. P. W. West gave a paper on a technique for sulfur dioxide determination, and literally thousands of preprints were in circulation before the published version appeared the following year.

The Federal Act of that year merely authorized research and technical assistance by the U. S. Public Health Service; in essence, it permitted Federal researchers to inform states and local communities that they had a problem (if that issue was in doubt), but it carefully enjoined Federal Government against exerting any legal power to abate the problems that were dis-covered. Nevertheless, there were ringing declarations that, with the superior knowledge that would be developed in the Federal laboratories, the problem was almost solved.

The first Federal initiative was to publicize an effort already underway. The Hi-Volume air sampler was a curious mating of the internals of commercial vacuum cleaners, an outer shell developed for the collection of radioactive particles, and a large adapter to hold a flat glass-fiber filter, the dimensions being set by the available size of the filter. It had been deployed initally by the Public Health Service under a contract with the Chemical Corps to learn whether particulate protein in the atmosphere presented sufficiently stable patterns to serve as an indicator of biological warfare. It is almost superfluous to say that it did not, but the Public Health Service, having got the rather large samples of airborne particulate matter, analyzed them for inorganic ingredients as well, and simply compiled the data. Such was the base for what is now called the National Air Surveillance Network. Unfortunately, this was so useful as a public relations device for the first few years of the program, when there was essentially no other output, that the Network and its methodology acquired an aura of tradition that probably, in the long run, slowed down the ultimate development of more desirable techniques of monitoring for particulate matter.

In the fall of 1955, sensing the need to do something, the Public Health Service convened a meeting of air pollution officials. The intent was undoubtedly good; I presume that it was to promote the exchange of information about the latest techniques of pollution control. However, it rapidly developed into a sort of beauty contest, with the traditional "smoke chasers" looking down their noses at the upstarts who were concerned about gaseous pollutants (and vice versa), and various other sorts of competitions springing up on the side. One session of the final day was devoted to air monitoring. A few months earlier, the Directors of the Los Angeles County Air Pollution Control District had declared that there should be set up within the County precisely 14 air pollution monitoring stations, each with an elaborate set of continuous monitors (although the legal declaration of excessive ozone had to be based on the rate of cracking of stressed strips of rubber) and had appropriated the money to equip them. The going cost of almost any continuous monitoring device was about $4,000 or $5,000, and the first set of equipment had just arrived. As a result, the representative of Los Angeles was obviously one up on everyone else. The representative from Detroit had also gotten several new monitors recently, and, although he did not have the prospect for as many, was able to maintain his position by disagreeing as to the types of monitors that were most efficient. Finally, out in the audience, the representative (and, I believe, sole member) of the air pollution staff of Chicago leaped to his feet and shouted, "You guys can say all you want to about $5,000 instruments, but what I gotta have in Chicago is something that costs $25 and can be operated by the Mayor's nephew!"

He was not far wrong, and he spoke for representatives of many city and state agencies nationwide. In fact, during my tenure with the U.S. Public Health Service, from 1955 to 1961, I made repeated efforts to acquire enough monitoring instruments to do some evaluation, and to be able to mount local studies on an experimental basis. I was invariably turned down on the grounds that such an effort would be altogether too expensive. As a result, it was well into the 1960's before there were any significant number of locally-owned air monitoring devices outside California, and, in fact, before the Federal Government itself entered the market in a significant fashion. Some time around 1957 I was assigned the problem of preparing a paper on future air sampling needs for oral presentation at a meeting. That particular paper seems to have gotten lost from my archives, which may well be merciful. However, I remember that I ultimately arrived at what seemed at the time ridiculously large numbers of instruments and ridiculously high costs. As the final step in the calculation, I computed a number of Hi-Vol samplers necessary to filter all the air in Los Angeles daily, pointing out that this might be the final solution of the problem there. I cannot now recall whether I pointed out the compensating problem from brush dust and electrically generated ozone. I do remember that I was taken to task for such unseemly levity. Needless to say, this proliferation did not in fact occur. On the contrary, I could truly write in 1971, "We are just emerging from a period in which there was too little competition to place the burden on the instrument manufacturers to improve their products" (4). A further indication is the fact that the first edition of Stern's monograph on air pollution, published in 1962, covers the field of automatic monitors rather completely in 20 pages. A single page is devoted to "sophisticated air monitoring instruments" with most of the space given to speculations about the applicability of gas chromatography. It is probably also indicative that the chapter was not judged dangerously out of date, despite the fact that the most recent reference was 1959, owing to production delays in the book. In fact, the whole subject of air pollution measurements occupied slightly in excess of 200 pages, including meteorological instruments, source sampling, and numerous other subjects not immediately directed to the automated measurement of pollutants.

It is also well worth noting that, with few exceptions, the instruments described were almost literal robot chemists. That is to say, with some necessary simplifications, they went through the same steps that a live chemist would use in an automated sequence. A few resembled nothing so much as enormous cuckoo clocks. Others functioned in a continuous flow system.

Nevertheless, it is probably fair to say that the successful instruments of that period were simply mechanized versions of laboratory operations. The only obvious exceptions were the non-dispersive infrared analyzers and the various photoelectric particle counters. They had the advantage that their performance did not deteriorate because of a night on the town or an argument with a spouse, but the disadvantage that they were incapable of feeling embarrassed by a run of bad data. In all honesty, many of the early instruments appeared to have been assembled by a committee of the less competent members of the Plumber's Union and the electronics trade, occasionally with some very bad optics added. The complex plumbing made for very slow response and long inherent integration times. The result was normally a device requiring as much manpower as obtained the same information by manual sampling. The only difference was that instrument technicians were somewhat more expensive than chemists.

The early 1960's saw the initiation of a number of state air pollution programs where none had existed before. Most were based on the Federal model, which is to say that the state agency was given the uncontested power to make measurements and inform local communities if they had air pollution problems. By the mid-1960's this pattern had changed. The Federal Agency began to gain enforcement power, which forced the states to establish enforcement agencies to escape undue Federal intervention in local affairs. The controversy over the role of automobiles in photochemical smog had been laid to rest, and the manufacture of air monitoring devices was suddenly perceived as profitable. My paper in 1971, quoted earlier, continues, "[We] are now seeing a population explosion of new instrument makers. Now the problem is one of deciding in the midst of an embarrassment of riches. There is clearly a need for those of us in the atmospheric monitoring profession to have a consumer's union of our own. I applaud the efforts by the Federal air pollution staff, and only wish it could be expanded. An unfortunate percentage of the instruments available today display poor design, poor construction, and/or poor performance."

To do justice to the monitoring instrument manufacturers, the current generation of monitoring instruments are remarkably good. A few months ago I participated in setting up and running a small selection of instruments to study indoor-outdoor pollutant variations in an asthma hospital. The nitrogen oxides monitor, a comparatively new device, was plugged in and ran with absolute stability for a month--on the nitric oxide channel, at least. The infrared carbon monoxide analyzer dated from about 1970, produced no usable data for the first week of operation, and drifted in a random fashion over the entire month. I am told that the new models are far more dependable.

Having reached the early 1970's, it is necessary to double back and examine happenings in the parallel fields of legislation, control technology, and atmospheric chemistry. In many respects the driving force was the continuation of urban sprawl. In many areas this had become so acute that gains in emission levels per source were overwhelmed by the proliferation of sources. For example, despite claims from local agencies to the contrary, it is my impression that the mean pollution levels have changed little; the standard deviation has decreased, so that there are no longer extremely bad or extremely good days, and the area covered by the pollution has increased enormously. It is probably appropriate to consider that air pollution has been successfully democratized. At the same time, the much greater mobility of the population has given city dwellers a chance to experience non-urban air, which, while it is no doubt less clean than it was in an earlier day, is still significantly cleaner than in that of the urban centers. This has led to a crisis of expectations that probably will not be met in this century.

One result was the Clean Air Act Amendments of 1971. These mandated a quite arbitrary 90 percent reduction in emissions from new cars on a specified timetable of about five years, the setting of the National Ambient Air Quality Standards, and their achievement during about the same five year period. Unfortunately, this approach ignored the fact that, if the automobile was indeed the primary source of pollutants, the impact of truly clean cars manufactured beginning (say) 1976 would not become truly apparent until about 1982. This essentially unworkable situation was compounded by two unfortunate decisions by the auto industry.

I do not have a pipeline into the highest offices of the Detroit auto makers. I learned long ago that it is impossible by outside observation to distinguish between conspiracy and stupidity. The facts are, however, that the industry gave little sign to taking the Federal law seriously for two or three years. By that time, their declaration of being unable to meet the legislative deadlines was indisputably true.

In addition, when serious research was started, almost all emphasis was placed on the catalytic afterburner. This device has the great advantage of allowing the engine operation to be optimized for performance and/or economy; it also allows the manufacturers to continue to use an engine with which they are thoroughly familiar. Philosophically, it has a close resemblance to promoting street sweepers as a basic design improvement on the horse. Practically, the result has been to vacate the field of innovative engine design to the Europeans and Japanese.

There have been other undesirable consequences. The present automobile is almost too complex to be successfully maintained,

and with this complexity has come decreased reliability. A fouled
spark plug can dump enough uncombusted gasoline into the converter
to change it briefly into a furnace that is a threat to itself,
to the attached car, and to the surroundings. There is a still
unresolved question as to whether the auto may now have become a
threatening source of sulfuric acid aerosol in the urban canyons.
(On the other hand, I recently learned of work suggesting that
man is such an excellent source of ammonia that acid aerosol is
little threat to him. It may be very hard on guinea pigs, but
not on humans.) Now we are told that, while the catalytic converter has allowed optimum tuning, and hence has improved fuel mileage,
the next step of nitrogen oxides control will undo that benefit.
Meanwhile, heavier cars, more safety equipment, the addiction to
automobile air conditioning, to automatic transmissions and to
power accessories has brought us to a level of fuel consumption
unparalleled in history, just at a time when the true finiteness
of the world petroleum supplies has really become apparent.

We now have demands for cars with better gas mileage; meanwhile the opinion makers assure everyone that the American public
will not tolerate lightweight cars of slow acceleration characteristics, especially those requiring the manual shifting of gears.On
another front, consumer groups demanding clean power plants are
clashing with consumer groups decrying the threatened increases
in electrical costs. Some techniques of removing sulfur from
power plant effluents have been shown effective and reasonably
dependable. On the other hand, they both lead to sizable increases
in cost and generate large volumes of useless wastes, primarily
calcium sulfate. In Denver, and I suspect in many other northern
cities, the highest airborne particulate levels occur in the few
days immediately following a snow storm. The citizens demand
increased street sanding to make roads navigable as soon as
possible after a snow, and reject increased taxes for more prompt
street sweeping once the things dry out. I cannot regard this
combination of circumstances as being totally independent of one
another.

If these opposing forces are not enough, consider a few
others. Whatever the final solutions, the path from where we are
to a stance of greater sanity is a difficult one, and will require
much study and, in particular, a great deal of environmental
sophistication. Yet hardly a week passes that I do not receive a
phone call from unemployed environmental scientists inquiring if
I know of any jobs. The new Federal budget calls for a decrease
in air pollution research within EPA. There is a shortage of
fuel, and yet coal production in Colorado is at a low ebb, and I
understand that small industries there are unable to buy coal.
We need cleaner cars on the road, and yet I personally save
hundreds of dollars annually on license fees by driving old cars.
(In fact, it may properly be said that the drivers of new cars

are subsidizing my use of the road.)   In short, our tax laws favor
the retention of old and dirty equipment over its replacement.

Overall, we may sum up the present situation as almost totally
schizophrenic. As an amateur theologian, I am forced to note that
some of the difficulty stems from a real lack of public goals
beyond material possessions. The overall result is legislative
irresolution and a great deal of petty bickering and mutual name-
calling between those emphasizing eating our cake and those
espousing having it too, with each being accused of failing to
deliver his part of the impossible accommodation.

There have been disgraceful scenes. For example, I personally
witnessed several repetitions, in Colorado, of a scenario in which
the Federal EPA set a deadline for the enactment of a particular
class of emission regulations, and simultaneously launched a
study, presumably by experts, of the optimum form of those regulations. We duly scheduled hearings at the latest possible date on
a proposed regulation of our own, and invited the EPA to give us
the benefit of their wisdom. At the hearings a lone EPA representative showed up to explain lamely that their study had slipped
six months, and they had no idea what we should do, but we would
certainly be sued if we failed to do it. At the same time, my
friends in EPA (and I still have some) tell me that much of the
problem is the continued addition of responsibilities and funds
to their burden without authorization of additional personnel to
insure that the duties are discharged or the funds wisely spent.
There are apparently a number of cases where a good laboratory
scientist is completely prevented from doing research by being responsible for 30 or 40 contracts with industry, many of them for
jobs better done inhouse where intermediate results can be
identified and acted on.

The public is confused and behaving accordingly. They believe
that the environmentalists saved the world from the supersonic
transport in 1971, when in fact the environmentalists had little
to do with that decision. The SST, it now appears, has only
minimal environmental impact; it is simply uneconomical and
wasteful of energy. The environmentalist heroes who claim credit
for the defeat of the SST and the impending doom of underarm
sprays are the environmentalist villains who are preventing us
from having nuclear power--also a completely unsupportable
statement.

From this entire discussion, I hope it is obvious that the
primary needs of the times are a less consumptive life style and,
perhaps, even a willingness to tolerate minor physical discomforts.
It may even be necessary to schedule ourselves a little less
tightly so that it is possible to make our appointed rounds
without vehicles capable of accelerating from a standstill to 60

miles per hour in eight seconds. It may be necessary to learn to
use a manual transmission of four or five speeds forward. This
raises the arresting thought that, at some time in the future, it
may be necessary to have a minimum intelligence level in order to
drive an automobile. We may need to learn to wear heavy clothes
in the winter and light clothes in the summer. As an ultimate
deprivation, housewives may be forced to learn to start with flour
and other basic ingredients to produce baked goods, instead of
buying them premixed. The mere thought of Americans relearning
that different varieties of cake taste different, and none of them
taste particularly like cotton candy if properly prepared, is
amost frightening in its revolutionary implications.

The contemporary scene is sufficiently absurd that it is easy
to burlesque it as I have done. It is even easier to call for a
revolution--or perhaps more accurately a devolution--into less
consumptive life styles that feature more creative labor and fewer
spectator sports. I have personally moved in that direction, and
am enjoying the results. On the other hand, I would certainly not
join those prophets calling for the establishment of this sort of
change by fiat and the establishment of a dictatorial form of
socialism or communism. We are here discussing the demise of
entire industries, the decimation of others, and the creation or
resurrection of still others. Under the best of circumstances
serious errors will be made. However, marching the whole country
to its new orientation in lock step is the thing that Federal
Government can do best, and it is precisely the way of maximum
hazard. I have a mental picture of some unknowledgeable
bureaucrat decreeing, "We must get back to nature! Everyone must
build himself a log cabin." Within the next 60 days, the U.S.
would be totally deforested, and shortly thereafter we should have
seen the last of a few thousand species of birds, together with
squirrels, deer, porcupine, and a large variety of other biota.

No, the way to the future is the hard way of shifting the
priorities of people, not governments, of voting with pocketbooks
as well as at the polls, of refusing to put into practice everything that we know how to do, and of constant vigilance over our
environment and our people. We need monitoring techniques that
have not yet been conceived. We need holistic approaches to
legislation that sets realistic balances between reductions in
energy use, achievements of energy efficiency, production,
employment, quality of product, quality of life, and protection
of all elements of the environment. (I have nightmares about
"solutions" to air pollution problems at a cost of irretrievable
soil pollution, and the like.) Most of all, we need somehow to
achieve a mutual trust among the various segments of our society,
together with behavior that warrants such trust.

Given the necessary support, and adequate encouragement, the scientific community can develop, deploy, and use the necessary monitoring equipment to chart our way. For guidance as to how to get the people to follow this lead, I suspect I must return to my hobby of theology.

REFERENCES

1   EVELYN, J.: Fumifugium in The Smoake of London, Two
     Prophecies. Maxwell Reprint, New York (1969).

2   LODGE, J.P., Jr.: Preface to The Smoake of London, Two
     Prophecies. Maxwell Reprint, New York (1969).

3   LODGE, J.P., Jr.: An atmospheric scientist views environ-
     mental pollution. Bull. Am. Meteor. Soc. 50 7 (1969) 530.

4   LODGE, J.P., Jr.: When NOT to use on-line air pollution
     instruments. Instrumentation Technol. (1971) 28.

THE STRATEGY FOR CLEANING UP OUR WATERS

Robert L. Collin

Rochester, New York

ABSTRACT

The United States is embarked on a national program to clean up its waters and eliminate pollutant discharge from all man-made sources. The purpose of this program is to protect not only human health and welfare from the direct impact of polluted water but also to protect the delicate web of life on which man's survival depends. The strategy for accomplishing this purpose is based on an attempt to blend a number of complex components including technological capability, economics, scientific measurement, and judgments about safety; all brought together by the political decision process. The strategy is being implemented through numerous daily decisions of government officials who are under pressure from industry, the general public, and special interest groups.

The program is working - our waters are getting cleaner. However, as we put more effort into the program some of its weaknesses and oversights are becoming apparent. As we accomplish the easier goals potential road blocks to continued success will begin to appear.

INTRODUCTION

After more than two centuries of viewing streams, rivers, and lakes as convenient dumping grounds for society's wastes the United States has embarked on the great national task of cleaning up its waters and eliminating from them all pollution by man-made

activity. Scientists working on the detection and measurement of
pollutants in water should have some appreciation for this national
program, its ramifications, its costs, and its potential problem
areas. Above all, the scientist should appreciate that a national
water pollution abatement program cannot be solely a program based
on scientific measurement and scientific criteria of hazard. It
must partake of economics, value judgments, and the compromises
that have to be made among millions of personal and institutional
goals that lie at the heart of the political process.

I would like to discuss the basis of our present water
pollution control program and in particular some of its administrative, economic, and political aspects - and some of its problems.
I will try to concentrate on that nebulous but critical area where
the technical and the value judgment processes come together.

THE NATIONAL PROGRAM

In 1972 Congress passed a series of amendments to the then
existing Federal Water Pollution Control Act which has provided
the vehicle for the national efforts in water pollution control.
These amendments (Public Law 92-500) had as their objective "to
restore and maintain the chemical, physical, and biological
integrity of the Nation's waters" (4). They superceded a system
which had been put into effect in 1965 and had shown itself to be
unworkable.

The old (1965) law was based on the idea that all waters had
a certain assimilative capacity and that water quality standards
could be set for any particular reach of water based on the desired
use of these waters. The addition of pollutants was allowed (since
water could supposedly assimilate some waste) as long as the water
quality standards were maintained. However, once the standards for
that particular receiving body were exceeded then the polluter that
caused the problem could be forced to reduce his pollution until
the standards were again attained. The 1965 law had a number of
bad features but its main failing was the extreme difficulty of
relating, and proving in court, that effluent from a particular
polluter caused the violation. In addition, Congress was beginning
to realize that pollutants are seldom "assimilated" in the sense
that they disappear or change into innocuous components. In many
cases toxic substances, heavy metals, and inorganic ions persist in
the environment and are merely shifted from one place to another in
the aqueous ecosystem which includes the water, the sediments, and
the plant and animal life existing in both.

Congress made an abrupt departure from past regulatory efforts
with the Amendments of 1972. The idea of an assimilative capacity

was thrown overboard and replaced by a primary goal which stated "...that the discharge of pollutants into navigable waters be eliminated by 1985" (5). I have underlined the word "eliminated". This is a clear statement that says no addition of pollutants to water is really acceptable. It sets us on the course of doing away with all pollution of the nation's waters, and by 1985. No one can rightly claim that the statement is vague or wishy-washy.

This goal of eliminating man-made pollution completely has been criticized as unrealistic, unattainable, and economically and socially unsound. It may very well be all of these but it has the virtue of defining a national purpose in unequivocable terms, in terms that can be used as a tool by government administrators to move forward in a firm manner on all aspects of water pollution. This goal is indeed the fulcrum upon which administrative leverage is currently exerted to control water pollution. Further, the law's definition of "pollutants" is very broad. "The term 'pollutants' means dredged spoil, solid waste, incinerator residue, sewage, garbage, sewage sludge, munitions, chemical wastes, biological materials, radioactive materials, heat, wrecked or discarded equipment, rock, sand, cellar dirt and industrial, municipal, and agricultural waste discharged into water" (8).

To move toward the goal of eliminating pollutants Congress specified an interim goal, "... that wherever attainable, an interim goal of water quality which provides for the protection and propagation of fish, shell fish, and wildlife and provides for recreation in and on the water be achieved by July 1, 1983" (6). Congress also specified a regulatory strategy by which this 1983 goal was to be attained. A permit was required for all sources of pollutant discharge to waters, both public and private. The permit system was to be administered by the Environmental Protection Agency (EPA) and was to be used to set all polluters on a compliance schedule by specifying effluent limitations with attainment dates that would allow the 1983 goal to be met in two stages. The first stage, up to July 1, 1977, requires industry to install "the best practicable control technology" which is to be based on the technology of the current leaders in the field. Municipal sewage treatment plants are to apply at least secondary (biological) treatment of their wastes before discharge into receiving waters. The second stage, up to July 1, 1983, pushes industry to achieve the "best available technology economically achievable". This technology is to be defined by EPA but it is to be based on the best existing technology in each industry.

Much of the administrative burden has been turned over to the states after they had developed their own programs that met EPA standards. As an extension of the 1965 Act, states are also required to complete the task of establishing water quality

standards for all waters in the state. These standards take
into account differences in natural water quality, differences in
flow characteristics, and the best use to which various bodies of
water are to be put; such as a source of drinking water, fish
propagation, and recreational use. In its promulgation and
periodic updating state water quality standards must reflect the
attempt to attain the 1983 goal of waters suitable for fish
propagation and body contact recreation. Under no conditions can
the "best use" be as a receptacle for pollutants as could have
occurred under the 1965 law. These state water quality standards
are supposed to play a significant role in attaining the 1983 goal.
It was foreseen that in many cases the "best practicable treatment"
as defined by EPA would be insufficient to bring a stream up to
its state water quality standards. If this were the case then
the permitted effluent limitations for a pollutant source must be
reduced even further so that the water quality standards are not
contravened. That is, the water quality standards of the states
exist as a floor below which water quality must under no condition
fall as we move to meet the 1983 interim goal.

The Act also recognizes that many pollutants such as sediment
and runoff from agricultural and urbanized lands do not enter a
watercourse necessarily from a well-defined source, such as a pipe.
These "non-point" sources are also to be controlled by broad
management plans drawn up by states on a regional or state-wide
basis.

Toxic substances such as chlorinated hydrocarbons and heavy
metals were seen to pose special threats to man and aquatic life
that could require immediate action. EPA was required to identify
such toxic substances and has been given special powers to establish
effluent standards for them immediately. The standards may include
an absolute prohibition on discharge.

To date, the implementation of the toxic substances part of
the Act has lagged seriously, partly because the Act specifies
that industry has only one year to meet the required effluent
limitations once they have been promulgated under the Act's toxic
substances section. Many feel this is an impossible timetable to
meet in most instances. Effluent limitations have been written for
only six substances - DDT, aldrin/dieldrin, benzidine, toxaphene,
endrin, and PCBs - and these limitations for all except PCBs, apply
only to manufacturers and packagers. Possibly the Toxic Substances
Control Act of 1976 which attempts to keep hazardous chemicals
from being produced in the first place may be a better approach to
this important problem.

The 1972 Act is exceedingly complex but at the same time it
is comprehensive. I have merely touched on some of its more

important points. Its virtue is that it gives the government
enforceable powers to move in the direction of water pollution
elimination with definite goals and attainment dates. It is a
clean-up program with teeth. It is working and for the first time
in our history this nation's waters are getting cleaner.

EFFLUENT LIMITATIONS, STANDARDS, CRITERIA, AND COSTS

Some of the implications and interrelations of the Act can
be seen by examining the process by which effluent limitations are
established in practice. Permits for industrial and municipal
point sources must express the allowable amount of pollutant in
numerical terms, e.g., 2 lb of zinc per day. And a cleanup
schedule must be given that shows how the course will comply with
the prescribed figures over the life of the permit. These numbers
and how they are derived are of critical importance to the success
of the pollution control program. If the numbers are too high and
if they do not force the polluter to make a significant reduction
in pollution over a period of time we are not going to clean up
our waters. On the other hand if they are too low or if the compliance schedule is too tight the only way the permit conditions
can be met is for the industry (or the municipality) to go out of
business. The determination of the middle ground must be the aim
of the people who administer the program. And the administrators
are forced to this middle ground by a number of opposing pressures.

Many industries, but not all, will fight any attempt to
develop a pollution abatement program that requires them to reduce
their use of a free service - in this case, the dumping of an
unlimited amount of waste into a water course. Most industries
will really dig in their heels when costs of required pollution
control begin to climb and they feel that it will be cheaper to
spend money on lawyers to carry their case to court rather than on
getting their daily zinc output down from 3 to 2 lb per day. Enlightened industries that understand their long-term existence
depends to a great extent on their management of waste and the
acceptance of this management program by the general public go to
some length to gauge public opinion. They will resist efforts to
clean up pollution that they feel go beyond what the public is or
will be demanding. Their stake in correctly gauging the future
temper of the public is as great as that of the politician. However, their perspective must be over a longer time span than that
of an elected official who may be focussing on a 3 or 6 year
election date target. It is not surprising that the "enlightened"
industry and the "enlightened"politician might have different views
of water pollution control requirements.

On the other side there is a clear public demand now for clean water and both industrial polluters and government administrators and politicians are prime targets for public wrath when fish kills and damage to aquatic ecosystems occur. Environmental groups, armed with their own lawyers, are watching over the shoulders of government administrators to detect indications of any violation of the spirit or the letter of the law. The Natural Resources Defense Council has concentrated considerable effort on the water pollution control program and by instituting a number of lawsuits has been able to stiffen the backbone of the EPA and alter the course of the program. An EPA lawyer has been quoted (2) as saying "We'd never do anything around here if it weren't for the environmentalists". The courts and the threat of court litigation keep the administrators responsive to the letter and the intent of the law. An administrator can become adept at insulating himself from Congressional reproach but as long as all concerned parties to an issue have access to the courts the administrator is forced to be responsive to the law.

The numbers in the permits are derived by a complex synthesis of scientific water quality criteria, technological capability, and economics. Many industry-wide effluent limitations based on "best practicable" and "best available" technology have been written by EPA. This procedure is fairly straight-forward because the limitations are determined by what the average of the best and the best of plants within a given industry are able to achieve in practice. These industry-wide limitations are based on amount of pollutant per unit of production and to translate these numbers into figures for a permit that will be applied to a specific plant the plant's production level must be known. This is often something a plant would prefer to keep flexible and in tune with market demands.

Where industry-wide limitations have not been written the process becomes more complex and in a sense the limitations must be worked out on a negotiation basis with each particular plant. EPA and state representatives do work with industry on an informal basis, and they also listen to arguments of both industry and other interested parties in the more formal public hearing procedures. EPA clearly has an upper hand in these negotiations and it is up to the polluter to justify why he cannot meet any limitation requirement or clean-up schedule. EPA is not going to go around shutting down industry on a major scale to achieve effluent limitations. The political process would put a quick stop to that. Indeed the procedures established in the Act with intermediate goals and permits which contain schedules for compliance indicate that the intent of Congress was generally to push industry to clean up its effluents but not push it beyond the point

where it would have to shut down. A few plants have closed
blaming EPA regulations but in most cases these were older plants
which were economically marginal and would probably have been out
of business in a short time even without pollution abatement
requirements.

But what happens where, because of a concentration of several
polluters on a segment of waterway or for some other reason, the
EPA prescribed effluent limitations are not sufficient to maintain state water quality standards? One might think that in this
situation, which turns out to be fairly common, industries would
be forced to shut down on a grand scale because there would be
no technological way for pollutants to be sufficiently reduced to
meet prescribed legal standards. In general this does not happen
because of the manner in which state water quality standards are
established.

The setting of state water quality standards illustrates the
complex interplay of the technical, judgmental, and economic
forces that come together in water pollution control. First the
"best use" of each stream segment must be decided so that each
segment can be classified into one of a small number of best use
categories. To some degree the best use reflects the current use
of each particular segment. For example, water which is used as a
drinking water source would be classified for that use - AA or A
in New York State. It is highly unlikely that water now grossly
polluted would be put in the "drinking water" use category. In
other cases "best use" reflects natural stream conditions that
would allow upgrading to that use with technologically achievable
controls. Water with high flow year round would be classified for
body contact recreation and fishing use, and water with low or
intermittent flow or a high level of natural pollutants would be
classified for non-contact recreational uses, and so on. Thus use
classification of water segments is based to a great extent on a
reflection of what has been achieved or what is achievable in
practice without too much dislocation.

The second part of the process involves prescribing water
quality standards for each use classification category. These
standards are to be set on the basis of public health criteria or
criteria based on the propagation of aquatic species. The criteria to be used are the ones appropriate for the particular "best
use". The standards just like the classification have a legal
basis - they are written into law - but the criteria upon which
they are based have no such standing. The criteria are the points
in the process where scientific knowledge is brought into play
together with value judgment.

A water quality criterion is defined as "a designated concentration of a constituent, that when not exceeded, will protect an
organism, organism community, or a prescribed water use or quality
with an adequate degree of safety" (10). Based on a survey of
scientific data EPA has assembled a collection of criteria for a
large number of pollutants (10). In developing legal standards
a state is allowed to appropriately modify these criteria to take
account of local conditions and these conditions are to reflect
actual and projected uses, background, presence or absence of
sensitive species, temperature, weather, flow character, and
synergistic or antagonistic effects of combinations. Clearly a
state has considerable leeway to ensure that standards reflect
site-specific conditions and in particular that they reflect the
technological ability of industry to meet the standards at not
unreasonable cost. Again it would be political suicide for a
state government to promulgate legally binding standards that
would force a widespread shutdown of industry in that state.
Because of the way they are derived, state water quality standards which are based on scientific and judgmental criteria also
reflect the realities of technology and economics just as do the
industry-wide effluent limitations based directly on technology
and economics.

In discussing economic factors I have stressed the costs
shouldered by an industry in cleaning up its pollution. But there
is the other side of the economic coin - the costs borne by the
users and potential users of the water because of the presence of
pollutants. In recent months New York has witnessed the closing
of a commercial fishing industry on the lower Hudson River and
the suspension of a sports fishing industry on Lake Ontario
because of water pollution. Both of these are worth looking at in
a little detail.

Polychlorinated biphenyls (PCBs) have been discharged into
the Hudson River from two plants of the General Electric Company;
one at Fort Edwards and one at Hudson Falls. PCBs are not very
soluble in water and they tend to concentrate in the sediments.
Bottom dwelling organisms ingest the PCBs and pass them up through
the food chain where they eventually concentrate in fish. The
concentration of PCBs in the Hudson River water is very low and
close to or below the detectable limit but fish caught downstream
as far as 150 miles below the discharge point have been found with
PCB concentrations averaging 31 ppm. The FDA standard for PCBs
in edible fish has been set at 5 ppm and hence the fish from the
Hudson River have been considered unsafe to eat. In February 1976
Commissioner Ogden Reid of the N.Y.S. Department of Environmental
Conservation banned commercial fishing for most fish species from
the river because of human health hazards. The ban caused severe
problems and financial loss to the commercial fishing industry.

Action by the N.Y.S. Department of Environmental Conservation has forced General Electric to drastically clean up its operations. The legal basis for state action was not connected with the permit system or with state water quality standards. General Electric was operating within the terms of its discharge permit and no state standards had been set for PCBs because neither the state nor EPA had adequately foreseen the problems that would become associated with such substances as PCBs which are persistent in the environment and toxic to man and nature in small quantities. However the sections of the Hudson where the fishery is located are classified either A or B, and for both categories the best use includes fishing. The state was able to show in an administrative hearing that the addition of PCBs to the Hudson River made the river unsuitable for its best use, fishing, because the PCBs made the fish from the river inedible. Thus the classification for the Hudson River had been contravened by the addition of PCBs. The blame for this rests partly with General Electric for adding the PCBs to the water but partly also with the state for issuing a permit, which out of ignorance put no restriction on the discharge of PCBs from the plant.

A similar episode occurred in Lake Ontario with the recent discovery that certain species of fish (among which are the Salmonidae) contained up to 0.97 ppm and on the average around 0.2 ppm mirex, a compound which has been found to cause cancer in rats and mice. The FDA limit for mirex in edible fish flesh is 0.10 ppm. Again the amount in Lake Ontario water itself is generally undetectable but it can be found in core samples of sediment taken in a wide swath across the southern part of Lake Ontario. A major source of mirex is suspected to be the Hooker Chemical Company plant at Niagara Falls on the western end of the lake where mirex was produced in large amounts prior to 1967. Last summer a ban was instituted on the possession of a wide variety of fish from the lake and its tributaries. This ban dealt a severe blow to the one industry in New York State that appears to be enjoying explosive growth - the sport fishing industry. The immediate future of this industry which has been estimated to have a potential value in the hundreds of millions of dollars is now in doubt.

There is a cost that must be borne to clean up water pollution but the demise of a commercial fishing industry on the Hudson River and a sport fishing industry on Lake Ontario, both within a period of 2 years, shows us that there is a cost to society in continuing water pollution. The costs are there in one form or another. They are paid either by the user of the water in damages to health, losses in recreational opportunities, or business losses; or they are paid by the polluter in his efforts to abate pollution. The balance that is struck will depend on the pressures brought to bear on government administrators in the short term and on elected representatives in the long term. Both groups are susceptible to public opinion and it is to be hoped that this opinion will be based on facts and scientific data carefully measured and interpreted.

SOME PROBLEM AREAS

Although the program to clean up the nation's waters that began in 1972 is showing success there are a number of concerns that might require changes in the water pollution control strategy in the near future. There are still technical and value judgment problems with control of pollutants that persist in the environment and act in very small concentrations. There is a continuing argument over costs and benefits of any particular action as well as on the overall economic efficiency of a program based on government regulation. We will face major political decisions when we attempt to eliminate pollution from general land run-off. We are just beginning to face up to the question: Once we remove pollutants from the water what do we do with them? These are not all the problems we are likely to face but they are a sample and I would like to touch briefly on each in turn.

Toxic Pollutants

Problems connected with the introduction of toxic substances into our environment are going to be with us for some time. The American Chemical Society estimates that about 1.8 million chemicals have been formulated and that about 250,000 new ones are being added each year. Of all known chemical compounds only about 3000 have been adequately tested for carcinogenesis. The number tested for teratogenicity and mutagenicity is undoubtedly less. Our ignorance about the effects of non-naturally occurring chemicals on man and on the environment is abysmal.

Technical problems include difficulty in the detection of small amounts of a toxic substance in the environment. For example, the EPA criterion for an upper limit on mirex to ensure aquatic ecosystem protection is 0.001 pp 10^9. However, commonly used techniques for detection have a lower limit for mirex in water of around 0.01 pp 10^9. In addition, there are great technical problems in establishing scientific data on damage to man and ecosystems. Some of this involves extrapolation to low doses and transfer of data from one species to another. I am pleased to see that this subject will be treated in a separate paper at this Conference by Professor Weiss.

Benzidine is a good example to illustrate current procedures for limiting discharge of toxic substances through the permit process. EPA has prescribed effluent limitations for six substances under its powers derived from the toxic pollutant section of the Water Pollution Control Act (7) and one of these is benzidine. Benzidine has been linked to bladder cancer among workers in the chemical industry and quantitative data have been assembled from studies on rats treated with benzidine. Two technical problems are involved. First, the extrapolation of cancer incidence data to much lower concentrations than were actually used in the experiments and second, the transfer of the extrapolated data from rat to man. Usually a linear extrapolation to zero effect at zero concentration is used in the absence of any indication that a repair mechanism exists. To account for the transfer of data from rat to man EPA assumed a safety factor of 100, a not uncommon assumption. The next step was to make a value judgment on what risk was acceptable. The assumption was made that a cancer incidence of one in one million for people drinking benzidine-containing drinking water was an acceptable risk. (It would be interesting to know about this decision by an administrator in EPA.) Translated into effluent limitations this came to a maximum allowable discharge of 1 lb of benzidine per day into a moderately large river flowing at 10,000 cubic feet per second.

The scientific assessment of the potential for damage from chemical compounds is essential for intelligent action to abate pollution. But even when we know accurately what damage to expect and in what species, a decision must be made on what level of risk is acceptable both over the short and the long term. As pointed out in a recent penetrating discussion of risk and safety by William Lowrance (3) two steps are involved in assessing safety. The first is the technical, objective problem of measuring risk and the second is the personal and social value judgment as to the acceptability of that risk. Even if the first step can be surmounted the second is not capable of any simple or neat solution in our present society. Problems connected with value judgments and how these are made and by whom will always be with us, but they will increasingly be the subject of public controversy as our awareness of toxic effects increases.

Economic Efficiency

The national water pollution control program based on a zero-discharge goal and effluent limitations enforced by a permit system has encountered continuous criticism from economists (1). Their argument runs along the following line. The present system,

in as far as it does not explicitly consider costs and benefits
of each increment of pollution elimination, leads to a less than
optimal allocation of resources. A more efficient allocation
would take place if the market mechanism were allowed to operate
free of government permit regulation with effluent charges levied
so that the costs of pollution to society were built into the
price of the product in the same way that the costs of raw
material, labor, and capital are built into the price. That is,
properly derived economic incentives in principal are a more
efficient way of allocating scarce resources than are regulatory
procedures. And indeed when the costs of cleaning up our waters
have been estimated at many hundreds of billions if not trillions
of dollars such questions of economic efficiency must be given
close attention. There are administrative hurdles and problems
in calculating costs and benefits that so far have prevented the
general acceptance of effluent charges in this country. But
building economic incentives into the current regulatory program
has been broached recently by the EPA Deputy Administrator John
Quarles (9). In referring to necessary extensions of the 1977
deadline for industrial discharges Quarles suggested that Congress
should consider some sort of economic incentive, such as an
effluent charge, to encourage polluters to achieve compliance
as quickly as possible. Questions of economic incentives and
overall economic efficiency will undoubtedly surface more and
more as the easier pollution problems are corrected and as it
becomes more expensive to make further incremental improvements
in water quality. We might very well end up with some amalgam
of the regulatory and economic incentive approach (probably
anathema to an economist!) before too many years have passed.

Non-Point Sources

I have focussed most of this discussion on our efforts to
clean up the so-called "point sources" of pollution where the
discharge is from a clearly defined outfall pipe. However, a
major source of water pollution is the general water run-off from
streets in urbanized areas, from sediment run-off on disturbed
land, from individual septic tank run-off, and from run-off of
fertilizer and pesticides from agricultural land. Elimination of
such diffuse water pollution sources demands strict controls
over how and where we build and how we make use of land. The
issue of the control of land use by government is a particularly
sensitive one because we have been attuned to the tradition that
once a person has a deed to a piece of land what he does with
that land should pretty well be left up to him. At least, if
there are to be government controls the current political thinking
is that these controls with few exceptions should be in the hands

of the local unit of government and certainly not in the hands of
federal or even state officials. But if we are really serious
about cleaning up our waters on a national scale it would seem
that federal land use standards of some type must be developed.
Such a role of the federal government is bound to engender
intense controversy.

Are We Asking the Right Questions

One of the drawbacks with government, as with most institu-
tions, is an apparent inability to take action unless clearly
defined and well isolated problems are perceived. Cleaning up
water is one such isolated problem. But we all know that everything in this world is connected to everything else and that
there are no problems that can be completely isolated from all
others. When we take industrial or municipal sewage and run it
through a treatment plant we may be able to remove the pollutants
and discharge pure water but we know that we have not eliminated
the pollutants. They are still around. We have merely moved
them from one place to another and altered their form. In the
case of sewage treatment plants we have produced a sludge which
contains the original pollutants slightly modified and concentrated. What we do with this material has important environemntal
implications. If we incinerate it, which is common practice, we
run the risk of air pollution - we clean up the water at the
expense of dirtying the air. If we spread the sludge on the land
we run the risk of contaminating food crops with toxic materials
that have entered the sewers from industry, households, and steet
run-off.

Another striking example of the problems in a piece-meal
approach to environmental control is the phenomenon of acid rain.
Strict air pollution control laws have resulted in a great
decrease in particulate matter leaving stacks from coal and oil
burning furnaces. These particulates formerly trapped sulfur
dioxide and caused it to settle out in the neighborhood of the
stack. Now the sulfur dioxide is dispersed into the air where it
is carried many hundreds of miles in the course of which it is
oxidized to sulfur trioxide. This in turn is washed out of the
air in the form of a sulfuric acid rain. Pristine wilderness
lakes that exist at high elevations in the New York State
Adirondack Park, far removed from sources of man-made pollution,
no longer support fish populations because of the increased acidity
of the water caused largely by sulfur dioxide emitted from factories
in Ohio and Indiana.

Because the focus of pollution abatement in this country is on air and water it is natural that the pollutants end up in a form that must be disposed of on land. More and more as air and water become cleaner the pollution problems we face will be connected with the management of solid pollutants that have been removed from the air and water and are looking for a place to go. As the interrelation of air, water, and land become more apparent to us the management of the residuals from our society will take on new meaning. New directions will appear as the questions we ask change to meet a more holistic perception of environmental management.

CONCLUSION

We are embarked on a massive program to clean up our waters after 200 years of neglect. Science and technology play a major role in this program but the decisions on goals and how to accomplish them are not solely scientific ones. Judgmental decisions in that arena of self-interest and social and economic struggle we call politics will play a fundamental part in the outcome. The role of the professional scientist is to contribute his expertise to the education of the public, its elected representatives, and government officials while realizing that the ultimate decisions of society must and will be made in the political and not the scientific arena.

LITERATURE REFERENCES

1 DORFMAN, R. and DORFMAN, N.S.: Economics of the Environment. W.W. Norton, New York (1972).

2 HOLDEN, C.: Mirex: Persistent pesticide on its way out. Science 194 (1976) 301.

3 LOWRANCE, W.W.: Of Acceptable Risk. Science and the Determination of Safety. W. Kaufman. Los Altos (1976).

4 Public Law 92-500. Section 101(a).

5 Public Law 92-500. Section 101(a)(1).

6 Public Law 92-500. Section 101(a)(2).

7 Public Law 92-500. Section 307.

8 Public Law 92-500. Section 502(6).

9 QUARLES, J.R.: The mid-course correction. Minor adjustment or major retreat? Talk presented to the Water Pollution Control Federation, Government Affairs Seminar, Washington, D.C. April 6, 1976.

10 U.S. Environmental Protection Agency: Quality Criteria for Water. (1976).

Discussion

FELDMAN - It's quite obvious that there's a dilemma between the
economic considerations and the desire for clean air and water and
so on. And this dilemma was very forcefully brought home to me
last week in Puerto Rico. I was in a limousine with twelve other
people, and the limousine should have had about eight. It was
air-conditioned with all the windows closed, and the native driver
noticed in his mirror that this lady had lit up a king-sized cigarette. So, he very forcefully told her that he could not allow smoking in the limousine because we were over-crowded, and he had to
keep the windows closed. But, when he saw her making motions to
put out the cigarette, he started screaming to her: "Don't put out
the cigarette. It's too expensive. By all means finish the
cigarette." I don't know myself which was the most important consideration.

WEST - I was glad to see Dr. Collin put in a plea for judgment.
I think, however, I would expand his plea to include the scientists,
because I think the scientists use very little judgment in permitting some of the regulations to be enacted. For example: in the
case of water, we can make a very simple case. Without contaminated water, we could not exist. Even without polluted water, we
could not exist on this earth, because water is simply H_2O. It
has to be contaminated with oxygen, otherwise we wouldn't have fish.
It must be contaminated with CO_2, otherwise we could not have
aquatic vegetation, including phytoplankton which starts part of
the food chain. These people say you cannot have trace metals in
the water. At a meeting at Forth Worth about four years ago, they
said they would shut down any plant they caught discharging things
like copper and zinc to the water. Well, if I lived on the Mississippi River, and the water was equivalant to distilled water, I
wouldn't enjoy South Louisiana. Because we have to have some copper in the water, otherwise we wouldn't have oysters. We have to
have some zinc and iron. Otherwise we wouldn't have shrimp,
snapper and flounder. And in fact, as ugly as it sounds, we've
got to have some sewerage. The most productive waters that we
have, anywhere around this country, at least, are those in the
Gulf, where the discharge of the polluted Mississippi River sup-
ports a tremendous aquatic life. You can see the contrast if you
go to the Bahamas, my favorite escape place, which has the most
beautiful waters in the world. Any idiot could look at that water
and say: "Well, that's the place to live." They could look at the
Gulf waters, and say: "No way should we have water like that."
Well, you could live in the Bahamas -- a few people could live.
The rest would starve to death because the productive waters are
those that carry the contaminated discharge of the Mississippi,
including sewerage, trace metals, and so forth.

When we talk about toxins, almost all toxins are also essential trace elements. For a few, there seems to be some argument, such as lead, cadmium, and mercury. We would even be wise to consider what optimum ranges exist, and maintain the quality of our water within this optimum range. Too little, we're in trouble. Too much, we're in trouble. But for most things, there is an optimum amount.

The subject of judgment is rather fascinating as the following examples will show. We all know we have to drink water. And still, I guess now it's about a month ago, there was a story about a girl who was emotionally disturbed because of the death of her mother, and for some strange reason this initiated a tremendous urge to drink water. She drank enough water that it killed her. Presumably the water was perfectly potable, but she just drank too much of it. And like anything else, when you say too much, you mean too much.

I had a neighbor who ate too many hotdogs one time. He took his family on a picnic and ate so many hotdogs that he died. So you can do two things. You can set up regulations and follow them blindly. This, I think, can get you in an awful lot of trouble. Somewhere along the line, if you set up ranges and you administer with judgment, I think we'd be much better off.

COLLIN - I don't have any problem with what you are saying. But I think you're missing the point of the argument. I think that everybody will admit that natural water is not pure H_2O. On the other hand, there's the other extreme of the situation where you have rivers catching fire. You have the situation where fish cannot propagate. You have situations where levels of Mirex and PCB's are accumulating in fish, high enough so that they are probably dangerous to the health of people that eat them. This is in a place like Lake Ontario where the water we are drinking this morning comes from. They collect in the sediments, in plant life, and in the fish (in the fatty tissues of the fish). I've got some real worries about what we're doing to Lake Ontario, which is a major supplier of drinking water in the northeast part of the United States and Canada.

So I think there are two sides to the picture. What we're trying to do now on a national level is not to end up with distilled water everywhere, but it's to cut back on the major pollutants that we know are going into the water and to try to ask questions about some of the things that we are not sure are harmful but might be harmful. We're trying to move in a direction of reducing the amount of pollution that is going into the water. I think that it's that movement to reduce the amount that is the important aspect of our present situation. To do that I think, you've got to have administrative tools. You've got to have legal powers.

They may not be written in a way that makes a scientist happy, but on the other hand, they have to be written in a way in which they can be used effectively. The primary goal of the Federal water pollution amendments is a case in point. It may not make sense from a scientific point of view but I think there's one fact that you've got to admit: it does give the Federal government power, tremendous power, to move against the gross polluters of our waters.

WEST - I complimented you on the very fact that you made that point of having that judgment in there. And I was supporting you observation on that. But I do think it is always a matter of importance that we recognize that there are ranges and that there is no such thing as purity. And unfortunately, there is the connotation for contamination and pollution -- the connotations are completely evil. No one wants to eat anything that's contaminated or drink anything that's contaminated or breathe anything that contaminated, But that's when the connotation gets us in trouble. And so, my plea is to use judgment and keep within a reasonable range.

LODGE - I must say I agree with both sides here. I think perhaps, however, that we get a little carried away with the notion that the way nature had things before we started stirring them up is invariably optimum. I'm reminded of the saying that an optimist is sure that this is the best of all possible worlds, and the pessimist is afraid that the optimist is right.

EUROPEAN ASPECTS OF ENVIRONMENTAL
RESEARCH AND LEGISLATION

F. Geiss, Ph. Bourdeau and M. Carpentier

Commission of the European Communities

Brussels and Joint Research Centre, Ispra

ABSTRACT

The mechanisms of legislative activities in environmental matters for the European Communities are explained. Lists of accepted "directives" and some others under discussion illustrate these activities. An overview on both the direct and indirect action of the EC Environmental Research Programme is given. Three selected projects are described in some detail: 1) The isotopic lead experiment (ILE), which aims at assessing the pathway of environmental lead through the environment to man, 2) the EURASEP project, using the NIMBUS G satellite-borne CZCS sensor for remotely measuring of coastal pollution and 3) the EC data bank on environmental chemicals (ECDIN).

When I was asked to present a paper at this conference, I was somewhat at a loss because techniques of measuring and detection of pollution in Europe are not far different from those in the United States. However, we do have other problems created by the presence of more borders, states and nations. These do not have any impact on the types of methods of measuring pollutants, but it will probably be of interest here to start with the legislative mechanism in the environmental area in Europe. This will be followed by a description of some of the work being done in specific areas.

Nine member-states form the European Communities (EC):
Belgium, Denmark, Germany, France, Luxemburg, Ireland,
Italy, the Netherlands and the United Kingdom. It has
an executive "Commission" with headquarters in Brussels.

In 1973 the EC adopted an environmental policy, the
purpose of which was "to help to bring expansion in the
service of man by procuring for him an environment providing the best conditions of life and reconcile this
expansion with the increasing and imperative need to
preserve the natural environment." The promotion
throughout the Community of the harmonious development
of economic activities and a continuous and balanced
expansion - which constitute the paramount purpose of
the European Economic Community (EEC), - is not con-
ceivable without an effective campaign to combat pollution and "nuisances" or an improvement in the quality
of life and the protection of the environment. These
objectives are among the fundamental tasks of the Community and led to this "Environment Programme", which
is now in its second period (1977-1981).

The primary aim of this program is to set quality
standards and to prepare regulatory directives to accomplish them. These regulatory directives have to be
applicable in all member-states and existing legisla-
tions in this field have to be "harmonized". In a
simplified manner the mechanisms implied can be described as follows: The initiative for new environmental
regulations lies both with the member-states and the
Commission. Initiatives taken by the Commission
generally lead to directives which after adoption by the
Council of Ministers of the Member-states, through the
appropriate national legislation mechanisms become
identical national regulations or laws in all member-
states. If a member-state initiates a legislative procedure in an environmental sector, it will "notify"
this intention to the Commission, who then decide whether
they want to launch a directive in the same field or not.
Some major directives and similar texts adopted by the
Council of Ministers or under preparation, are:

Water Pollution

Adopted:
- Quality of surface waters for abstraction of drinking
  water,
- Quality of fresh and sea bathing water,

- Reduction of water pollution by certain dangerous substances discharged into the aquatic environment*,
- The Commission is a signatory of the Paris Convention
  on the prevention of marine pollution from land-based
  sources (North-East Atlantic),
- The Commission is a signatory of the so-called Barce-
  lona Convention on the Protection of the Mediterrani-
  an against pollution and dumping from ships and air-
  craft,
- Directives on pollution from wood pulp mills and from
  TiO_2 industry.

Under preparation:

- Directives relating to quality of water for human
  consumption, on waters supporting freshwater fish,
  water for agricultural use, industrial waters,
- Directives for the exchange of information on the
  quality of surface fresh water**,
- Directives on the protection of underground waters
  against dangerous substances.

Atomospheric Pollution

Adopted:
- Directives concerning exhaust gases from motor vehicles, sulphur content of liquid fuels and the exchange of monitoring data.

Under preparation:

- Directives for air quality with regard to lead,
- Health standards for SO_2, particulate matter.

Waste

Under preparation:

- 3 Directives concerning dangerous wastes, waste oil
  and PCBs,

---

*This very important directive introduces a system of
prior authorizations for the discharge of a number of
dangerous substances into the aquatic environment. It
provides for the setting of limit values and quality
objectives of substances on a black and a grey list.
It includes sea water.
** This proposal is aimed at exchanging information between the pollution surveillance and monitoring networks.

Chemicals

Adopted:
- A series of directives on detergents.

Under preparation:
- A series of directives on paints and pesticides,
- A directive on the marketing of new chemicals*,
- Directive on the surveillance of dangerous chemicals in industrial processes (including manufactured, by-products, additives). As a consequence of the Seveso accident, notification would have to be made to the authorities, containing information on substances and preparations involved, assessment of immediate and delayed risks (toxicity, etc.), plant location, manufacturing processes, premises, safety equipment, notification of accidents.

The Commission funds oriented research to support these legislative activities, both through extra-mural contracts and research in their own Joint Research Centre at Ispra (Italy). The funds for these external contracts represent about 5-10% of the total expenditures on environmental research in the member-states. In comparison with research carried out at national levels, the joint execution of Community projects in the enviromental field is beneficial in many respects:

- to orient research towards the most pressing needs of the program.
- To focus on particularly difficult problems the joint effort of a series of specialized laboratories in the various member-states.
- to carry out major projects throughout the territory of the Community in order to obtain results which are statistically significant, e.g., epidemiological surveys or field studies must cover the greatest available range of variation in subjects of biotopes and in types of environment,
- to facilitate scientific collaboration with other countries, e,g, with the United States in matters such as epidemiological surveys and toxicity testing of environmental chemicals.

The research areas, where work is contracted externally are:

---

*It requires that every new chemical be tested for its effects on man and environment. Dossiers to be compiled by the manufacturers. This directive will parallel, to a certain extent, the U.S. Toxic Substances Control Act.

1) research aimed at criteria, i.e. exposure-effect relationships (heavy metals, organic micropollutants,
   asbestos, air quality, marine pollution, new chemicals, noise),

2) research on environmental information management
   (data bank project ECDIN on environmental chemicals),

3) reduction and prevention of pollution,

4) management and improvement of the environment.

The accent is placed on exposure-effect relationships.

The environmental research program of the EC Joint
Research Centre concentrates on selected topics within
this framework of R & D. It is also developing a
general EC potential for the manifold applications of
remote sensing for environmental and resources management. It provides a focal point for the processing
and archiving of remote sensing data and carries out and
coordinates community-wide pilot operations in this
field. The major projects are:

I.   Particle formation and transport of pollution,
     Pathway of automotive lead.

II. Eutrophication of lakes,
     Remote sensing of coastal pollution (project EURASEP).

III. ECDIN, a data bank on environmental chemicals,
     Analysis of organic micropollutants in water.
     Impact of heavy metals from fossil-fuelled power
     plants.

IV. Application of remote sensing to monitoring of
     soil moisture (project TELLUS).

We have chosen four projects out of these programs
to illustrate the forms of collaboration in the EC-
countries.

1. The Isotopic Lead Experiment (ILE)

In order to assess the fraction of the lead present
in the human blood, originating from automobile exhaust,
a large scale experiment has been started in Northwest
Italy. Here in 1975 the overall isotopic ratio of lead
in environmental samples was about 1.18 for 206/207.

In the region of Piedmont with the capital Turin (25,000 km^2, 5 million inhabitants) all antiknock lead in petrol is being substituted, for 2 years, by a lead (originating from an Australian mine) with the isotopic composition 206/207 = 1.04. The pathway of this natural tracer through the environment to man will be followed (isotopic lead analyses in food, particulate matter in atmosphere, blood and urine). Throughout the running period of the experiment, samples are taken periodically from the same individuals in the case of adults (some 200) and from changing groups of infants with two samples (distant in time) per individual. In May 1977 the environmental isotopic ratio of the lead showed already a sharp drop of the isotopic ratio to 1.08, whereas blood-lead seems as expected, to follow at a much lower speed.

2.   Remote Sensing of Coastal Zone Pollution Project
     EURASEP

   In 1978 NASA will launch the satellite NIMBUS-G, carrying i.a. the Coastal Zone Colour Scanner (CZCS is a scanning radiometer which will measure color and temperature of coastal zone waters with a resolution of 800 x 800 m.) It has 4 channels measuring reflected sunlight in the visible region and 1 in the thermal infrared (10.5 - 12.5) to determine the temperature of the water surface. It is anticipated that, after appropriate data processing, this sensor will be in the position of remotely measuring concentrations of chlorophyll, suspended matter and yellow substances. During the flight of the satellite several teams of marine biologists and model-lists will measure the sea truth, the levels of certain substances to be correlated with the sensor signals of the satellite, at test sites in the Mediterranean (Italy, France), North Sea and Channel (Germany, the Netherlands, Belgium, France, the United Kingdom), the Irish Sea, the Belt Area in the Baltic Sea (Denmark, Germany) and possibly Greenland (Fig. 1).

   A pre-launch flight experiment using the so-called Ocean Colour Scanner (OCS), A CZCS simulator, was successfully accomplished in July 1977 along the North Sea coastline of France, Belgium and the Netherlands, with eleven research teams from the member states collaborating in the test area for sea truth measurements. These activities are being coordinated by scientists of the EC Commission.

Figure 1.   European Communities EURASEP Project (Nimbus-G).
Coastal Zone Color Scanner Test Sites.

3. Analysis of Organic Micropollutants in Water

During the years 1971-1976 twelve European counties
(four of them not belonging to the Community) coordinated their efforts for the methodological improvement of
analytical methods for organic compounds present in
great variety in most ground and surface waters (COST
Project 64b).

In the framework of this activity more than a thousand
different compounds have been encountered and identified
in European waters. It is assumed that this is only a
small fraction of the whole. Part of these compounds are
of natural origin, but most of them originate from human
activities (industrial and municpal effluents, oil spills,
washout from soil, rainout from the atmosphere).

The approach adopted was to consider in detail sampling and sample treatment, separation and detection,
GC-MS coupling reference data collection and data processing or automatic structure identification, respectively.

It is intended to further pursue this collaboration
along the previously adopted lines with a new accent on
LC/MS coupling and on linking with ECDIN.

4. Environmental Chemicals Data and Information Network (ECDIN)

Some 20,000 to 30,000 chemicals are manufactured by
the chemical industry in sizeable quantities. Since no
complete collection of data necessary for the assessment
of the environmental impact of these chemicals existed,
in 1973 the EC started a pilot project of an environmental chemicals data bank ECDIN. It covers:

- All chemicals with an annual production of more than
  500 kg,
- all highly toxic chemicals produced in any quantity or of natural
  origin,
- the metabolites and degradation products of these two groups.

During the pilot phase the number of compounds for which data
are collected, is restricted to 3000 - 5000 compounds.

For each chemical a record of more than 100 data fields is established. The data are organized in 10 major "categories":

    I. Identification, nomenclature,

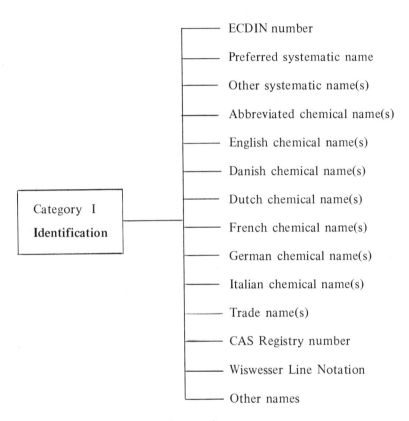

Figure 2a

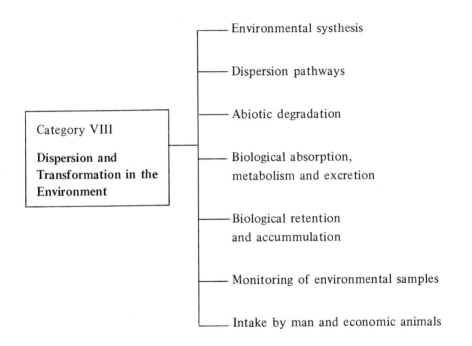

Figure 2b

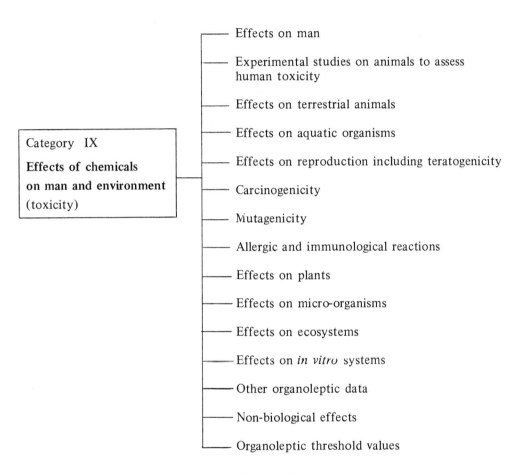

Figure 2c

II.   Chemical structure information,

III.   Physical and chemical properties,

IV.   Chemical analysis data and methods,

V.   Supply, production and trade,

VI.   Transport, handling, storage,

VII.   Use Patterns and disposal,

VIII. Dispersion and transformation in the environment,

IX.   Effects of the environment (toxicity, etc.),

X.   Regulatory data.

   To illustrate the nature of the data fields, the structure of
the categories I and VIII and IX are shown (Fig. 2a,b,c).

   Data fields may contain five text or structured data according
to the nature of the data stored.

Of course the system is computerized and as the project was conceived as a network from the onset, one of the early tasks was to
develop a format for the exchange of data between network partners.
Development of this format and the software necessary to handle it,
was one of the tasks performed under contract. Much of the data
collection and software development, particularly in relation to
chemical structure handling, is also carried out under contract.
Physically, the data bank is located at the Ispra Establishment
of the EC Joint Research Centre in Italy, where the task of assembling the collected data for input to the retrieval system is
executed.

   The recent accident at Seveso demonstrates the necessity of a
data bank such as ECDIN.

Discussion

FILBY - I notice you made no reference to radioactive materials. Is this covered by a separate agency such as Euratom?

GEISS - In 1968 the former independent agencies ("Commissions") Common Market, Coal and Steel Community and Atomic Energy Community (Euratom) were merged into the Commission of the European Communities, now functioning on the basis of the three treaties.

The Euratom radioprotection standards of 1960 have been updated in 1976. The legislative transfer mechanism is the same as for the non-nuclear environmental protection standards.

ETZ - This one effort that will monitor by satellite the surface ocean waters: what are these gelbstoffe, or yellow matters that you wanted to track down?

GEISS - These are degradation products of humic acids. We don't want to extract them, but just try to remotely measure their concentration at the surface.

ETZ - And what is the significance of their presence in concentrations higher than background concentrations?

GEISS - In case of pollution coming from land, their concentration is higher around estuaries. Their concentration depends on the nature of the river and the coastal zone.

ETZ - What did you see in the mass spectrum of the emission of these aerosols from these pine trees, or fir trees? These aerosols from pine trees, I believe, are the same substances that you see here over the Blue Ridge Mountains in the haze over these forested areas. (See Figure 3).

GEISS - The emitted gases and solids are photochemically speaking the most reactive compounds that are known. They are likely to be transformed in the course of a couple of minutes into nuclei. The compounds you saw are just the ones that came out from a branch.

ETZ - By squeezing the branch?

GEISS - No, just by putting a hood over them and sucking out the air.

ETZ - What about the quantities of these pinenes and other natural emissions?

GEISS - On a world-wide basis biogenic emissions by far exceed anthropogenic emissions into the air. The area where this study on particle formation due to the joint impact of both types of emis-

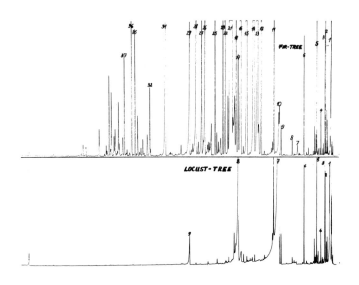

a) the Fir-Tree

1	isoprene	22	m/e = 136
12	tricylene	23	γ-terpinen (?)
13	α-pinene	25	m/e = 136
14	camphene	26	camphor
15	β-pinene	28	m/e = 152
16	myrcene	31	m/e = 154
18	β-ocimene (?)	35	β-caryophyllene
21	limonene + m/e = 154	36	caryophyllen (?)
		37	humulene (?)

b) the Locust-Tree

7	3-hexene-1-01
8	3-hexene-1-yl acetate

Figure 3. Organic emissions from trees, studied in
the framework of an EEC research project on the joint impact of biogenic and anthropogenic emission on particle
formation in the air. Living branches in a ballon swept
with an air stream, enriched on a Tenax column and separated on a 50 m OV 101 GC glass capillary column. Identification by MS computerized library search.

sions is being done (Po Valley in Italy) is protected by high moutains around (3000 to 4000 m) and is therefore characterized by very low wind velocities. This leads to a very stagnant "photochemical soup" which, at some irregular intervals, is flushed by a mountain breeze (Föhn). From these "zero conditions" the cycle starts again.

This is, of course, only a flash shot on this research project.

CHANEY - Is the sampling that you do in this valley done in a sort of a routine basis or do you just do it whenever you want?

GEISS - This particle formation project is now starting. Samples will be taken on a periodical basis. We will have a sampling station in a cable car, going up a thousand meters in order to get the vertical profile. With a series of sensing and remote sensing instruments, we hope we can identify the major parameters involved. The focus is on organic emissions.

COLLIN - You mentioned a number of environmental directives that presumably have the force of law in the various Common Market countries, but I didn't get a feel for what these involve. I wonder if you could take one and sort of tell what you said in a particular directive. Maybe the one on PCBs.

GEISS - I expected this question. The directive on PCBs says that it cannot be drained anywhere. There is no restriction on use. There are national laws on use, but there is community wide law now on the disposal of them.

COLLIN - Zero disposal into the water.

GEISS - Zero disposal into the water.   Another one on pesticides is for example banning of ten pesticides: immediate banning of DDT and three others. For the other six a deadline of three of four years. They are specified.

COLLIN - Was there a timetable for the PCBs or was that immediate?

GEISS - I think its immediate. I would have to check.

THE WHOLE ANIMAL AS AN ASSAY SYSTEM

Bernard Weiss

Department of Radiation Biology & Biophysics
University of Rochester Medical Center
Rochester, New York 14642

ABSTRACT

Advances in our ability to detect and measure environmental
contaminants and their tissue products parallel our new emphasis
on incipient functional impairment and long-range risk estimates
as indices of toxic effects. Death and overt pathology no longer
are credible criteria by which to set environmental standards. By
turning to functional criteria, however, we provoke a new set of
vexing problems, namely, how to define toxicity. Among the
factors that blur such a definition are the following: How do we
subsume adaptive and compensatory mechanisms? How do we take
account of non-specific subjective complaints? How shall we
incorporate the responses of susceptible populations such as the
very young, the very old, and constitutional predispositions to
adverse responses? Behavior exemplifies many of these problems
of definition because it encompasses so many functions and because
toxic end points are merely arbitrary stages in a continuum. It
is an indispensable criterion, however, because it is the whole
animal that responds--that dies, or grows weak or becomes irritable
and depressed. The impact on its welfare is what defines an agent
as a pollutant.

REPORT

Although some criteria of toxicity may parallel some criteria
of pollution, they are not locked into correspondence. A substance
must offend us in some manner to acquire the epithet of pollutant,
but its offensiveness can range from threats to health to putrid
odors to repugnant esthetics. A fast food outlet may embody all

53

three characteristics, but the toxicologist's concerns typically embrace only the first.

The futuristic technology to which this symposium is addressed challenges the traditional tenets of toxicology. To a non-chemist, extrapolating from the pace of recent advances, you seem almost at the stage where you can pick up an individual molecule with a pair of tweezers. What this progress implies for environmental health science, besides a few more problems with the Delaney clause, is not, I hope, a new set of tools desperately seeking applications. E. K. Marshall, the renowned pharmacologist, who himself originally trained as a chemist, once told me of a letter from an investigator proudly announcing the acquisition of a mass spectrometer and requesting Marshall to suggest some suitable problems for it. It would be a catastrophe should the fruits of this advancing technology be dissipated in the cause of fashion. My somewhat anomalous role in this symposium is to set these advances in a framework reflecting many of the current issues that assail toxicology and the environmental health sciences.

The ultimate definition of toxicity is death. Step back from the brink and the criterion blurs, especially if function, not structure defines our measures, and incipient rather than overt toxicity our criteria. One criterion of functional integrity is behavior, a summarized, integrated reflection of the organism's capacity to handle its environment. This overwhelming virtue also incurs disadvantages. Since the central nervous system coordinates its accessory and sub-system less like an authoritarian maestro than a central government permitting local options, compensatory mechanisms, which exist in abundance, may mask deficiencies in one or more such systems. Every bodily system maintains intrinsic redundancies, however, so behavior offers no unique problems in principle.

Compensatory mechanisms are among the factors that preclude an unambiguous definition of functional impairment, and, by extension, incipient toxicity. Consider, for example, the physiological impact of organophosphate pesticides. These compounds inhibit the enzyme acetylcholinesterase, which is required to sweep the neuorotransmitter acetylcholine from the nerve terminals where it is released. How much of an exposure is toxic? Reducing brain acetylcholinesterase by 75% may fail to provoke any functional consequences because the enzyme is so abundant that adequate reserves still exist. Moreover, even a degree of inhibition that flares into impairment may represent only a temporary state. Animals adjust to chronically low acetylcholinesterase levels rather quickly, perhaps in part because acetylcholine turnover diminishes (12). Are agricultural and cropduster pilots suffering toxicity when blood cholinesterase is 30% of normal, but their ability to function indistinguishable from normal?

Such questions can be multiplied endlessly, because, even now, our ability to detect pollutants and chemical changes induced by pollutant exposure far exceeds our ability to define a clear toxic end point. Distinguishing between a healthy adaptive response and a toxic progression is one of the challenges now posed to toxicology and environmental health science. Behavior, as a reflection of the total organism's functional capacity, illustrates many of these enigmatic problems.

Heavy metal toxicology is a good place to begin, and methyl-mercury a potent illustration of the complexities we face. By now, you need no introduction to methylmercury as an environmental hazard or as a source of vehement debate. It sliced into our awareness after Minamata (11) and erupted again with poisoning episodes in other parts of the world, notably Iraq (2). Methyl-mercury is primarily a central nervous system poison, destroying brain tissue. A current debate about its toxicity now inflames Canada, where certain Indian communities, because of high fish consumption, shelter individuals whose hair and blood levels exceed values frequently considered to represent the threshold of toxicity. The trouble is that many indices of methylmercury toxicity (Table I) are subjectively defined; for example, numbness and tingling around the mouth and extremities, a symptom called paresthesia by neurologists. Here is a clear instance of adequate guides to exposure--hair and blood concentrations--but flimsy correlations with the toxic end point because the latter is too ambiguous. A similar ambiguity prevails in assessing lead toxicity because hematologic indices of exposure may only disclose compensa-tory or trivial responses, with actual impairment coming at higher body burdens. And, even here the impairment may be largely incipient rather than clinically overt.

Laboratory research can draw from, amplify, and refine clinical criteria, however. Many human victims of methylmercury poisoning experience constriction of the visual field, so that they cannot detect objects in the periphery. An experimental arrangement that explored this phenomenon in monkeys was based on the structure of the visual system (Evans, Laties, and Weiss, 1976). The peripheral field is largely represented by retinal elements called rods, which are sensitive to low luminance, while the central field is represented by cones, which function at higher luminance levels and provide the precise discriminative mechanisms we need for tasks like reading. If a monkey is required to distinguish geometric forms at near-threshold levels of illumination, peripheral rod vision provides the sensitivity, and, if the peripheral field does not function because of damage to the corresponding brain areas, the monkey will not perform well on the task. This is, in fact, an early sign of methylmercury intoxication. Armed with such data, we can proceed to ask questions about the source of the impairment --which visual system structures have failed, where the poison is

TABLE I   Indices of Methylmercury Toxicity

SENSORY
      Paresthesia
      Pain in limbs
      Visual disturbances (visual field constriction)
      Hearing disturbances
      Astereognosis

MOTOR
      Disturbances of gait
      Weakness, unsteadiness of legs; falling
      Difficulty manipulating objects
      Thick, slurred speech (dysarthria)
      Tremor

OTHER
      Headaches
      Rashes
      "Mental disturbance"

localized, in what form, and, ultimately, how much is required to produce an impairment.

Asking "how much?", however, neglects biological variability, a contribution of inherited endowment, acquired predispositions, and life stage. Let me discuss an example of the dilemmas posed. Food additives now are the subject of intense debate, particularly as instigators of behavioral toxicity. A pediatric allergist named Ben F. Feingold proposes that synthetic colors and flavors, plus some natural constituents, induce behavioral disorders in certain children intolerant to them, and that such reactions are responsible for a significant proportion of the population of children labelled as "hyperactive," "hyperkinetic," or suffering from "minimal brain dysfunction" (5). Whatever the merits of the hypothesis, based primarily now on parental testimony and clinical impressions, it illuminates a problem so far neglected. Although such additives are subjected to conventional toxicity testing, freedom from carcinogenesis or other pathology surely does not imply an absence of effects on behavior--which are not examined in this field (4). Given the vast number of current additives (Table II) and how some are combined (as in synthetic pineapple flavor, Table III), you quickly appreciate the dimensions of the problem. It is a chemical morass, particularly when you consider how transformations produced in food processing and by our own metabolic pathways may multiply the number of unique substances.

TABLE II   Intentional Food Additives

Class	Number in Use
Preservatives	33
Antioxidants	28
Sequestrants	45
Surface active agents	111
Stabilizers, thickeners	39
Bleaching and maturing agents	24
Buffers, acids, alkalies	60
Food colors	34
Non-nutritive and special dietary sweeteners	4
Nutritive supplement	117
Flavorings - synthetic	1610
Flavorings - natural	502
Miscellaneous: yeast foods, texturizers, firming agents, binders, anticaking agents, enzymes	157
Total number of additives	2764

TABLE III   Components of Synthetic Pineapple Flavor

Agent	Percent of Mixture
Allyl caproate	5.0
Isopentyl acetate	3.0
Isopentyl isovalerate	3.0
Ethyl acetate	15.0
Ethyl butyrate	22.0
Terpinyl propionate	2.5
Ethyl crotonate	5.0
Caproic acid	8.0
Butyric acid	12.0
Acetic acid	5.0
Oil of sweet birch	1.0
Oil of spruce	2.0
Balsam Peru	4.0
Volatile mustard oil	1.5
Oil cognac	5.0
Concentrated orange oil	4.0
Distilled oil of lime	2.0

The question of, "how much?," is further distorted by the
fact that we do not even need chronic exposures for a chronic
effect. A single or limited exposure may be enough to determine
a toxic effect far in the future long after the agent disappears
from tissue. Carcinogenesis is a cogent example of such a process.
But behavior may serve as well. For example, mice exposed to
methylmercury prenatally may not demonstrate the aftermath until
months later, or even in advanced age (10). Longevity is one of
the most sensitive indicies of toxicity, which is why lifetime
studies are mandatory for assessing chemicals with wide communal
import--food additives, for example. But even this criterion may
be blemished. Suppose that the consequence of intoxication is not
a reduction of lifespan but a diminution in functional capacity
(13), so that the organism in advanced age is less able to perform
complex tasks, strenuous exercises, or to endure stress or
privation. The typical environment of the laboratory rodent
imposes few demands other than the ability to eat and drink. And,
although our own environment may not be as circumscribed, it, too,
demands minimum vigor, and attenuates the impact of individual
differences in functional capacity. It still is novel to set
the question of toxicity in such a context.

Clear definitions of toxic end points are hampered also by
neglect of variables that, by custom, are relegated to insignifi-
cance. One of these is nutrition. Consider the current dilemma
in lead research. An effective animal model, useful at least for
heuristic purposes, is produced by feeding lead to nursing dams.
The rat or mouse pups are then exposed through maternal milk. But
the procedure is riddled with logical hazards. Most experimenters
have tended to feed the dam such high lead concentrations that,
either because of taste or because of illness, food and water
consumption decline, less milk is available for the pups, and they
develop more slowly than their controls (1). Maternal behavior
is disturbed as well, another source of growth retardation and
variable development. A milk diet, moreover, predisposes animals
toward elevated absorption of some metals, at the same time that
nutritional inadequacy promotes selective imbalances of others
(7,8). After weaning, the pups then move to a diet typically
prepared for optimal breeding, and containing a superabundance of
minerals and vitamins. For example, the amount of iron may contain
five times the Recommended Daily Allowance of iron. Most of us are
unable to determine what our experimental animals eat, even when we
prescribe a formulation from the supplier. Unresolved issues of
nutrition suffocate lead research in ambiguity.

We might find it simpler to resolve judgments of quantity,
however, if we could unambiguously agree on a definition of
functional impairment. What are we to conclude if we expose a
trained animal subject to a recognized toxin and find its performance to improve? Levine (9) trained pigeons to peck a response key

in special behavioral test chambers. This behavior was maintained
by lifting a magazine filled with grain into feeding position for
a few seconds after certain pecks specified by the computer program
that controlled the experiment. Under one arrangement, the pigeons
earned access to grain only if they paused at least 20 sec between
pecks. Pigeons typically peck at such a high rate that they earn
rather few rewards under this contingency. After exposure to
carbon disulfide, a potent neurotoxin, their response rates, as
shown in Figure 1, dropped far enough to earn them a substantial
rise in rewards. Do we postulate this change as performance
impairment, accept it at face value as performance enhancement,
simply view it as an index of central nervous system responsive-
ness or adopt some other strategy? By measuring behavior controlled by other contingencies, Levine was able to conclude that
response rate lowering comprised a rather uniform consequence of
carbon disulfide exposure--a more intelligible finding than one
piece of information in isolation.

     The title of my lecture, improvised in haste, emphasizes
the whole animal, because the whole animal is our ultimate concern
and final criterion: its development, its longevity, its functional
integrity, its quality of performance, its feelings. Sometimes
these constitute our only measures. The farm families in Michigan
who consumed PBB-tainted food complain mainly of subjective ail-
ments. An agent like ionizing radiation may simply "age" the
organism more rapidly so that it acquires the diseases of senility
in middle age. Early lead exposure may enhance predisposition to
diseases and disturbances ranging from mental retardation to kidney
dysfunction, to nospecific early mortality (6). So how do our concerns intersect? One obvious decussation is in specifying a
pollutant or a toxin. For example, we now know that PCB's
represent health hazards, a conclusion we can confirm by feeding
the commerical product to animals. Its constituents are so complex,
however, that only refined chemical analysis of both environment and
tissue enables us to delineate the dose-effect relationships that
precede a search for mechanisms. Although TCDD is the major
contributor to PCB toxicity, it is not the sole component that
threatens health.

     A further partnership is bound to emerge in the near future.
The products of coal combustion are not as well-characterized as
one might assume, given coal's long industrial history, but the
products of coal gasification and liquefaction are even more
tenuous, because, as commercial scale evolves from pilot and
demonstration plants, the nature and balance of products and
effluents will shift. Expensive technologies need to be erected
to protect both workers and the general public. But protect
against what? Which constituents and at what levels? And by

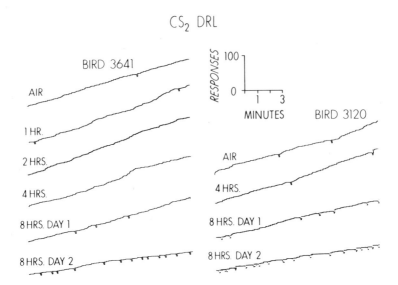

Figure 1.   Performance changes in pigeons induced by carbon
disulfide exposure (2 mg/1) Each peck on the response key
incremented the recording pen, so that each tracing is a
cumulative record of performance. The steeper the slope, the
higher the pecking rate. Each time a pigeon earned a reinforcement (4 sec access to mixed grain), the pen deflected downward.
These records demonstrate that exposure reduced response rate
and increased the frequency of reinforcement (9).

which criteria of toxicity? These issues are not trivial and
academic exercises. They are essential precursors to estimating
risk. And if you once thought that exercise to be rather academic,
simply remember saccharin.

      Supported in part by grants ES-01247 and ES-01248 from the
National Institute of Environmental Health Sciences, grant
MH-11752 from the National Institute of Mental Health, and in part
by a contract with the U.S. Energy Research and Development
Administration at the University of Rochester and has been
assigned Report No. UR-3490- 1201.

REFERENCES

1   BORNSCHEIN, R. L., MICHAELSON, I. A., FOX, D. A. and LOCH, R.:
     Evaluation of animal models used to study effects of lead on
     neurochemistry and behavior, Biochemical Effects of Environ-
     mental Pollutants. Ann Arbor Science Publishers. Ann Arbor,
     Michigan (in press).

2   CLARKSON, T. W., AMIN-ZAKI, L. and AL-TIKRITI, S.: An
     outbreak of methylmercury poisoning due to consumption of
     contaminated grain. Fed. Proc. 35 (1976) 2395-2399.

3   EVANS, H. L., LATIES, V. G. and WEISS, B.: Behavioral
     effects of mercury and methylmercury. Fed. Proc. 34 (1975)
     1858-1867.

4   FAO/WHO EXPERT COMMITTEE ON FOOD ADDITIVES: Toxicological
     Evaluation of Some Food Colors, Enzymes, Flavour Enhancers
     Thickening Agents, and Certain Other Food Additives.
     World Health Organization. Geneva (1975).

5   FEINGOLD, B. F.: Why Your Child is Hyperactive. Random
     House. New York (1975).

6   HENDERSON, D. A.: A follow-up of cases of plumbism in
     children. Australian Annals of Medicine. 3 (1954)
     219-224.

7   KELLO, D. and KOSTIAL, K.: The effect of milk diet on lead
     metabolism in rats. Environ. Res. 6 (1973) 355-360.

8   LEVANDER, O. A.: Nutritional factors in relation to heavy
     metal toxicants. Fed. Proc. 5 (1977) 1683-1687.

9   LEVINE, T. E.: Effects of carbon disulfide and FLA-63 on
     operant behavior in pigeons. J. Pharmcol. Exper. Therap.
     199 (1976) 669-678.

10   SPYKER, J. M.: Behavioral teratology and toxicology.
     Behavioral Toxicology (Weiss, B. and Laties, V. G. Eds.).
     Plenum. New York (1975) 311-349.

11   TAKEUCHI, T.: Biological reactions and pathological changes
     in human beings and animals caused by organic mercury contam-
     ination. Environmental Mercury Contamination (Hartung, R.
     and Dinman, B. D. Eds.). Ann Arbor Science Publishers.
     Ann Arbor, Michigan (1972) 207-222.

12   WEISS, B. and HELLER, A.: Methodological problems in evaluating
     the role of cholinergic mechanisms in behavior. Fed. Proc.
     28 (1969) 135-146.

13. WEISS, B. and SIMON, W. Quantitative perspectives on the long-
     term toxicity of methylmercury and similar poisons. Behavioral
     Toxicology (Weiss, B. and Laties, V. G. Eds.) Plenum.
     New York (1975) 429-435.

DISCUSSION

COLEMAN - With some of the behavioral modifications that you are interested in, how many animals does it take to acquire statistical significance? Is it a megamouse sort of situation? Or is it small?

WEISS - It's rather difficult to perform megamouse experiments with behavior, because I can envision laboratories stretching from Rochester to Buffalo, although I'd rather have them extend from San Francisco to Los Angeles. Megamouse experiments are employed for risk assessment. We are concerned in behavioral toxicology not with gross risk assessment, but with subtle functional changes -- with incipient toxicity. We in part can overcome the deficiencies of small numbers of animals by employing stable baselines, making it easier for us to detect small displacements, small perturbations from the baseline. We're stuck with the fact, too, that behavior is no more a unified concept than health. Behavior is comprised of many different functions. We choose those that we believe have some relevence to the substance whose toxicity we've been asked to gauge. We picked vision for methylmercury because of the history of methymercury poisoning in Japan and Iraq. For other substances we choose other functions. We determine dose-response relationships because they provide clues to mechanisms. We abjure the large scale behavioral study because it contributes so much variability that it doesn't make sense to commit that many resources, and because we're not interested in gross effects. We're interested in early covert effects, perhaps at a stage when an intoxication is still reversible.

HEINRICH - Does the reversibility, itself, enter your criteria? An effect which is not reversible would be more serious than a temporary one?

WEISS - Yes, I suppose you might say we're interested in the entire continuum. In studies, say, with the visual system, it may be possible to detect, with appropriate techniques, incipient toxicity before the structures that underlie the behavior itself have been destroyed, that is, before actual cellular damage has occurred. Often, intoxications are heralded by the appearance of rather nonspecific complaints: feeling irritable, or tense, or not well, or fatigued, or not eating properly. Clinicians look upon these as nonspecific symptoms, often to be ignored. But psychology has a long history of measuring precisely those kinds of subjective variables, not from our toxicologic standpoint, but from the standpoint, say, of the efficacy of psychotherapy. It's a technology that the environmental health sciences will have to employ in answering those kinds of questions and providing guidelines for exposure.

FILBY - What would you propose that an agency like FDA do to its present testing procedures to take into account the effect that you're measuring or trying to measure? I know that's very difficult and you probably can't be very specific.

WEISS - FDA, in fact, is trying to take those variables into account. Also, the Toxic Substances Control Act, as you know, specifies behavior as one of the variables, one of the end points to be
taken into account in determining adverse effects. At the National
Center for Toxicological Research in Arkansas, funded jointly by FDA
and EPA, a behavioral toxicology unit is now under construction. FDA
laboratories in Washington also have some behavioral expertise.
The Environmental Agency, the National Institute for Environmental
Health Sciences, and, of course, many chemical manufacturers, now
are building behavioral toxicology laboratories for precisely the
purposes i've outlined. But remember that I'm using behavior to
illustrate for you some of the problems in defining exactly what
we mean by toxicity. As soon as you give up death as the end
point, many problems emerge.

TORIBARA - Could you comment on extrapolating animal data to humans?
A brief comment with saccharine in mind.

WEISS - You're not being sweet to me, Taft. All I can tell you
about saccharine is that it's not very useful for assisting in
weight reduction. Since it makes unpalatable foods palatable, we
know that people will eat just as much, if not more. In fact,
Monsanto, in the 1950's, showed that if you sweeten cellulose
with saccharine, cows will eat it.

    You can extrapolate from animals to humans not directly, of
course, on the basis of dose, but on the basis of function, taking
into account similarities and differences. For example: in rats
and mice, the target areas within the brain for methymercury are
different than they are in the primate, which is precisely why we
chose monkeys for our work with methylmercury. For the production
of carcinogenesis, there do not seem to be any distinctive species
differences. Duration of life, of course, is an important component which is why extrapolation from the animal studies to humans
is not as ridiculous as many might think.

HEINRICH - In the case of asbestos, for instance, the time from
the contamination to the outbreak may be twenty or thirty years,
How licit is it to extrapolate from an animal that lives two or
three years?

WEISS - It certainly is a warning, isn't it? If an animal acquires
a disease or a functional impairment in that short a period of time,
and you know that the human is going to be exposed for perhaps
forty times that duration, then you begin to worry about the process.
That's certainly one of the reasons that humans appear to be more
sensitive.

COLEMAN - One other thing that concerns extrapolation and maybe reflects my ignorance of the way that primate brains are formed, is the difference between control centers in a human brain that has a neocortex, especially if they control behavior. Is there a suitable model for human behavior in an animal?

WEISS - We can train animals to perform many complex tasks. For example, Dr. Laties in our laboratory trains animals to count. Pigeons can be trained to peck on one key eight or nine times, before pecking on a second key once in order to gain access to grain. That is surely a complex performance for the pigeon. Rats can also be trained to perform extremely complex tasks. And I don't think we've even demanded, say, 20 percent of what we can expect from primates. Psychologists now are teaching chimps to "talk", which is an extremely complex performance. We've attempted in the experimental analysis of behavior not to examine specific behaviors, but to work with principles, and with functions such as the ability of animals to discriminate signals in the environment. Behavioral toxicology is built on a long history of behavioral pharmacology, which first exploded in the 1950's with the intro- duction of tranquilizing drugs. We have a lot of guides that tell us what to do, a lot of performances that we can use as baselines. We understand what happens to them when they're challenged. We understand from this long history how impairments of the central nervous system may be reflected or expressed as behavioral im- pairments. We are not working in a vacuum without a history.

LODGE - One of the things that strikes me very strongly and has done so every time we've had an interaction between the analysts and the medical profession is that --

WEISS - I'm a psychologist.

LODGE All right. The behavioral or human element of this is that each thinks that he's got problems and the other doesn't. I think most of the people analyzing the atmosphere figure if they could just actually nail down what in the devil is in it, that the toxi- cologists and so on could quickly tell them how dangerous it is. And I have the feeling that the medical and behavioral people have the feeling that once they nail down the symptomatology, that the problem will be solved because obviously those chemists know what's in the air. And of course, both are pure myth. I am not at all sure looking at your ozone chamber that the rats are breathing the ozone concentration that's in the chamber. And in many other cases, I suspect that the chemistry is at least as badly out of control as the medical aspects. Let this be a warning on both sides. We ain't got the answers either.

WEISS - It's not a marriage of convenience; it's a marriage of ignorance.

LODGE - A pooling of ingnorance I think.

ETZ - What are the chances that we all adapt to these various environmental stresses over the period of generations, and that a new species in essence evolves from all of this.

WEISS - I don't know about you, but -- [adopting primate posture].

SESSION II:

MORE FAMILIAR PRINCIPLES

ANALYSIS OF ATMOSPHERIC POLLUTANTS OF POSSIBLE IMPORTANCE IN HUMAN CARCINOGENESIS

Eugene Sawicki

Environmental Protection Agency
Environmental Research Center
Research Triangle Park, N.C.

ABSTRACT

Genotoxic properties and analytical methodologies of a large variety of inorganic and organic gases, vapors and particulates have been discussed. The relevance of this information to the human condition is examined. With the prime importance of mutagenesis in the key processes of life, knowledge about the chemical and radiation background of our environments has become an absolute necessity for our survival with grace. Any chemical or radiation which affects DNA could be expected to affect one or more of the various carcinogenic pathways in living tissue. These are some of the reasons for our intense interest in all genotoxic pollutants.

INTRODUCTION

In the "good old days" a few small groups worked and/or lived in highly polluted environments. Not too many people cared, mainly because they didn't recognize the danger of the situation and because they faced relatively greater dangers from the much better known infectious diseases of the day.

Since those days the killing and crippling diseases of that day, such as yellow fever, rabies, smallpox, polio, measles and the various plagues have diminished in importance while yesterday's trickle of toxic new chemicals entering our environments has become a flood, so that some of the biggest health problems of our day are the chemical diseases.

It is only recently that the importance of the role of mutagenesis in the key processes of life has been realized. Extensive
evidence has been accumulating which indicates that mutagenesis is
the vital spark, which initiates chemical changes in DNA that force
changes on all living things. Thus, mutagenesis is postulated as
playing the key role in evolution, carcinogenesis through somatic
mutation, inborn errors of metabolism through inherited germ cell
mutations, atherosclerosis through mutation of a smooth muscle
cell in an arterial wall, teratogenesis (in some cases through
mutation of an embryonic cell), and aging through an inexorable
accumulation of metabolic mutations leading to a decreasingly
efficient operation of the organism.

Thus, our contacts with carcinogens do not just mean we either
get cancer or we don't. In the same way a cancer patient's contacts with highly mutagenic anticarcinogens does not just mean
s(he) is cured, alleviated temporarily or there is no effect.
Thus, there can be an accumulation of various toxic effects and
various metabolic mutational errors, all these gradually being
accumulated over the years. Under this type of gradual battering,
the organism becomes less and less efficient with time. And so we
can rot gradually from our contacts with carcinogens, even though
we may never get cancer.

When a toxic chemical can be metabolized or hydrolyzed in
living tissue to a highly active electrophilic cation, then nucleophilic groups present in tissue chemicals and biopolymers can be
attacked. The wide variety of reactions and reaction products are
only limited by accessibility of the two diverse reacting groups
to each other and the relative rates of the various reactions. The
result of these types of reactions could be any one phenomenon or
any combination of phenomena as shown.

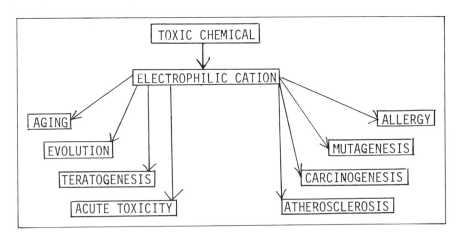

The reactive cation could react with key biochemicals, and affect key cells and cause specific and general types of acute toxic effects. The attack of the cation on the genetic apparatus could result in an accumulation of metabolic mutations over the years leading to premature debilitation during the latter stages of aging. Since this type of cation is very reactive and there are a large number of chemicals in tissue with nucleophilic groups, acute toxic effects of non-specific and specific types could also take place, the type depending on where the cation is formed and its half-life. The cation acting as an hapten could combine with a protein or a polypeptide and thus, acting as an antigen, induce the formation of antibodies or of sensitized cells and be responsible for clinical allergic manifestations. The reactive cation could react with appropriate bases in the DNA of a germ cell eventually leading to inherited inborn errors of metabolism. In a somewhat similar fashion the genetic apparatus in a somatic cell could be affected and a carcinogenic pathway could be initiated which under further provocation could lead to a cancer. The smooth muscle cell in an arterial wall could be affected by such a cation present in the blood stream. This postulated type of mutation could lead to a thickening of the arterial wall and the resulting atherosclerosis. The embryo in a pregnant woman could be affected teratogenetically by the transplacental effect of the toxic chemical, its proximate or ultimate metabolite or the final cation. And last but not least such a cation could affect the evolution of the species.

Looking at the cancer picture from an overall viewpoint it would seem that in the vast majority of cases neither genes nor environmental agents by themselves cause cancer. Most likely, genetic susceptibility and environmental contacts are both necessary to the genotoxic process. Based on animal and human carcinogenesis experiments it would appear that the varying inter-play between genes and various genotoxicants (many times with one carcinogen playing a key role) will determine whether an individual will get cancer or will suffer other as yet unmeasurable effects. Somewhere between the no adverse effect and the carcinogenic effect of a mutagen there is a continually decreasing inability of an individual's homeostasis (conditioned by his genotype and the continually increasing number of genotoxic contacts to maintain equilibrium) and of the individual's ability to survive gracefully in his microenvironment.

Using this definition of disease as a state of individual homeostatic abnormalities one could consider an individual's life as a continuum of health-disease states, varying in diverse ways at one and various times in various parts of the body between robust good health to just short of death and these various equilibria continually shifting and changing under the continual

impacts from the environment. The temporal effect of these insults results in a gradual decrease in the resilience of these
reversible phenomena with the degree of reversibility of these
equilibria gradually decreasing to a point of irreversibility
and eventual breakdown of some bodily function(s).

In considering the possibly dangerous activity of an atmospheric carcinogen or a mutagen on human beings, the composition
of the microenvironment of the individual in terms of other
carcinogens and mutagens and cofactors and antifactors would need
to be known. In this sense knowledge of the chemical and radiation background of the high risk individual is necessary in terms
of cigarette smoking, cosmetic use, drug intake, and miscellaneous
radiation and chemical contacts.

INORGANIC POLLUTANTS

Confining the discussion of the analysis of inorganic gases
to the genotoxic aspects of air pollution, real time monopollutant
monitoring for the inorganic gases has proved to be of some value
in kinetic, control and trend studies but from the genotoxic viewpoint this type of analysis is too expensive and gives too little
information to justify expansion to other pollutants.

The monotoxicant approach in environmental carcinogenesis
studies has led to an inordinate amount of controversy and para-
dox. This approach was the natural outcome of the necessarily
simplified determination of the carcinogenicity of a chemical
through long term treatment of a highly inbred mammal with large
amounts of the chemical. The contact with huge amounts of a carcinogen followed by a very high incidence of cancer (50 - 100%)
is a rare human event. Usually human beings are in contact with
a wide variety of chemicals and radiations.

Since it takes quite an extended contact with a carcinogen
conglomerate before cancer results, time-integrated studies are
most logical and desirable. Polypollutant assay is the desirable
way to lead us out of the controversial situations we're now
seeing in epidemiological studies.

The desirability for a thorough knowledge of the chemical and
radiation backgrounds of our environments is based on the failure
to satisfactorily derive a relation between contact with one
chemical or mixture and a human cancer. Examples of the complexity
of the human cancer problem are the carcinogenic activities of
arsenic and benzene in humans as shown by epidemiological studies
and the negative results obtained for animals. Other doomed monopollutant correlation studies have been the atmospheric BaP-lung

cancer studies and the Zn/Cd ratio in soils versus gastric cancer studies. The continuing controversy as to the carcinogenicity and anticarcinogenicity of selenium is another area where the monotoxicant approach is breaking down. There are many other human carcinogens and here again it would seem that other factors are involved in the carcinogenicity of these chemicals; examples of some of these are aflatoxin and liver cancer, cigarette smoke and lung cancer, vinyl chloride and angiosarcoma, asbestos and mesothelioma, cadmium and prostate cancer, etc. None of these epidemiological correlations are completely satisfactory, since the majority of people in contact with these carcinogens never get cancer and those who get cancer are also in contact with other types of carcinogens and genotoxicants.

Chlorine, nitrogen dioxide, ozone and sulfur dioxide are some of the prominent genotoxic inorganic gases. Large accidental spills of chlorine into the atmosphere have been recorded, e.g. two large ones at chemical plants in Louisiana in early 1977. This halogen would readily react with olefins and other unsaturated compounds to form halogenated compounds, a few of which are known to be mutagens and carcinogens, but most of which have never been tested for mutagenicity or carcinogenicity. Thus, chlorine is a precarcinogen and a premutagen. Its analysis is not too satisfactory. Colorimetry with methyl orange (68) or 4-nitroaniline (24) can be used. The latter method has a possibility of differentiating chlorine from bromine.

Nitrogen dioxide concentrations in non-urban, urban and highly polluted areas are available (55). Since this compound is a precursor of nitrous acid which can readily react with secondary aliphatic amines to form carcinogenic and mutagenic nitrosamines, it can be considered a premutagen and a precarcinogen. It has been suggested that nitrosamines found in factory areas are produced from the atmospheric reaction of secondary amines with nitrogen oxides (12).

Nitrogen oxides can be determined by a variety of commercial instruments utilizing real-time monitoring and chemiluminescent detection (7). Methods utilizing 24 hour sampling and colorimetry are also available (46,50). The latter method has the least interferences but the commercial chemiluminescent instruments are superior for obtaining a large amount of data quickly and with minimal difficulty.

Concentrations of ozone have been reported in urban, non-urban and highly polluted areas (55). Ozone shows a wide variety of genotoxic effects. It has been reported as accelerating aging, causing lung tumors in tumor-susceptible mice, and being a clastogen, mutagen, precarcinogen and a premutagen. Thus, lipid peroxidation damage occurs in the lungs of ozone-exposed rats (15). One of the products is malonaldehyde which is known to be carcinogenic to Swiss female mice (61). Ozone is best determined with a commercial real time monitor utilizing chemiluminescence for detection (47). This method is based on the gas-phase chemiluminescent reaction of ozone with ethylene to produce an excited species which emits visible light. Calibration methods used with the method include gas phase titration with NO or excess ozone and ultraviolet assay of the ozone standard.

Concentrations of sulfur dioxide have been reported in urban, non-urban and highly polluted atmospheres (55). Sulfur dioxide is a genotoxic gas which has cocarcinogenic and premutagenic properties. When inhaled with BaP (104 ug/m^3), sulfur dioxide (26,200 ug/m^3) caused a carcinogenic effect in rats and hamsters greater than obtained with BaP alone (41). The premutagenic properties of sulfur dioxide stems from the ready formation of mutagenic sulfite from SO_2.

This gas can be analyzed through real time monitoring by flame photometry (18) or gas phase fluorescence (60) or through 24 hour sampling followed by colorimetry (2). The flame photometric method is essentially a method for the measurement of sulfur-containing air pollutants through the coloration of a flame resulting from their combustion. Usually there are no interfering sulfur compounds in the air since SO_2 is usually the sulfur species overwhelmingly present.

Some of the genotoxic vapors found in polluted atmospheres include arsenious oxide, bromine, elemental mercury, dimethylmercury, nickel tetracarbonyl, nitrous acid and selenium dioxide.

Arsenious oxide is stated to be a human carcinogen. A three-fold increase in respiratory cancer has been reported among 8000 smelter workers exposed to this chemical (44). It can be measured by atomic absorption, colorimetry (silver diethyldithiocarbamate), neutron activation or spark source mass spectrometry. Vapor pressure studies of arsenious oxide and selenium dioxide indicate that these compounds would be present mainly in the vapor form in the atmosphere (6). As yet, analytical data does not seem to bear this out. This could be due to the collection of As (III) and Se (IV) as arsenite or selenite on a filter.

High concentrations of atmospheric bromine could be found in industrial areas where this halogen is manufactured or utilized. Industrial accidents would be one source of elevated concentrations of the compound. Bromine could be considered a precarcinogen and a premutagen since it would react readily with olefins and other unsaturated compounds in the atmosphere to form alkylating agents, some of which would certainly be carcinogenic. The

carcinogenic, mutagenic and teratogenic activity of the halogens
need to be explored because of their high reactivity and because
of the potential mutagenicity of the halogenated derivatives they
would form in polluted environments.

Bromine can be determined by a quenchofluorimetric method
with fluorescein as the reagent (6) and by a colorimetric procedure
with o-tolidine as the reagent (21). Chlorine would be an interference in the latter method.

Elemental mercury is stated to be carcinogenic (19) and is also
cocarcinogenic (5) in that it enhances the carcinogenicity of some
carcinogens. This element can be determined through atomic absorption (39) or atomic emission (11). In the latter method collection,
characterization and determination of particulate mercury, methylmercury (II) chloride type compounds, elemental mercury and
dimethylmercury has been achieved using sequential selective
absorption tubes for separation and a dc discharge spectral
emission type detector for analysis of the derived volatilized
mercury at $\lambda 253.65$ nm.

Dimethylmercury is a volatile metabolic product derived from
mercury and its compounds. It is a clastogen, mutagen and a
teratogen. There is an increase in the frequency of chromosome
breaks reported in lymphocytes from Swedish fishermen and their
families exposed to dimethylmercury from ingested fish (63). Methylmercury compounds are mutagenic to rats, drosophila and certain
plant cells, and cause neurological disturbances and death in
human adults and central teratogenic anomalies in newborn infants
as well as extensive teratogenic malformations in human infants (66).

The inhalation of nickel tetracarbonyl by rats induces
metastasizing pulmonary tumors (67). The vapor is probably deposited
as a finely subdivided nickel powder which is then metabolized
into a carcinogenic pathway. The compound can be analyzed by the
wide variety of methods available for nickel. Recently the vapor
has been analyzed by a chemiluminescent procedure (64).

The genotoxic properties of nitrous acid are essentially
those of nitrite, which will be discussed.

Nitrous acid in the atmosphere is probably formed from nitrogen dioxide and is stabilized to some extent by the olefins (10,52).
Nitrous acid is believed to form the major proportion of nitrous
fumes from some sources while nitrogen dioxide in moist air can
be slowly converted to nitrous acid (51). Analysis of these two
chemicals is based on the solubility of nitrous acid and the in-
solubility of NO_2 in water. It was estimated that nitrous acid
is about 35 times as volatile as water at 10^0C. Utilizing the

differential absorption method it was found that over a 3 month period atmospheric concentrations of NO_2 and HONO at one location in southern England varied between 7.5 and 39ug/m^3 and 0.8 and 21ug/m^3, respectively (52). The highest proportion of HONO was found in air which had passed over the industrial regions of W. Europe.

The evidence for the presence of substantial amounts of nitrous acid in polluted atmospheres is meager. Confirmation by other means is needed before the analytical methodology for the compound and the problem its presence would cause in the analysis of NO_x could be taken seriously. If nitrous acid is present in polluted atmospheres, then all methods for NO_2 would have to be modified.

A differential impinger sampling method was used to collect nitrous acid and nitrogen dioxide. The nitrous acid diazotized sulfanilamide to form a diazonium salt which was then coupled with N-1-naphthylethylenediamine to form an azo dye absorbing at λ550 nm. Nitrous acid has also been produced in the free atmosphere of a reaction chamber (4).

Inorganic anions of various types have been shown to have genotoxic properties. They would be found in particulate samples and would be analyzed after filter collection by the appropriate procedure.

Arsenate and arsenite would be collected in the particulate fraction and could be analyzed by the procedures used for arsenious oxide. Arsenate has been reported to be carcinogenic, in that increased rates of respiratory and skin cancer have been reported for factory workers in contact with it (30,33). Attempts to induce cancer in animals treated with arsenic have ended in failure. This data would seem to indicate that arsenic is either a cocarcinogen or other cofactors or carcinogens are involved in the carcinogenicity of arsenic. Sodium arsenate is mutagenic in a bacterial system (53) and teratogenic with Swiss-Webster mice (31).

Indications are that chromate is carcinogenic, clastogenic and mutagenic. Workers in the chromate-producing industries are exposed to an excessive risk of lung cancer (34). Potassium dichromate in chronic and acute poisoning evokes a significant increase in the frequency of cells with chromosome aberrations in the bone marrow of rats (9). Chromates are also mutagenic in bacteria (70). In a screening test for metal mutagens or carcinogens, chromate has been found to give a positive result (62). The test indicates that if a metal is a carcinogen and/or a mutagen, it interferes with ability of DNA molecules to correctly select the bases that they need for proper replication.

Chromate can be determined through atomic absorption (38) or through colorimetry with o-dianisidine (37) or s-diphenylcarbazide (1) as the reagent.

Nitrate is considered to be a precarcinogen (16) and a pre-mutagen (36) in the sense that it could be one of the precursor molecules necessary for nitrosamine formation. The presence of nitrate in cigarette tobacco results in the presence of frameshift mutagens in cigarette smoke condensate.

Although there are a very large number of methods for the determination of nitrate, the ion chromatographic procedure (49) appears to be the method of choice.

Nitrite and many of the other water-soluble anions found in polluted airborne particles could also be analyzed by ion chromatography. Nitrite could also be determined colorimetrically or fluorimetrically. One method of this type would be treatment of $1,N^6$-ethenoadenosine with alkali, followed by the nitrosation reaction to form the fluorescent $1,N^6$-etheno-2-aza-adenosine (69).

Nitrite can be a mutagen, clastogen, precarcinogen or a pre-mutagen. The mutagenic action of nitrite has been demonstrated in many microorganisms, e.g. Salmonella typhimurium (54) and in FM3A cells, a C3H mouse mammary carcinoma cell line (40). Severe chromosomal aberrations were produced in the latter cells. It has also been shown that concurrent administration of nitrite and a secondary amine can induce lung adenomas in Swiss mice (27). Mice treated with nitrite and a primary or secondary amine or an alkylurea show a definite increase in mutation frequency (17).

Atmospheric selenate and selenite can be determined through atomic absorption, gas chromatography (through derivative for-mation of Se (IV) with 5-nitro-o-phenylenediamine to form the thermally stable and volatile 5-nitro-1,3-benzoselenadiazole, which is extracted into toluene, separated by GC and, finally, determined with a microwave emission spectrometric detector), neutron activation and spectrophotofluorimetry (conversion of selenium and its compounds to the dioxide which is reacted with 2,3-diaminonaphthalene to form the selenadiazole, F370/517).

The genotoxic properties of selenium are highly controversial (22,23,59). In dispute are the anticarcinogenic, carcinogenic, clastogenic and teratogenic properties of this essential element.

The preferred method for the determination of atmospheric sulfate is ion chromatography (49). Sulfite could also be deter-mined in this fashion. The main problem in the determination of atmospheric sulfate, sulfite, nitrate and nitrite is artifact formation.

Colorimetric methods can be used for the determination of sulfate. Although they have interferences, automated methods are available. X-ray fluorescence can be used for the determination of sulfate, but assumption has to be made that all sulfur is sulfate. Other problems are the use of highly trained personnel, complicated instrumentation, liquid nitrogen and the continual use of calibration and correction factors. However, automated analysis can be used and the filter can be analyzed directly.

Since sulfate is a good leaving group in the alkyl sulfates, these compounds are fairly powerful alkylating agents. Thus, the ultimate carcinogen formed from N-2-fluorenylacetamide in rat liver appears to be the sulfuric acid ester of N-hydroxy-N-2-fluorenylacetamide. This follows from the report that the hepato-carcinogenicity of N-2-fluorenylacetamide is inhibited when the sulfate pool is depleted (72). These results appear to indicate that the sulfate group can play a cocarcinogenic role under some conditions. Potentially dangerous would be the situation where this group could readily leave an organic compound under physio-logical conditions leaving behind an electrophilic cation adjacent to appropriate DNA nucleophilic groupings.

If sulfate conjugation is necessary for the activation of N-hydroxy-N-2-fluorenylacetamide and other carcinogenic aromatic amines, then knowledge about (i) the concentrations of sulfate in polluted atmospheres and in the environment and (ii) the meaningful (from a carcinogenesis viewpoint) portals of entry for sulfate becomes vital for the prevention and control of this problem.

Sulfite is a mutagen which exerts its effect by directly modifying DNA bases in such a way that base-pair errors arise at subsequent cell divisions (48). Some reasons have been presented for considering the toxicity and mutagenicity of atmospheric sulfur dioxide and sulfite to human beings as unproven (3).

Quite a few metallic compounds have been found to have some type of genotoxic activity, Table I. Beryllium compounds have been reported to be an etiological factor in the development of some cases of industrial lung cancer (65). Compounds of beryllium, cadmium, chromium (II), cobalt, copper, lead, manganese, and nickel are believed to be carcinogens and/or mutagens on the basis of a screening test wherein they were found to interfere with the ability of molecules of DNA to correctly select the bases that they need for proper replication (62).

Table 1. Analysis and genotoxic properties
of particulate metallic compounds

Metal	Analysis[a]	Genotoxicity[b]
Beryllium	ES, SPF	C, M
Cadmium	AA, ES, FAAS, NA	C, Cl, M, T
Chromium (II)	NA	M
Cobalt	NA	C, M
Copper	NA	M
Iron	AA, NA, XRF	C, CC
Lead	AA, ES, FAAS	C, M
Magnesium	ES, NA	CC
Manganese	AA, NA, XRF	C, M
Mercury	AA, AE, NA	CC
Molybdenum	AA, NA	C, M
Nickel	AA, NA	C, M
Silver	AA, NA	C
Titanium	AA, NA, SSMS, XRF	C, CC
Vanadium	AA, NA, SSMS, XRF	C

[a] AA = Atomic absorption, AE = Atomic emission,
ES = Emission spectroscopy, FAAS = Flameless
atomic absorption spectroscopy, NA = Neutron
activation, SPF = Spectrophotofluorimetry,
SSMS = Spark source mass spectroscopy,
XRF = X-ray fluorescence.

[b] C = Carcinogen, CC = Cocarcinogen,
Cl = Clastogen, M = Mutagen, T = Teratogen

Smelter workers with at least one year of exposure to cadmium oxide dust and fumes have a higher risk of prostatic cancer than the general population (13).

The supposed carcinogeniticy of iron stems from the report that the intramuscular injection of iron-dextran resulted in the induction of sarcomata in rabbits (29).

In a review of epidemiological studies on nickel, IARC concluded that the studies demonstrated conclusively an excess risk of cancer of the nasal cavity and lung in workers in nickel refineries, which is most likely due to some form(s) of nickel (35). However, the affected individuals are in contact with many other chemicals so it would seem likely that other chemicals are also involved in this process of carcinogenesis.

ORGANIC POLLUTANTS

Although the search for organic compounds in outdoor and indoor atmospheres has been of low priority through the years, many hundreds of these compounds have been discovered. The composition of the atmosphere and its genotoxic components have been discussed to some extent (55-57).

Some of the organic genotoxicants which have been found in polluted atmospheres are shown in Table II. Some of the main methods of analysis are shown. There are many other methods of analysis available for these compounds.

In our investigations of the ambient atmosphere we have found approximately 50 carcinogens and a total of about 200 genotoxic materials. More thorough investigations would disclose many more compounds of this type.

The halogenated organic compounds are a ubiquitous family present in polluted atmospheres. Many of these compounds are produced industrially in huge amounts. We have found approximately 70 halogenated aliphatic compounds and 60 halogenated ring compounds in the atmospheric samples which we have collected in highly polluted areas in California, Louisiana, New Jersey, Texas, West Virginia, etc. Many of these compounds have never been tested for carcinogenic or mutagenic activity.

Fairly extensive epidemiological evidence has been gathered which indicates that some of the compounds in Table II are human carcinogens. These include vinyl chloride (angiosarcoma), bis-chloromethyl ether (lung cancer), benzene (leukemia), asbestos (mesothelioma), cigarette smoke (lung cancer), coal tar pitch fumes (bladder, lung and skin cancer), and 2-naphthylamine (bladder cancer).

Other pollutants are highly suspect. Dimethyl sulfate is carcinogenic to rats when administered by inhalation and has been associated with cancer of the lung in humans exposed to this industrial pollutant (20). Vinyl bromide is very closely similar structurally to vinyl chloride so this chemical should be considered as having definite potential as a human carcinogen. Chloroprene, or 2-chlorobutadiene, is used in the manufacture of neoprene and is postulated as causing skin and lung cancer in workers in contact with the material (45).

A look at the relative amounts of organic genotoxicants in the particulate and vapor fractions of air would be educational. Since this field has had such a low relative priority, not much data is available to make such a meaningful comparison. However, a preliminary look at the relative weights of the total particulates as collected by a high volume sampler and the organic vapors as collected by a Tenax cartridge method in a few areas indicates that a reordering of priorities may be necessary. The data in Table III show that the collected organic vapors were present in 4 orders of magnitude larger amounts than the total airborne particulate material. In addition, some of the carcinogens in the vapor fraction are present in 5 to 6 orders of magnitude larger amounts than are the carcinogens in the particulate fraction. These aspects need much more thorough investigation.

Another aspect that needs a much more thorough examination is the composition of the vapors in environmental mixtures which have been shown to be carcinogenic to man. Thus, particulates in cigarette smoke and in coke oven effluents have been blamed for the human cancers, but the vapors could be the dangerous factor or could contain some of the dangerous factors.

In our spot checking for atmospheric organic vapors we have found that every polluted area which we examined had enough carcinogens at high concentrations in the atmosphere to indicate that a much more thorough investigation was needed. This applies to established industrial areas. Such a thorough investigation would be desirable for areas where new giant industrial complexes are or will be built.

The preliminary data in Table IV emphasizes the necessity for thorough studies. The key carcinogen in this area which needed study was believed to be dimethylnitrosamine, but as the Table indicates other pollutants may be present in high concentrations which may be just as important or even more important in terms of genotoxic activity. The last two compounds in the Table have been selected for long term carcinogenicity studies by the National Cancer Institute's Bioassay Operations Program. The data in this Table and other data that we have obtained indicates unequivocally

Table II. Some genotoxic organic air pollutants

Pollutant	Analysis[a]	Genotoxicity[b]
	Gases	
	C_1	
Cyanogen chloride + 14° [c]	SP	PC, PM
Formaldehyde - 20°	SP	C?, Cl, M
Methylamine - 6°	GC, 1C	PM
Phosgene + 8°	SP	?
	C_2	
Azomethane + 20	GC	C
Diazomethane - 23°	SP, GC	C, M
Dimethylamine + 6°	GC, 1C	PC, PM
Ethylene - 102°	GC	PM
Ethylene oxide + 11°	GC-MS	C?, M
Vinyl chloride - 14°	GC-MS	C, Cl, M
	C_3	
Propylene - 48°	GC	PC, PM
Trimethylamine + 4°	GC, 1C	PC, PM
	C_4	
1-Butene - 7°	GC	PC, PM
Cis-2-Butene + 4°	GC	PC, PM
trans-2-Butene + 0.4°	GC	PC, PM
Perfluoroisobutene - 7°	GC, MS	?

Table II (Cont'd)

Pollutant	Analysis	Genotoxicity
	Vapors	
	C_0	
Hydrazine + 114^0	CL, GC, SPF	AC, C, M, T
	C_1	
Carbon tetrachloride + 77^0	GC-MS	C, M
Chloroform + 62^0	GC-MS	C, M
Methyl hydrazine + 87^0	CL, GC, SP	C, T
Methyl iodide + 43^0	GC-MS	C
	C_2	
Acetaldehyde + 21^0	GC, SP	M
Aziridine + 57^0	GC, HPLC, SP	C, Cl, M
Bis-chloromethyl ether + 104^0	GC-MS, HPLC	C
1-Bromo-2-chloroethane + 107^0	GC-MS	C
Chloromethyl methyl ether + 59^0	GC-MS, HPLC	C
1,1-Dimethylhydrazine + 63^0	CL, GC	C, T
1,2-Dimethylhydrazine + 81^0	GC	C, T
Dimethyl sulfate + 188^0d	GC-MS	C, Cl, M
Ethylamine + 16.5^0	GC, IC	PM
Ethylene bromide + 132^0	GC-MS	C, M, T
Ethylene chloride + 84^0	GC-MS	C, M
Ethylene sulfite	GC-MS	M
Ethylene sulfate	GC-MS	C
Ethyl iodide + 72^0	GC-MS	C, M
Trichloroethylene + 87^0	GC-MS	C, M
Vinyl bromide + 16^0	GC-MS	C, M
Vinylidene chloride + 32^0	GC-MS	C, M

Table II (Cont'd)

Pollutant	Analysis	Genotoxicity
C_3		
Acrolein + 53^0	SP	M
1,2-Dibromo-3-chloropropane + 196^0	GC-MS	C, M
Dimethylcarbamoyl chloride	SP	C
Dimethylformamide + 153^0	GC-MS	PC, PM
Methyl ethylamine + 35^0	GC, IC	PC, PM
Propylene oxide + 35^0	GC-MS	C, M
C_4		
1-Chlorobutadiene + 68^0	GC-MS	C, M
2-Chlorobutadiene + 59^0	GC-MS	C, M, T
Diethylamine + 56^0	GC, IC	PC, PM
Piperazine + 146^0	GC-MS	PC, PM
C_5		
2-Methyl-1-butene + 39^0	GC-MS	PM
2-Methyl-2-butene + 38^0	GC-MS	PM
C_6		
Aniline + 184^0	GC-MS	PC
Benzene + 80^0	GC-MS	C, Cl
4-Methyl-1-pentene + 54^0	GC-MS	PM
Phenol + 182^0	GC, SP	CC
Particulates		
Alkanes[d] $>300^0$	GC-MS, TLC-SPF	CC
Alkenes[e] $>300^0$	GC-MS	PC, PM
Aza arenes[f] >300	LC-SP	C, M
Imino arenes[g] >350	TLC-SPF	C, M
PAH[h] $>380^0$	GC-MS	C, M

Table II (Cont'd)

Miscellaneous

Pollutant	Analysis	Genotoxicity
Asbestos	EM, LM	C, CC
Cigarette smoke tar	GC-MS	C, M
Coal-tar-pitch volatiles	GC-MS	C, M
2-Naphthylamine + 306°	GC-MS	C, M, PC

[a]CL = Chemiluminescence, EM = Electron microscopy, GC = Gas chromatography, HPLC = High pressure liquid chromatography, IC = Ion chromatography, LC = Liquid chromatography, LM = Light microscopy, MS = Mass spectrometry, SP = Spectrophotometry, SPF = Spectrophotofluorimetry, and TLC = Thin-layer chromatography.

[b]AC = Anticarcinogen, C = Carcinogen, CC = Cocarcinogen, Cl = Clastogen, M = Mutagen, PC = Precarcinogen, PM = Premutagen, T = Teratogen.

[c]Temperatures after name of each pollutant are boiling points in degrees Centigrade.

[d]Examples are aliphatic hydrocarbons ranging from C28 through eicosane, b.pt. 343°, which is found in the particulates fraction, down through hydrocarbons like n-dodecane, b.pt. 216°, which would be found in the vapor fraction. These compounds are cocarcinogenic in that they can convert a noncarcinogen (essentially an incomplete carcinogen) to a carcinogen or can convert a weak carcinogen to a strong one (32).

[e]Example is 1-hexadecene which can be transformed to the carcinogenic 1,2-epoxyhexadecane by hepatic microsomes (71) and can probably undergo the same transformation in polluted atmospheres.

[f]Example is the carcinogen, dibenz(a,h)acridine, b.pt. ~500° (58).

[g]Example is the carcinogen, 11H-benzo(a)carbazole, b.pt. 450° (8).

[h]Example is benzo(a)pyrene, b.pt. 496°.

Table III. Atmospheric organic vapors[a]

Pollutant	ng/m^3	Carc. Activity
CH_2ClCH_2Br	25,000	+
CCl_3CH_3	25,000	?
$CHCl_3$	74,000	+
$CHCl=CCl_2$	93,000	+
CH_2Cl_2	375,000	?
Benzene	1,550,000	+
Toluene	2,600,000	?

and many more[b].

[a]Collected on a Tenax GC cartridge 0.25 mile downwind from KIN-
BUC chemical dump near Edison, N. J. on March 25, 1976.

[b]Identified organic vapors (total) = 138,000 ug/m^3. Estimated
total organic vapors = ⁓400,000 ug/m^3. This compares with an
average of about 50 ug/m^3 of airborne particulates in the New Jersey
urban areas. About 5 ug/m^3 of this would be the benzene-soluble
fraction and about 0.002 - 0.005 ug/m^3 would be BaP in the particulate fraction. These aspects need much more thorough
investigation.

Table IV.   Atmospheric Carcinogens - The Unknown Factor[a]

Pollutant[b]	ng/m^3	Carc. Activity
Dimethylnitrosamine	36,000	+
2,3-Dichlorobutane	270,000	?
2-Methyl-3-chloro-1-propene	400,000	?
1-Chloro-2-methyl-1-propene	670,000	?

[a]Samples collected on a Tenax GC cartridge in Baltimore, Md. in February, 1976 and then analyzed by GC-MS-COMP.

[b]Highest values obtained in this study.

that monopollutant monitoring for a "key" carcinogen is unsatis- factory for the delineation of the carcinogenic risks in an area. In addition, the attempts to correlate the extensive data obtained from expensive monopollutant monitoring systems with health effects have been failures.

When we mull over reports that we have many real time mono- pollutant monitors (over 100 different monitors for sulfur dioxide) and very little data on the numerous atmospheric genotoxicants, we certainly need to improve our pollutant relevance, our ordering of pollutant priorities.

The crucial fact is that the generation of numbers which are not needed and not used (or used in hopeless correlations) raises the bill without contributing significantly to the value of the service. For most situations, would not a few more sophisticated polypollutant assays in the highly polluted areas of the country be better than grinding out vast quantities of monopollutant numbers and air quality indices that ignore the toxic chemicals in the air?

There is a particularly cruel relationship between man-made
evolution and waste. The evolution of methodology, machine,
instrument and scientist is proceeding at a rapid rate so that we
seem to be swimming in (and sometimes against) an inexorably rush-
ing stream of constantly changing methodologies, instruments,
chemicals and specialists. And yet the embarassing question arises
whether we're contributing too many tombstones to an enormous
cemetary of obsolescent literature, methods, instruments, data and
specialists.

Since the simple domino theory of carcinogen   cancer
does not apply for most people, we know that many other types of
genetic and environmental factors are involved. To understand and
prevent this problem we must know the genotoxic composition of our
atmospheres and our other environments. To do this we will have
to depend on computerized polypollutant methods of assay and information on the areas of the country with a high rate of chemical
production, use, pollution and/or cancer.

There is a financial limitation to our analytical capabilities
in collecting the numbers necessary for an understanding of human
environmental carcinogenesis. This lack of meaningful numbers is
the insurmountable problem that epidemiologists face in attempting
to correlate the concentrations of atmospheric pollutants with
health problems. Consider that over 99% of the polluted areas in
the country have never been examined for organic pollutants and of
those examined, over 99% of the organic compounds present have
never been identified.

Thus, to determine the composition of highly polluted indoor
and outdoor atmospheres, the GC-MS-COMP method is the most for-
midable for polypollutant assay. Before this is done, the pollutants must be collected. The sampling method will vary dependent on
whether the pollutant is a gas, a non-polar vapor, a polar vapor,
a high boiling vapor or a particle of some particular size.

For the analysis of gases, air samples can be collected in a
glass sampling bulb followed by separation on Chromosorb 102 with
a cryogenic gas chromatograph (25). About 12 to 18 inorganic and
organic gases can be determined. However, a complementary method
is needed that can collect, separate and analyze approximately 100
organic gases. Examples of compounds which would be collected in
this fashion would include vinyl chloride, perfluoroisobutene,
phosgene, dichlorodifluoromethane, chlorodifluoromethane, methyl-
amines, etc. A cartridge method would be preferred here.

The window presently available looks out upon about 125
organic vapors boiling between approximately 15^0 to about 215^0.
The cartridge for this group utilizes Tenax GC as the collecting

agent. What is needed is to subdivide this group into the hydrocarbons and the polar compound by collecting on two different
cartridges, each specific for its group.

Charcoal has been used for C_7 to C_{20} compounds (28). Reproducibility and routine usage might be difficult with a carbon ad-
sorbent. Another difficulty would be the collection of nitrogen
oxides and ozone. These pollutants could oxidize and/or nitrosate
some of the other material collected in the cartridge.

A fourth type of cartridge is needed for the higher boiling
organic vapors.

It is felt that with these four cartridges, capillary gas
chromatography, mass spectrometry and the INCOS system, approximately
five hundred to eight hundred organic gases and vapors could be
determined eventually in highly polluted indoor and outdoor atmospheres.

It is frustrating and sad to be only able to collect organic
vapor samples batchwise for 2 days near Baton Rouge in one area containing chemical companies and petroleum refineries and to find 12
dead cows in a field nearby and air difficult to breathe and then
to collect in another area 30 miles away crosstown near a chemical
waste incinerator and to find 6 more dead cows in an adjacent field
and an acknowledgement of an additional 6 more dead cows in this
field previously. The factors that are disturbing are (i) the low
priority of this type of problem so that one can't go back
immediately and do a thorough study of this contaminated area
where the lung cancer rate is so high, (ii) the restricted range of
the Tenax cartridge for just some of the air pollutants, (iii) the
possible toxic effects of the air on the people in this and nearby
areas, (iv) the possibly contaminated milk if milk cows graze in the
area, and (v) the possibly contaminated beef.

Let us briefly discuss organic particulates. There are a wide
variety of sampling techniques that can and are being used to collect
airborne particles. Four types are shown in Table V to demonstrate
the range in the amount of material each one would collect in an
average urban area. Of course the amounts shown represent the total
24 hr. sample, except for the personal sampler which is for an 8
hr. run. In most cases adequate amounts of sample are obtained only
with the 2 high volume samplers.

Another problem is the extraction of organic material from the
particulates. Usually benzene (14) or cyclohexane (42, 43) is used
with Soxhlet extraction for 4 to 8 hours. With preliminary cleanup

Table V. Atmospheric samplers[a] - 24 hour operation

Type	M^3	mg Particulates	mg Organic[b]	ug $SO_4^=$	ng BaP	ng BcACR
Trichotomous[c]	50,000	6000	500	500,000	100,000	5000
HI-VOL[d]	2,000	236	20	20,000	4,000	200
Dichotomous[e]	20	2	0.2	200	40	~2
Personal[f]	<1.	<0.1	<0.01	<10	<2	<0.001

[a] All values estimated from the HI-VOL values

[b] Essentially the material extracted by benzene

[c] Cutoffs at 1.7 and 3.5 um. Sampler developed by Battelle of Columbus, Ohio

[d] EPA recommended sampler

[e] Cutoff at 2 um

[f] NIOSH developed sampler. All values are for 8 hr operation

and separation of the benzene extract (5.8% of the particulates)
into various cuts - 22 alkanes, 36 arenes, 2 benzocarbazoles, 11
neutral oxygenated arenes, 19 aliphatic acids, 13 aromatic acids,
6 phenols and 45 aza arenes have been determined (14). With
cyclohexane as an extractant and GC-MS-COMP for separation-analysis
approximately 100 arenes can be determined in air samples (42)
or in coke oven emissions (43). Most of these arenes are only
identified as to ring structure. The problems in this type of
characterization have been discussed (56).

Another problem which we noticed over 15 years ago is the in-
solubility of most of the particulate organic compounds in benzene
or cyclohexane. After extraction with hot benzene approximately
4 times as much organic material could be extracted from the
residue with dimethylformamide at room temperature. This has been
shown in recent work also (73).

For the first step in investigation toward development of the
most thorough organic extraction system three extractions would be
necessary e.g. (i) cyclohexane, (ii) dimethylformamide, and (iii)
water. Ultrasonic extraction for 10-15 min at or below room
temperature would be preferred (26). Freeze-drying would then be
used to obtain the residues (73). Following well known fractionation
procedures and GC-MS-COMP of the fractions it should be possible to
determine 700 to 1000 compounds in airborne particulates, especially
with the use of capillary GC columns, HPLC and GC-IR. This type of
analysis would mainly be useful where the organic particulate mass
was very much elevated.

Thus, with the development of the appropriate instrumental
packages and procedures approximately 2000 atmospheric gases,
vapors and particulates could be determined. Eventually, dependent
on the type of pollution the appropriate sampling-analytical procedure could be chosen for a thorough look through the appropriate
portion of the picture window to determine how serious the problem
is.

We would like to emphasize very strongly that a thorough look
at the air pollution profile of an area would indicate possible
chemical disease problems in that area. This would be the first
step of a multidisciplinary investigation of the genotoxic problems
in the area.

Acknowledgement: Much of the organic vapor data discussed
in this paper has been obtained through the brillant investigations
of Dr. E. Pellizzari and his staff at RTI. The opinons expressed in
this paper are those of the authors and not necessarily those of EPA.

REFERENCES

1   ABELL, M.T. and CARLBERG, J.R.: Simple reliable method for determination of airborne hexavalent chromium. Am. Ind. Hyg. Assoc. J. 35 (1974) 229.

2   ADAMS, D.F. et al: Tentative method of analysis for sulfur dioxide content of the atmosphere (colorimetric). Health Lab. Sci. 6 (1969) 228.

3   ALARIE, Y. Rebuttal-Health effects of atmospheric sulfur dioxide and dietary sulfites. Arch. Env. Health 31 (1976) 110.

4   ANONYMOUS: Chem. Eng. News (October 27, 1975) 18.

5   ARRHENIUS, E.: Publication of the Department of Cell Physiology. Wenner-Gren Institute. Stockholm, Sweden, p. 13.

6   AXELROD, H.D., BONELLI, J.E. and LODGE, J.P., Jr.: Fluorescence determination of bromine: application to the measurement of bromine aerosols and airborne particulates. Env. Sci. Technol. 5 (1971) 420.

7   BAUMGARDNER, R.E., CLARK, T.A. and STEVENS, R.K.: Comparison of instrumental methods to measure NO_2. Env. Sci. Technol. 9 (1975) 67.

8   BENDER, D.F., SAWICKI, E. and WILSON, R.M., Jr.: Characterization of carbazole and polynuclear carbazoles in urban air and in air polluted by coal tar pitch fumes by thin-layer chromatography and spectrophotofluorometry. Int. J. Air Water Poll. 8 (1964) 633.

9   BIGALIEV, A.B., ELEMESOVA, M.S. and BIGALIEVA, R.K.: Chromosome aberrations induced by chromium compounds in somatic cells of mammals. Tsitol. Genet. 10 (1976) 222.

10 BOURBON, P., ALARY, J., ESCLASSAN, J.F. and ALENGRIN, F.: On the determination of NO_2 and NO_2H in the atmosphere. Pollution Atm. 16 (1974) 39.

11 BRAMAN, R.S. and JOHNSON, D.L.: Selective absorption tubes and emission techniques for determination of ambient forms of mercury in air. Env. Sci. Technol. 8 (1974) 996.

12 BRETSCHNEIDER, K. and MATZ, J.: Nitrosamines in the atmospheric air and in the air at the places of employment. Arch. Geschwulstforsch. 42 (1973) 36.

13 BUELL, G.: Some biochemical aspects of cadmium toxicology. J. Occup. Med. 17 (1975) 189.

14 CAUTREELS, W. and CAUWENBERGHE, K.V.: Determination of organic compounds in airborne particulate matter by gas chromatography - mass spectrometry. Atm. Env. 10 (1976) 447.

15 CHOW, C.K. and TAPPEL, A.L: An enzymatic protective mechanism against lipid peroxidation damage to lungs of exposed rats. Lipids 7 (1972) 518.

16 CORREA, P., HAENSZEL, W., CUELLO, C., TANNENBAUM, S. and ARCHER, M.: A model for gastric cancer epidemiology. Lancet 11 (1975) 58.

17 COUCH, D.B. and FRIEDMAN, M.A.: Interactive mutagenicity of sodium nitrite, dimethylamine, methylurea and ethylurea. Mutation Res. 31 (1975) 109.

18 CRIDER W.L.: Hydrogen flame emission spectrophotometry in monitoring air for sulfur dioxide and sulfuric acid aerosol. Anal. Chem. 37 (1965) 1770.

19 DRUCKREY, H., HAMPERL, H. and SCHMAHL, D.: Carcinogenic action of metallic mercury after intraperitoneal administration to rats. Krebsforsch 61 (1957) 511.

20 DRUCKREY, H., PREUSSMAN, R., NASHED, N. and IVANKOVIC, S.: Carcinogenic alkylating substances. I. Dimethyl sulfate-carcinogenic effect on rats and probable cause of occupational cancer. Z. Krebsforsch 68 (1966) 103.

21 ELKINS, H.B.: The chemistry of industrial toxicology. 2nd ed., Wiley and Sons, Inc. New York (1958) 305.

22 EXON, J.H., KOLLER, L.D. and ELLIOTT, S.C.: Effect of dietary selenium on tumor induction by an oncogenic virus. Clin. Toxicol. 9 (1976) 273.

23 FROST, D.V.: The two faces of selenium - Can selenophobia be cured? CRC Critical Reviews in Toxicology (October, 1972) 467.

24 GABBAY, J., DAVIDSON, M. and DONAGI, A.E.: Spectrophotometric determination of free chlorine in air. Analyst 101 (1976) 128.

25   GIANNOVARIO, J.A., GROB, R.L. and RULON, P.W.: Analysis of
     trace pollutants in the air by means of cryogenic gas
     chromatography. J. Chromat. 121 (1976) 285.

26   GOLDEN, C. and SAWICKI, E.: Ultrasonic extraction of total
     particulate aromatic hydrocarbons (TpAH) from airborne parti-
     cles at room temperature. Int. J. Environ. Anal. Chem. 4
     (1975) 9.

27   GREENBLATT, M., MIRVISH, S. and SO, B.T.: Nitrosamine studies:
     induction of lung adenomas by concurrent administration of
     sodium nitrite and secondary amines in Swiss mice. J. Nat.
     Cancer Inst. 46 (1971) 1029.

28   GROB, K. and GROB, G.: Gas-liquid chromatographic-mass
     spectrometric investigation of C_6-C_{20} organic compounds in an
     urban atmosphere. J. Chromatog. 62 (1971) 1.

29   HADDOW, A., ROE, F.J.C. and MITCHLEY, B.C.V.: Induction of
     sarcomata in rabbits by intramuscular injection of iron-
     dextran. Brit. Med. J. i (1964) 1593.

30   HILL, A.B. and FANNING, E.L.: Cancer from inorganic arsenic
     compounds. I. Mortality experience in the factory. Brit. J.
     Med. 5 (1948) 2.

31   HOOD, R.D. and BISHOP, S.L.: Teratogenic effects of sodium
     arsenate in mice. Arch. Env. Health 24 (1972) 62.

32   HORTON, A.W., ESHLEMAN, D.N., SCHUFF, A.R. and PERMAN, W.H.:
     Correlation of cocarcinogenic activity among n-alkanes with
     their physical effects on phospholipid micells. J. Nat. Cancer
     Inst. 56 (1976) 387.

33   IARC Monographs on the Evaluation of Carcinogenic Risk of
     Chemicals to Man: Some inorganic and organometallic compounds.
     International Agency for Research on Cancer. Lyon 2 (1973)
     48-73.

34   IARC Monograph on the Evaluation of Carcinogenic Risk of
     Chemicals to Man: Some inorganic and organometallic compounds.
     International Agency for Research on Cancer. Lyon 2 (1973)
     100-125.

35   IARC Monography on the Evaluation of Carcinogenic Risk of
     Chemicals to Man: International Agency for Research on Cancer.
     Lyon 11 (1976) 75-112.

36   KIER, L.D., YAMASAKI, E. and AMES, B.N.: Detection of
     mutagenic activity in cigarette smoke condensates. Proc.
     Nat. Acad. Sci. USA 71 (1974) 4159.

37   KNEEBONE, B.M. and FREISER, H.: Determination of chromium
     (IV) in industrial atmospheres by a catalytic method. Anal.
     Chem. 47 (1975) 595.

38   KNEIP, T.J., et al: Tentative method of analysis for chromium
     content of atmospheric particulate matter by atomic absorption
     spectroscopy. Health Lab. Sci. 10 (1973) 357.

39   KNEIP, T.J. et al: Tentative method of analysis for elemental
     mercury in the working environment by collection on silver
     wool and atomic absorption spectroscopy. Health Lab. Sci. 12
     (1975) 158.

40   KODAMA, F., UMEDA, M. and TSUTSUI, T.: Mutagenic effect of
     sodium nitrite on cultured mouse cells. Mutation Res. 40
     (1976) 119.

41   KUSCHNER, M.: The causes of lung cancer. Am. Rev. Resp. Dis.
     98 (1968) 573.

42   LAO, R.C., THOMAS, R.S., OJA, H. and DUBOIS, L.: Application
     of a gas chromatograph-mass spectrometer-data processor
     combination to the analysis of the polycyclic aromatic hydro-
     carbon content of airborne pollutants. Anal. Chem. 45 (1973)
     908.

43   LAO, R.C., THOMAS, R.S. and MONKMAN, J.L.: Computerized gas
     chromatographic-mass spectrometric analysis of polycyclic
     aromatic hydrocarbons in environmental samples. J. Chromatog.
     112 (1975) 681.

44   LEE, A.M. and FRAUMENI, J.F., Jr.: Arsenic and respiratory
     cancer in man. An occupational study. J. Nat. Cancer Inst.
     42 (1969) 1045.

45   LLOYD, J.W., DECOUFLE, P. and MOORE, R.M., Jr.: Background
     information on chloroprene. J. Occup. Med. 17 (1975) 263.

46   MARGESON, J.H., BEARD, M.E. and SUGGS, J.C.: Evaluation of
     the sodium arsenite method for measurement of NO_2 in ambient
     air. J. Air Poll Control Assoc. (1977) in press.

47   MCKEE, H.C.: Collaborative testing of methods to measure air
     pollutants.   III. The chemiluminescence method for ozone:
     Determination of precision. J. Air Poll. Control Assoc. $\underline{26}$
     (1976) 124.

48   MUKAI, F., HAWRYLUK, I. and SHAPIRO, R.: The mutagenic
     specificity of sodium bisulfite. Biochem. Biophys. Res.
     Commun. $\underline{39}$ (1970) 983.

49   MULIK, J., PUCKETT, R., WILLIAMS, D. and SAWICKI, E.: Ion
     chromatographic analysis of sulfate and nitrate in ambient
     aerosol. Anal. Letters $\underline{9}$ (1976) 653.

50   MULIK, J., FUERST, R., GUYER, M., MEEKER, J. and SAWICKI, E.:
     Development and optimization of a twenty-four hour manual
     method for collection and colorimetric analysis of atmospheric
     NO_2. Env. Anal. Chem. $\underline{3}$ (1974) 333.

51   NASH, T.: Chemical states of nitrogen dioxide at low aerial
     concentration. Ann. Occup. Hyg. $\underline{11}$ (1968) 235.

52   NASH, T.: Nitrous acid in the atmosphere and laboratory
     experiments on its photolysis. Tellus $\underline{26}$ (1974) 175.

53   NISIOKA, H.: Mutagenic activities of metal compounds in
     bacteria. Mutat. Res. $\underline{31}$ (1975) 185.

54   Eisenstark, A. and ROSNER, J.L.: Chemically induced
     reversions in the cysC region of salmonella typhimurium.
     Genetics $\underline{49}$ (1964) 343.

55   SAWICKI, E.: The chemical composition and potential genotoxic
     aspects of polluted atmospheres. Presented at the Workshop
     for the "Investigations on the Carcinogenic Burden by Air
     Pollution in Man." Hanover, Germany (October 22-24, 1975).

56   SAWICKI, E.: Analysis of atmospheric carcinogens and their
     cofactors. (Rosenfeld, C. and Davis, E. Eds.: Environmental
     Pollution and Carcinogenic Risks). IARC Scientific Publication
     No. 13, INSERM, Paris (1976) 297-354.

57   SAWICKI, E.: The genotoxic environmental pollutants. Presented
     at the Symposium on Management of Residuals from Synthetic Fuels
     Production. Denver, Colorado (May 25, 1976).

58   SAWICKI, E., MEEKER, J.E. and MORGAN, M.J.: The quantitative
     composition of air pollution source effluents in terms of aza
     heterocyclic compounds and polynuclear aromatic hydrocarbons.
     Int. J. Air Wat. Poll. $\underline{9}$ (1965) 291.

59   SCHROEDER, H.A., FROST, D.V. and BALASSA, J.J.:   Essential
     trace elements in man: selenium.   J. Chron. Dis. 23 (1970)
     227.

60   SCHWARZ, F.P., OKABE, H. and WHITTAKER, J.K.:   Fluorescent
     detection of sulfur dioxide in air at the parts-per-billion
     level.   Anal. Chem. 46 (1974) 1024.

61   SHAMBERGER, R.J., ANDREONE, T.L. and WILLIS, C.E.:   Anti-
     oxidants and cancer.   IV.   Initiating activity of malonaldehyde
     as a carcinogen.   J. Nat. Cancer Inst. 53 (1974) 1771.

62   SIROVER, M.A. and LOEB, L.A.:   Infidelity of DNA synthesis
     in vitro: Screening for potential metal mutagens or
     carcinogens.   Science 194 (1976) 1434.

63   SKERFVING, S., HANSSON, K., MANGS, C., LINDSTEN, J. and
     RYMAN, N.:   Methylmercury-induced chromosome damage in man.
     Env. Res. 7 (1974) 83.

64   STEDMAN, D.H. and TAMMARO, D.A.:   Chemiluminescent measure-
     ment of parts-per-billion levels of nickel carbonyl in air.
     Anal. Letters 9 (1976) 81.

65   STOECKLE, J.D.:   Beryllium disease.   Science 183 (1974) 449.

66   SU, M. AND OKITA, G.T.:   Embryocidal and teratogenic effects
     of methylmercury in mice.   Toxicol. Appl. Pharmacol. 38
     (1976) 207.

67   SUNDERMAN, F.W. and DONNELLY, A.J.:   Studies of nickel
     carcinogenesis: Metastasizing pulmonary tumors in rats
     induced by the inhalation of nickel tetracarbonyl.   Am. J.
     Clin. Path. 46 (1965) 1027.

68   THOMPSON, C.R. et al: Methods of air sampling and analysis.
     American Public Health Association.   Washington, D.C.
     (1972) 282.

69   TSOU, K.C., YIP, K.F., MILLER, E.E. and LO, K.W.:   Synthesis
     of 1,N^6-etheno-2-aza-adenosine (2-aza-e-adenosine):   A
     new cytotoxic fluorescent nucleoside.   Nucleic Acids Res.
     1 (1974) 531.

70   VENITT, S. and LEVY, L.S.:   Mutagenicity of chromates in
     bacteria and its relevance to chromate carcinogenesis.   Nature
     250 (1974) 493.

71   WATABE, T. and YAMADA, N.: The biotransformation of 1-hexa-
     decene to the carcinogenic 1,2-epoxyhexadecane by hepatic
     microsomes. Biochem. Pharmacol. 24 (1975) 1051.

72   WEISBURGER, J.H. and WEISBURGER, E.K.: Biochemical formation
     and pharmacological, toxicological and pathological properties
     of hydroxylamines and hydroxamic acids. Pharmacol. Rev. 25
     (1973) 1.

73   WITTGENSTEIN, E., SAWICKI, E. and ANTONELLI, R.L.: Freeze-
     drying in the recovery of organic material from extracts
     of air particulate matter. Atm. Env. 5 (1971) 801.

Discussion

GEISS - What is the separation scheme you usually apply for organic samples, as for an example the extract of particulate matters.

SAWICKI - Our particulate samples are collected on glass fiber filter with the help of a high volume sampler. The filters are extracted at or below room temperature with methylene chloride-cyclohexane (1:9, V:V) in a Polytron homogenizer for about 5 min. The polar material in the filtrate is removed with Corning controlled-pore glass powder. The PAH in the resultant filtrate are then separated and analyzed by HPLC or GC-MS-COMP. The polar residues are then extracted with other appropriate solvents to obtain the water-soluble organics, acids, bases and neutral oxygenated materials. The aliphatic hydrocarbons are determined in the original non-polar filtrate.

GEISS - Do you use a standard procedure for any sample? How do you do your screening to have an idea what you're looking for? You showed that you have hundreds of compounds potentially. They cannot be obtained from one chromatogram. There must be some standard procedure.

SAWICKI - Yes, what we look for depends on the way we sample. For gaseous substances like methyl bromide, vinyl chloride, vinylidene chloride, vinyl bromide and ethylene oxide, we collect in a cartridge containing Tenax GC. We consider these methods state-of-the-art but imperfect and we hope to expand and improve them.

If we want to examine the organic particles according to size then we use particle sizing samplers followed by our usual extracting procedures.

GEISS - Do you use liquid chromatographic techniques to cut the fraction somewhere?

SAWICKI - Yes, we do that. Although for the polynuclear hydrocarbons we use liquid chromatography to get a pure PAH fraction. We get rid of everything except the PAHs.

ROBILLARD - Two questions. First of all, do you classify the particulate matter on the basis of size before you do your extraction and analysis of the organics?

SAWICKI - In some cases we do. In most cases we don't. We feel that we're more interested in where there is a high atmospheric concentration of particulate matter, and of course the feeling is that, no matter what the size of the particle, if it's a carcinogen you don't want it in your body.

ROBILLARD - The chances of respiring a particle into your lungs is a function of the size of the particle.

SAWICKI - Right, but if the particle is "non-respirable" it can get into the stomach. The literature has described cases of stomach cancer in coal miners where it is believed that the miners breathe in large ferric oxide particles in the mine and at home but also inhale respirable particles containing PAH at home where they utilize coal to heat their homes.

ROBILLARD - Yes, that would come about more on the pollutants settling on food and utensils than actually breathing them, wouldn't it?

SAWICKI - That would be an additional source of contact. But the "non-respirable" ferric oxide particles are postulated as playing a key role here.

ROBILLARD - But does that pass through to the stomach just by inhaling it?

SAWICKI - Yes, but only after the mucociliary system rejects it and the particles are then swallowed. I think too many people are under the misconception that a "non-respirable" particle is unimportant physiologically. But approximately 40 million people suffer from hayfever for which the main aeroallergen is ragweed pollen, a non-respirable particle that is greater than 20 μ in diameter.

ROBILLARD - The second question is - do you have any information that would either confirm or deny the idea that organics in small quantities absorbed on respirable particles may be in the long run more toxic than just breathing the organic vapor?

SAWICKI - Most of these postulated effects refer to gases absorbed on particles. You're talking about SO_2, and mainly SO_2, I would imagine.

ROBILLARD - Just organics in general. Is there any information?

SAWICKI - Very little. A large amount of work has been done, however, with intratracheal instillation of benzo(a)pyrene and small dust particles of beryllium oxide, carbon, titanium dioxide, aluminum oxide, magnesium oxide and ferric oxide. When administered separately to hamsters, tumors were not obtained. When the BaP was given with the carrier dust, respiratory carcinogenesis took place.

Very little is known about what happens to vapors in the respiratory system. For our purposes we define a vapor as a gaseous pollutant which in the neat form in a bottle is a liquid or solid.

While respirable particles can penetrate the respiratory tract and lodge in the regions beyond the tracheo-bronchial tree, i.e. the alveoli, and thus can have long residence times (months), vapors in highly polluted atmospheres can penetrate and condense in the various areas of the respiratory tract in relatively huge amounts continually. What effect this will have on the respiratory defense system only further study and time will eventually tell.

ION-SELECTIVE ELECTRODES

Harvey B. Herman

Department of Chemistry
University of North Carolina at Greensboro
Greensboro, North Carolina 27412

ABSTRACT

The role of ion-selective electrodes in the detection and measurement of environmental pollutants is reviewed. The material is intended for the non-specialist in the area. First the theory of ion-selective electrodes is discussed. Included are sections on calibration curves, standard addition, selectivities and potentiometric titrations. Second a section is included on applications. Included are sections on reference electrodes, instrumentation and specific electrode types. The electrodes discussed are: glass type, silver sulfide type, fluoride, calcium, potassium, gas sensing and enzyme electrodes. They were chosen for their current and potential application to pollution analysis. A section on speculation about future applications of ion-selective electrodes concludes the paper.

INTRODUCTION

Techniques for the measurement of ionic and non-ionic species in water have undergone radical changes in the last generation. The 1960's saw a burst of activity developing instrumental methods using ion-selective (membrane) electrodes. Several thousand papers describing work with the newer membrane electrodes flooded the literature. Indirect applications to non-ionic species were also developed. An early review (1) of this subject strongly recommended the term "ion-selective electrode" to emphasize the point that the response of these electrodes was rarely, if ever, specific to only one ionic species. The recommended term rightly implies that it is still necessary for the analyst to understand his chemical system and potential interferences before using this or any other method.

103

Notwithstanding the caveat above, methods based on ion-selective electrodes have proven themselves in a variety of disciplines. This paper is intended to brief the non-specialist on recent developments in the analysis of pollutants, using ion-selective electrodes. A section on principles is followed by a discussion on experimental methods and specific examples in pollution analysis. As we will see the area is comparatively new and the future has much to offer.

THEORY

An ion-selective electrode can be represented by a model consisting of a membrane separating two solutions each containing a reference electrode (2). In practice one of the solutions contains the substance measured (analyte) and, in which, an external reference electrode is placed. The other solution, usually less visible, contains a reference solution and an internal reference electrode. Variation of the membrane potential is of interest to the anlayst as this can be related to changes in the analyte concentration. The measured difference in potential between the two reference electrodes allows calculation of the membrane potential by the equation:

$$E_{meas} = E_{ERE} - E_{IRE} + E_{mem} + E_j \qquad (1)$$

where E_{meas} is the measured potential, E_{ERE} and E_{IRE} are the potentials of the external and internal reference electrodes, respectively, E_{mem} is the membrane potential and E_j is a junction potential formed at the interface of two different ionic solutions.

If one is dealing with an ideal membrane electrode and reference electrode we can assume: one, that the membrane potential is given only by a function of the logarithm of the activity of the analyte solution and two, E_j will not vary with changes in the ionic composition of the analyte solution. Under these conditions Eq. 1 reduces to the familiar Nernst equation (3), at 25°C for cations,

$$E_{meas} = E^{O'} + 0.0591/n \cdot \log_{10} a_{analyte} \qquad (2)$$

where $a_{analyte}$ is the activity (not concentration) of the ion of interest, n is the cation charge, and E^O is the apparent standard potential of the cell. Frequently, concentration can be substituted for activity. As the activity or concentration of the analyte decreases all real electrodes reach a limiting potential where the Nernst equation is no longer followed. Possible causes will be discussed below but among them are finite membrane solubility or the presence of an interferent ion.

In the region where the Nernst equation is followed the most straight forward way, but not necessarily best way, to use an ion-selective electrode is by the direct measurement of potential (dir-

ect potentiometry). A calibration curve (said to be Nernstian if
the constant in Eq. 2 is $\pm 0.0591/n$ at 25°C) is constructed using
known solutions. Anions give the negative sign. A plot is prepared from the measured potential (y) and the log of activity or concentration (x) (3). The potential measured in the unknown solution
is used to read the unknown activity (concentrations) off the calibration curve.

However, the experimental variable most easily measured is concentration. Since the Nernst equation is a function of activity it
is often necessary for the plot to relate concentration and activity
by:

$$a_M = \gamma_\pm C_M \qquad (3)$$

where a_M is the activity of the ion, C_M is its concentration and
$\gamma_\pm$, the variable relating the two, is called the single ion activity coefficient. When the ionic strength (defined by

$$\mu = 1/2 \cdot \sum_i C_i Z_i^2 \qquad (4)$$

where C_i and Z_i are the concentration and charge of each ion in solution) approaches zero, $\gamma_\pm$ approaches 1 and a_M and C_M become inter-
changeable. When the natural ionic strength is very high or can be
made so by the addition of added salts, the activity coefficient is
relatively constant and concentration can be substituted for activity
without error in the final result. The curve, however is shifted so
that $E^{O'}$ now incorporates another constant, $0.0591/n \cdot \log_{10}\gamma_\pm$.

In other situations where the ionic strength is constantly
changing a considerable deviation from the Nernstian behavior is observed if concentration is used in the calibration curves. In this
case it is necessary to estimate $\gamma_\pm$ and calculate activity from concentration using Eq. 3. The modified Davies equation (4):

$$-\log \gamma_\pm = AZ^2[\mu^{\frac{1}{2}}/(1 + \mu^{\frac{1}{2}}) - 0.3\mu] \qquad (5)$$

where $\gamma_\pm$ is the single ion activity coefficient, μ is the ionic
strength (molar units), Z is the charge on the ion and A = 0.512, has
been found useful for this purpose. Activity coefficients calculated from this equation agree very well with the results of other experiments up to about 0.1M ionic strength.

Linear relationships, such as the Nernst equation, are not
always plotted on graph paper. Frequently the method of least
squares (5) (linear regression) is used to calculate the coefficients
(a and b) in the equation,

$$E_{meas} = a \log_{10}C_M + b \qquad (6)$$

The unknown concentration can be calculated by the coefficients by simple algebra. This method also gives us an estimation of the "goodness of fit" of the correlation with little further calculation. An experienced analyst can use this number to tell if anything is wrong with the procedure or electrodes.

In a "real" situation (as opposed to "ideal" laboratory conditions) such as encountered in pollution analysis, the solutions to be measured are quite different from normal standard solutions. In this case we can no longer be assured that the liquid junction potential will remain constant when switching from standards to the solution to be measured. The method of standard addition (6) has been used to minimize this problem.

The method of standard addition uses in essence a two point calibration curve. The electrode potential is measured in a known volume of the unknown solution. A known volume (usually small) of a standard solution is then added and the electrode potential again measured. The change in potential and concentration is used to calculate the unknown concentration. Thus,

$$C_x = \Delta C / (10^{\Delta E/k} - \frac{V_x + V_s}{V_x}) \tag{7}$$

where C_x is the unknown concentration, ΔC is the change in concentration, ΔE is the change in potential and k is 0.0591/n at 25°C. The change in concentration is given by,

$$\Delta C = C_s V_s / (V_x + V_s) \tag{8}$$

where C_s is the concentration of the standard solution, V_s is the volume of the standard solution and V_x is the volume of the unknown solution. Tables and graphs (7) have been provided as an aid in solving Eq. 7 for C_x.

If k cannot be assumed to be Nernstian a calibration curve needs to be prepared or a sequence of standard additions can be made. In the latter case it is necessary to use a calculator or computer (8) as the equations are non-linear and the multi-point calculation is more complicated. The results of this procedure are potentially more accurate as more points are taken and no assumption is made about Nernstian behavior.

Another perturbation on the ideal model of an ion-selective electrode is the presence of interferent ions. The Nernst equation, Eq. 2, with concentrations instead of activities, is modified to include the effect of other ions on the measured potential (9). Thus, for one cation interferent of the same charge,

$$E_{meas} = E^{0'} + 0.0591/n \cdot \log_{10}(C_{analyte} + K_{I,J}^{Pot} C_{interferent}) \tag{9}$$

where $C_{interferent}$ is the concentration of the interferent ion and $K_{I,J}^{Pot}$ is called the selectivity coefficient for the interferent. The latter term accounts for the observation that the interferent ion may have less (or sometimes more) of an effect on the measured potential than the analyte ion. These terms have been calculated for various ion-selective electrode models and may consist of partition coefficients, mobilities, formation constants and solubility products, depending on the model (10).

The user of ion-selective electrodes cannot be cautioned too strongly, not to ignore the effect of other ions on the measured electrode potential. For example, if the selectivity coefficient of an interferent ion is 0.1 then the concentration ratio of interferent to analyte should be less than 10 in order that the calibration curve be linear and no significant error in calculated concentrations be introduced. When the ratio is closer to 10 but when the correction is still small it is frequently possible to make reliable estimates of the sought-for ion. When the correction becomes too large, it is necessary to make changes in the method such as masking an interferent ion or to seek out a better solution to the problem.

Since the selectivity coefficients are so important, it follows that the analyst should have some idea of their values. In many cases they are known and in others they must be determined by the analyst. Recently (11) the selectivity coefficients have been shown to be to some extent concentration dependent and therefore should be used with caution and only in a qualitative sense. Several years ago (12) the methods in vogue at that time for the measurement of selectivity coefficients were summarized. Two of the methods using separate solutions with divalent cations are:

1. potential measurements at constant activity

$$\log K_{I,J}^{Pot} = (E_2 - E_1)/(2.3\ RT/2F) \qquad (10)$$

    where E_2 and E_1 are the potentials measured in the interferent and analyte solutions, respectively, and

2. Concentration ratio at constant potential

$$K_{I,J}^{Pot} = C_{analyte}/C_{interferent} \qquad (11)$$

These two methods can only give concentration independent selectivity coefficients if their calibration curves are parallel i.e., have the same slope. Otherwise the values calculated should be qualified so that the user will not apply them in non-appropriate situations. A more realistic picture is obtained when the calibration curves are run in the presence of expected levels of the interferent. In this case the selectivity coefficient is calculated from the ratio of activities at the point where deviation from Nernstian behavior is

apparent. It has been recommended (11) that this procedure be followed at several different concentration levels of the interferent.
It is also possible to use a non-linear regression procedure (13)
to evaluate the selectivity coefficients.

The concentration of an analyte can also be measured by potentiometric titration. The nature of ion-selective electrodes makes
them ideally suited for this type of application. The experimental
set-up is the same as in direct potentiometry except that a previously standardized reagent solution is added to the cell and the measured potential is followed throughout the course of the titration. A
potentiometric titration curve for silver ion with sulfide (probably
atypical) exhibits tremendous swings of potential (>1V) at the end-
point (14). Even if each potential measurement is grossly in error
(by ion-selective standards) it is still possible to estimate accurately the concentration of the unknown solution because the end
point volume is so easily and accurately determined. Of course
determinate error still is possible as in the example above when the
end point is observed slightly too soon.

In other more common situations where the end point change is
not as severe, replotting the data using first and second derivatives (15) gives better results. The first derivative, approximately the first difference, is a maximum for most titrations very
close to the equivalent point (assuming no other determinate error).
The second derivative goes through zero close to the equivalence
point and is frequently used to find the end point. Neither method
gives results which coincide exactly with the equivalence point but
are usually very close (16).

Gran (17) has described still another method to locate the end
point. The antilogarithm of the measured potential divided by the
calibration slope, $10^{\pm E_{meas}/k}$, is plotted versus volume of the reagent, and the volume intercept is taken as the end point. Frequently
a clearly defined linear region, away from the end point, is observed
even when no inflection point can be seen on the normal graph. Orion
Research, Inc. (Cambridge, MA 02139) markets "Gran Plot" paper which
can be used to facilitate the calculations needed for this method.
Several kinds of papers are offered with volume corrections built in.
The method of least squares, including volume corrections, has also
been used (18).

APPLICATIONS

Equipment

Reference Electrodes The measurement of ion-selective electrode
potentials requires a second electrode (external reference electrode)
to complete the cell. Ideally this electrode should give stable and
reproducible (or easily calculable) half-cell potentials. An impor-

tant part of the experimental design is the selection of a reference
electrode and its filling solution. Orion (19) frequently cautions
about the problems routinely encountered by faulty reference electrodes. Erratic potential readings are observed when the junction
between the reference electrode filling solution and the solution
to be measured is clogged. Furthermore even if the electrode is
working properly, changes in the liquid junction potential could
cause some measurements to be seriously in error. The standard
addition method (discussed above) was designed to help alleviate
this problem.

The two most common commercially available reference electrodes
are the calomel and silver-silver chloride electrodes. The choice
of either of these two electrodes is probably less important than
the selection of the type of junction and filling solution. The
sleeve type junction has been recommended (20) because of its fast
response and small, stable, junction potential. Filling solutions
are designed to minimize the junction potential and are called equitransferent. The sleeve type junctions have relatively high leak
rates which contribute to their fast response and stability but cannot be used if leakage of a significant amount of the filling solution
is intolerable. A double junction electrode can frequently circumvent
this difficulty. Alternatively, a design using another junction with
a lower leak rate could be used. The cracked bead electrode is particularly notable in this regard. For precise work, the fiber type
junction, which is used on many commercial calomel electrodes, is
not usually recommended.

The liquid junction potential, calculated using a simple model
(21), between a saturated solution of potassium chloride and $10^{-3}\underline{M}$
and $10^{-1}\underline{M}$ potassium chloride solutions is 4.0mV and 1.8mV respectively. This change in potential would be superimposed on the ideal
Nernstian change of a potassium ion-selective electrode/saturated
calomel reference electrode pair when transferred between those two
solutions. It can be calculated that this deviation from Nernstian
behavior would introduce a 9% uncertainty in the concentration of
potassium ion. This uncertainty (calculated by the simple model) can
be almost completely eliminated if an equitransferent filling solution is used.

Instrumentation A pH meter, specific ion meter, digital volt-
meter or potentiometer can be used to measure the membrane potential
of an ion-selective electrode. Instruments are manufactured by a
large number of companies. Whatever the ion meter device chosen, important specifications are precision, accuracy and input impedance.
Precision, in terms of repeatability, ranges from about 10mV to 0.1mV
or 0.1 to 0.0001 pH units. The accuracy, which is affected by input
impedance, calibration, standards, junction potentials, etc., is usually less than the repeatability.

The input impedance of an ion meter should be much larger than the resistance of the membrane electrode being used. Thus if one was using a silver sulfide pressed pellet electrode ($\sim$one megohm resistance) the input impedance of the ion meter selected could be 100 times smaller than if a glass type electrode ($\sim$100 megohm resistance) was being used.

The effect of low input impedance manifests itself as a decrease in the observed electrode slope. As long as this is not a significant fraction of the response it can be tolerated in some cases. Frequently, however, the membrane resistance changes with temperature and humidity, and the effect is not reproducible, causing an intolerable error. In these cases it is advisable to keep the meter/ electrode impedance ratio as high as practical.

The precision and/or accuracy of the ion meter will affect the accuracy of the calculated concentration. For direct potentiometry the relative error in concentration, $\Delta C/C_{analyte}$, can be calculated by

$$\Delta C/C_{analyte} = 2.3n/59.1 \cdot \Delta E_{meas} \qquad (12)$$

where ΔE_{meas} (mV) is the error in the measured potential. Thus an error of 1mV in E_{meas} gives a 3.8% error in concentration for a univalent ion (22). For most work with direct potentiometry then it would be prudent to have a meter with an error of <1mV unless the repeatability of measurements is known to be a good bit higher than that value.

Potentiometric titrations, in many cases, have substantially less error than direct potentiometry. The electrode, in this case, merely serves as an end point indicator and for many titrations the potential requirements are less rigid. For example, the end point potential change was approximately 1V in the titration of silver sulfide with silver nitrate (14). An error of 100mV in determining the end point volume would correspond to about an error of 1% in concentration. If this precision is satisfactory an ion meter with a resolution of 100mV could be used. In this case, we have noted earlier that, the end point is premature (about 2%) and more accurate potential measurements would be useless unless some sort of correction was made.

Electrode Types

Several commercially available electrodes have been found useful for pollution analysis. In this section I have chosen to limit discussion to seven easily available electrodes or electrode types. These and others are offered by companies like Orion Research, Corning and Beckman Instruments, and it is possible to construct many

different kinds oneself. Nevertheless it seems to me that these
electrodes are quite often used in practice and, for the most part,
their advantages and limitations are well understood by practitioners
of the ion-selective electrode art. It would therefore, serve a useful purpose to review their performance for the non-specialist who
might consider using them for pollution analysis. In a concluding
section ideas will be presented about possible future developments
in this area.

   Glass Type Electrodes A membrane potential forms when a thin
pH responsive glass wall separates two solutions which differ in
their hydrogen ion activity. An electrode, made from the glass membrane, responds to the logarithm of the hydrogen ion activity (or as
we have noted, sometimes concentration) over a very wide pH range.
Undoubtedly this is the most widely used ion-selective electrode.
An early design using Corning 015 glass (23) was found useful to
pH 11 or 12 where deviations from Nernstian behavior, alkaline
error, are observed. Later, other glasses were developed which
exhibited a much smaller error in alkaline solutions. An acid error
is also observed in strongly acidic media.

   Since the alkaline error is caused by response to other cations
it became apparent that electrodes sensitive to potassium or sodium
could be fabricated if this response could be accentuated. For
example, if the percentage of Na_2O is decreased and Al_2O_3 added to
the glass an electrode with a more favorable selectivity coefficient
for Na^+ over K^+ is obtained (9). Nernstian behavior to $pNa^+ \approx 5$ is
observed.

   Applications of glass type electrodes are too numerous to mention. They range anywhere from the measurement of the pH of seawater
at up to 6,000m (24) to the determination of the sodium ion content
of the effluent from power stations (25). The selectivity coefficient for potassium over sodium ion is not good enough for these electrodes to measure potassium ion in many practical situations. However,
sometimes correction can be made for the contribution of sodium ion
to the observed potential and the potassium ion concentration calculated by difference.

   Silver Sulfide Based Electrodes Silver sulfide powder subjected
to compression and heat becomes an electrically conducting pellet.
Electrodes have been constructed by sealing the pellet in an inert
tube and making electrical contact through a reference solution and
electrode, or by directly cementing a wire to the pellet. Factors
contributing to the success of this type of electrode are its good
electronic conductivity and its low solubility in water (26).

   The silver sulfide electrode responds to sulfide ion, silver
ion and mercury (II) which forms an insoluble sulfide. Nernstian
slopes (30mV/decade) are observed down to $10^{-7}\underline{M}$ sulfide. The res-

ponse time is considered to be excellent. Applications abound and
range from silver ion in photographic fixers (27) to sulfide in
natural water (3).

It is also possible to construct other sensors using silver
sulfide as the matrix but adding substantial amounts of other sulfides. A mixture of silver sulfide and lead, cadmium or copper sulfide is heat pressed into a conducting pellet. Two conditions important for proper operation of the electrode are that the silver
sulfide has the smaller solubility product and that the exchange
reaction is fast (28). Mercury (II) and silver ion interfere as
expected. Metals that form more insoluble sulfides than the analyte
also interfere.

Fluoride Electrodes A single crystal of lanthanum fluoride
sealed in an inert tube with an internal reference solution was originally shown by Frant and Ross (29) to make an excellent ion-selective electrode. The electrode responded rapidly to fluoride concentrations as low as $10^{-6}M$. Conduction through the membrane was
believed to be by fluoride ion (30).

Non-nernstian behavior is observed at both high and low pH.
At high pH interference is caused either by hydroxide penetration
of the membrane (29) or a reaction of the type (31)

$$LaF_3 + 3OH^- \rightleftarrows La(OH)_3 + 3F^-$$ (13)

The selectivity coefficient for hydroxide ion is strongly concentration dependent. At low pH formation of HF effectively removes fluoride ion from the solution. This can easily be corrected for, if
necessary, if the pH is known.

A large volume of papers utilizing the fluoride electrode has
appeared since the electrode was introduced. This should not be
surprising as the fluoride ion has historically been difficult to
determine. An early review (32) summarized the work up to that time
on measuring fluoride in potable water supplies. The soluble fluoride in rain, snow, fog or aerosols in 10ml samples has been determined by the fluoride electrode at 0.28 ppb level (33). Many other
papers too numerous to mention have appeared. It appears that this
electrode has been the most successful and easiest to use ion-selective electrode.

Calcium and Divalent Liquid Membrane Electrodes It is possible
to construct an ion-selective electrode using a liquid membrane as
the active element. One of the most successful of these types is one
marketed by Orion Research sensitive to calcium ions. They also make
a close relative which they call a divalent cation electrode. Their
original design using a porous hydrophobic millipore filter impregnated with active material has been replaced by disposable preloaded

cartridges. Both designs can be used for constructing different
liquid membrane electrodes.

The liquid membrane is normally made from a hydrophobic solvent in which is dissolved a salt consisting of a large organic ion
and, as its co-ion, the analyte ion. The materials used for the solvent and active site determine the selectivity of the membrane
electrode. The calcium and divalent cation electrode both use the
same salt, calcium didecylphosphate, as the exchanging site. They
differ in the choice of solvent used. Dioctylphenylphosphonate is
used for the calcium electrode (34) and the much less expensive
decanol is used for the divalent cation electrode. The latter
electrode has about equal selectivity for both calcium and magnesium.

The calcium and divalent cation electrodes have Nernstian response (29.6mV/decade), in the pH range 5.5 to 11, down to 10^{-5} and
$10^{-8}\underline{M}$ levels, respectively. The principle interferents are divalent
cations and sodium ion (35). Again, the literature is replete with
applications. Early references are in the book edited by Durst (1).
Considerations necessary for use of these electrodes in the potentiometric titration of calcium are discussed in a more recent reference (36).

Potassium Liquid Membrane Electrode It is not necessary to
have charged exchange sites in a liquid membrane electrode. Electrodes sensitive for many ions are being designed with electroneutral
carriers (37). For example, a solution of valinomycin, a ring
structure with alternating α-amino acids and α-hydroxyacids, in di-
phenylether forms a strong complex with potassium ion. A useful
potassium ion-selective electrode ($K_{K,Na}^{Pot}$ = 2.6 x 10^{-5}) (38) can be
prepared with this solution and the older Orion liquid membrane body.
This high selectivity can be compared with the much smaller value
noted for glass membrane electrodes. The electrode lifetime and
hydrogen ion selectivity are quite poor for this electrode. Even
so, applications have appeared for these electrodes in many disciplines. Among them are, potassium determination in sea water (39),
soils (40), and plasma (41).

Potentiometric Gas Sensing Electrode Several gas sensing
electrodes have been developed and are available commercially (42).
Among them are ammonia and sulfur dioxide electrodes. These electrodes normally make use of two membranes. The first membrane allows
passage of the analyte gas into an internal electrolyte. The second
membrane monitors changes in the concentration of some species in
the internal solution. For example, a glass membrane can monitor
pH changes if the gas is acidic or basic. The slope of the response
curve depends on the stoichiometric coefficients relating the given
gas molecule to hydrogen ion. For ammonia and sulfur dioxide the
observed slope is 60mV/decade and in opposite directions. The lifetime of the membrane is severely limited if the measured solution

contains wetting agents. Another design, the air gap electrode
(43), circumvents this difficulty by not immersing the gas sensing
electrode in the analyte solution. This electrode has been applied
to the determination of total inorganic and organic carbon contents
in water (44).

   Enzyme Electrodes   Enzyme catalyzed reactions are frequently
quite specific. Several workers (see ref. 45 for a review) have
found ways to couple enzymatic reactions with ion-selective electrodes. When an enzyme is confined at the surface of an electrode
a product produced (or reactant consumed) can be sensed by the
electrode. For example, a method developed for the analysis of
urea (46) is based on the production of ammonium ion by urease im-
mobilized in a gel layer on an ammonium ion-selective electrode.
The procedure requires very little of the expensive enzyme and the
electrodes can be used repeatedly. The response curves, in this
case, are non-linear because of substrate saturation at high urea
concentrations. Many other enzyme electrodes have appeared (45)
either with the enzyme confined at the electrode surface or in a
separate reactor in a flow system.

                            CONCLUSIONS

   Much of the activity in the field of ion-selective electrodes
has only taken place in about the last 15 years. We seem to be just
beginning to exploit the potential (no pun intended) of this method
in all fields of analytical chemistry, pollution analysis not except-
ed.

   Where do we go from here? The most obvious route is to extend
the use of electrodes already available to new analysis problems.
For example, the fluoride selective electrode (developed originally
in 1966 (29)) was recently established as an official AOAC method
for the analysis of feeds (47). Other examples could be cited of
an older electrode being newly applied to a real analysis need.
The second and less obvious route is to develop entirely new electrodes to solve existing or future problems. Liquid membrane electrodes and enzyme electrodes appear to show the most promise. With
liquid membrane electrode two parameters can be varied, the solvent
and the exchange site, both of which can affect the critical parameter selectivity. For example, Professor Rechnitz and the author
(13) developed a $CO_3^=$ electrode based on the solvent trifluoroacetyl-
p-butylbenzene. In this case the properties of the solvent were used
to enhance the selectivity of $CO_3^=$ over Cl^- and allow measurement of
total CO_2 in human serum (48). With enzyme electrodes, the enzyme
used, the internal sensor and electrode design can be varied so as
to arrive at favorable selectivity. Natural selection has worked in
our favor so that enzymes are inherently selective to begin with.
If a reaction could be devised so that one of the reaction products

could be sensed by an electrode then we are likely to have a new
and useful electrode. This is a very promising and currently
active area of research (49). Recently, a new method was described
(50) for determining urea using a separate enzyme (urease) reactor
and an ammonia gas electrode. In this case the design of the system
circumvented previous problems. In other cases new enzyme/sensor
couples will be developed. A recent example involves a novel enzyme
electrode method for the determination of nitrite based on nitrite
reductase (51). The reported advantages made the method simple,
valuable and economical for the routine analysis of nitrite ion.

SUMMARY

The introduction to the principles and practice of ion-selective
electrodes is intended to appraise the non-specialist of their utility. Electrodes are in some cases in routine use in the field of
water and air pollution analysis. Their use has been extended to
procedures involving automatic monitoring and continuous analysis.
Electrodes sensitive to non-ionic species have been and are continuing to be developed.

Many advantages can be cited for the user of ion-selective
electrodes. The measurement step is normally not the slow step in
the analysis as the response time of most electrodes is relatively
fast. Electrodes have a relatively long useful life (with certain
important exceptions) and require little or no maintainence. In
a surprising number of examples the electrodes are usable without
extensive pretreatment of the sample. The experimental paraphenalia,
unlike many other newer techniques, can be inexpensive to buy and
simple to operate.

Notwithstanding the advantages, let the non-thinking analyst be-
ware. All electrodes have some interference and should not be used
in situations where the results could be in substantial error or
even meaningless. As in many other areas, if you know what you are
about and aware of a method's limitations, you will find the use
of ion-selective electrodes only limited by your own ingenuity and
imagination.

ACKNOWLEDGEMENT

I would like to thank the North Carolina Science and Technology
Committee and the UNC-Greensboro Research Council for support of
this work. Portions of this manuscript were adapted from a more
extensive article by the author in the forthcoming book, "Water
Quality Handbook: Recent Advances in Analytical Techniques", H. B.
Mark, Jr. and J. S. Mattson, Eds., Marcel Dekker, Inc.

LITERATURE

REFERENCES

1 DURST, R. A., Ed.: Ion-Selective Electrodes. NBS Spec. Pub.
   314. U. S. Government Printing Office. Washington, D.C. (1969).

2 MOODY, G. J. and THOMAS, J. D. R.: Selective Ion Sensitive
   Electrodes. Merrow. Watford, England (1971) 1.

3 DURST, R. A.: Chap. 11 in ref. 1.

4 NANCOLLAS, G. H.: Interactions in Electrolyte Solutions. Elsevier. New York (1966) 15.

5 HERMAN, H. B.: in Electrochemistry, Calculations, Simulation
   and Instrumentation (H. B. Mark, et al., Eds.). Dekker. New
   York 2 (1972) 63.

6 Ref. 2, p. 52.

7 Orion Research, Inc.: Analytical Methods Guide. (1973) 8.

8 BRAND, M. J. D. and RECHNITZ, G. A.: Anal.Chem. 42 (1970) 1172.

9 RECHNITZ, G. A.: Chem. Eng. News. 45 (25) (1967) 146.

10 EISENMAN, G.: Chap. 1 in ref. 1.

11 BUCK, R. P.: Anal. Chim. Acta. 73 (1974) 321.

12 SRINIVASON, K. and RECHNITZ, G. A.: Anal. Chem. 41 (1969) 1203.

13 HERMAN, H. B. and RECHNITZ, G. A.: Anal. Chim. Acta. 76 (1975)
   155.

14 HSEU, T. M. and RECHNITZ, G. A.: Anal. Chem. 40 (1968) 1054.

15 BENSTON, M. L. and JAATEENMAKI, M. K.: Quantitative Chemistry.
   Van Nostrand. New York (1972).

16 CARR, P. W.: Anal. Chem. 43 (1971) 425.

17 GRAN, G.: Analyst 77 (1952) 661.

18 MacDONALD, T. J., BARKER, B. J., and CARUSO, J. A.: J. Chem.
   Ed. 49 (1972) 200.

19 Orion Research, Inc.: Guide to Specific Ion Electrodes and Instrumentation. (1969).

20 Orion Research, Inc.: Newsletter IV 3 and 4 (1972).

21 Orion Research, Inc.: Newsletter I 4 (1969).

22 ROSS, J. W., Jr.: Chap. 2 in ref. 1.

23 BATES, R. G.: Determination of pH. Wiley. New York (1973) 345.

24 PARK, K.: Science 154 (1966) 1540.

25 WEBBER, H. M. and WILSON, A. L.: Analyst 94 (1969) 209.

26 HSEU, T. M. and RECHNITZ, G. A.: Anal. Chem. 40 (1968) 1661.

27 VRBSKÝ, J. and FOGL, J.: Chem. Prům 22 (1972) 241.

28 KORYTA, J.: Ion-Selective Electrodes. Cambridge University
    Press. Cambridge (1975) 97.

29 FRANT, M. S. and ROSS, J. W.: Science. 154 (1966) 1553.

30 URE, R. W.: J. Chem. Phys. 26 (1957) 1363.

31 BUTLER, J. N.: p. 157 in ref. 1.

32 LIGHT, T. S.: Chap. 10 in ref. 1.

33 WARNER, T. B. and BRESSON, D. J.: Anal. Chim. Acta 63 (1973) 165.

34 ROSS, J. W.: Science 156 (1967) 3780.

35 WHITFIELD, M. and LEYENDEKKERS, J. V.: Anal. Chem. 42 (1970) 444.

36 HULANICKI, A. and TROJANOWICZ, M.: Talanta 20 (1973) 599.

37 MORF, W. E., AMMANN, D., PRETSCH, E., and SIMON, W.: Pure Appl.
    Chem. 36 (1973) 421.

38 PRODA, L. A. R., STANKOVÁ, V. and SIMON, W.: Anal. Lett. 2 (1969)
    665.

39 ANFALT, T. and Jagner, D.: Anal. Chim. Acta 66 (1973) 665.

40 BANIN, A. and SHUKED, D.: Agrochimica 15 (1971) 238.

41 MIYADA, D. S., INAMI, K., and MATSUYAMA, G.: Clin. Chem. 17
    (1971) 27.

42 ROSS, J. W., RISEMAN, J. H., and KRUEGER, J. S.: Pure Appl. Chem.
    36 (1973) 473.

43 RŮŽIČKA, J., and HANSEN, E. H.:   Anal. Chim. Acta 69 (1974) 129.

44 FIEDLER, U., HANSEN, E. H., and RŮŽIČKA, J.:   Anal. Chim. Acta
    74 (1975) 423.

45 GOUGH, D. A. and ANDRADE, J. D.:   Science 180 (1973) 380.

46 GUILBAULT, G. G. and MONTALVO, J. G.:   J. Amer. Chem. Soc. 92
    (1970) 2533.

47 TORMA, L.:   J. Assoc. Off. Anal. Chem. 58 (1975) 477.

48 HERMAN, H. B. and RECHNITZ, G. A.:   Anal. Lett.   8 (3) (1975)
    147.

49 RECHNITZ, G. A.:   Chem. Eng. News 53 (4) (1975) 29.

50 JOHANSSON, G. and ÖGREN, L.:   Anal. Chim. Acta 84 (1976) 23.

51 KIANG, C. H., KUAN, S. S. and GUILBAULT, G. G.:   Anal. Chim.
    Acta 80 (1975) 209.

Discussion

FILBY - One of the advantages of this type of technique is that this a response to a specific ion, take copper ion for example. There are problems associated with this in that it does respond only to the ion and not to other forms of copper. That can be an advantage in the sense that you can possibly use electrodes like this say for detection of copper complexes. There is a lot of interest right now in speciation of the trace metals.

HERMAN - Ion-selective electrodes can be used in a solution where complex ions are formed. If, for example, the fraction of free copper remains constant, as it would in some metal ion buffers, then the total copper concentration can be measured by calibration standards. If the fraction of free copper ion varies then, if the conditions are known, it still may be possible to estimate the copper concentration by calculations using the relatively simple formulas.

FILBY - What I was thinking of is more an electrode for a specific complex.

HERMAN - It is possible, in principle, to develop an electrode selective to one specific complex. The ones commercially available, since they are presumably already optimized for a given ion, would probably not be suitable. However, a liquid membrane could be designed to extract a specific complex as opposed to the free ion. This could be made the basis for a liquid membrane ion-selective electrode. The commonly used copper ion solid-state electrode would not be suitable for the application you suggest.

FILBY - Do you know if anything is being done on this?

HERMAN - I am not aware at this time of any specific studies. Vitamin B_1 is normally charged, but a neutral Vitamin B_1 complex is extracted into the liquid membrane and the electrode responds to changes in the Vitamin B_1 concentration.

TORIBARA - The people here may be interested in a short, but disastrous experience we had with a so-called ion-selective electrode. The problem was as follows: One of the physiologists was studying the effects of catecholamine drugs on smooth muscles. He noticed that he got some kind of an effect which was reversed by the addition of calcium.

He sent his observations to a journal, and a referee suggested that he check the obvious explanation that the catecholamine must be complexing the ionic calcium. Since we

had done a lot of work with calcium, he came to me and asked
whether we could determine if the calcium was being complexed.
I told him that it should be easy to determine that with our
calcium electrode. All we would have to do would be to measure
the potentials with and without the catecholamine. We did that
and the calculation indicated that we had a thousand times more
calcium in the presence of catecholamine than we had put in. We
decided that the electrode was not responding properly, and that
ended that experiment. The precaution here is that if we had not
paid careful attention to the sign of the potential we might
easily have decided that we had a very strong complexer. Obviously
the catecholamine was affecting the membrane of the electrode so
that it was no longer responding properly. Our experience points
out the importance of knowing if the electrode is operating
properly.

HERMAN – Development of new ion-selective electrodes has similar
pitfalls. As I briefly mentioned, I have worked with a new thiamine ion-selective electrode. We were concerned that the buffer
components were contributing to the electrode response. It was
possible to isolate the effects and, happily in this case, most of
the response was due to thiamine. In other cases this may not be
true, so the investigator should be on his guard against this potential problem.

WEE – At this time what is your lowest limit of detection with
some precision?

HERMAN – The answer to your question is dependent on which electrode is being discussed. For example, a sulfide ion-selective
electrode can be used to make free sulfide ion measurements down
to about 10^{-20} molar. The silver sulfide active material has a
very low solubility and measurements can be made at much lower
concentrations than in other systems. A typical lower limit for
other electrodes would be about 10^{-5} molar. There is no simple
answer to your question. It depends on the electrode and the
system being measured.

FELDMAN – Do you know of any other example besides sulfur?

HERMAN – I chose that one as a particularly good example.

ULTRATRACE METALS ANALYSIS BY ELECTROTHERMAL ATOMIZATION ATOMIC ABSORPTION - PRESENT TECHNOLOGY AND POTENTIAL DEVELOPMENTS

Marti Bancroft

Celdat Design Associates

Rochester, NY

ABSTRACT

The history of electrothermal atomization atomic absorption (ETAA) is briefly reviewed for background. Some principles of ETAA and its differences from more traditional flame AA are discussed to explain the advantages of this technique for ultratrace metals analysis. The present status of ETAA is also reviewed - ETAA is now routinely used for certain trace metals in a wide variety of samples. Some refractory elements (examples are V, Mo, Ba) are now readily analyzed, and there are a growing number of applications to solid samples and even indirect analysis of non-metallics. Some recent advances in temperature and power control methods have greatly improved sensitivity and repeatability of analytical conditions. Likely future developments of ETAA are in enhanced sensitivity, applicability to an ever-wider variety of samples, and multielement determinations. Some of the author's research on substrate effects, fill gas effects, and speciation is discussed where applicable to the future of ETAA.

The recent growth of electrothermal atomization (ET) in atomic spectroscopy is usually traced from Arthur S. King's publications in astrophysical journals in the first part of this century (1). Samples were sealed in this "King Furnace" and the furnace was subsequently slowly

raised to (and maintained at) the observation temperature. Analytical use of ETAA began with the work of
B.V. L'Vov in Russia, starting in 1960 (2,3). The
L'Vov furnace was used for ultratrace metals analyses -
desolvated samples were volatilized into the hot furnace by an electric arc to the sample-carrying electrode. Both the furnace and electrode were made of graphite and L'vov also used rether high pressures of inert
fill gases to prevent oxidation of the furnace. His book
in 1969 (4) contributed to the theoretical groundwork of
ETAA as an analytical technique. More university research on furnaces (5,6) followed L'Vov's impressive
sensitivity results (better than 10^{-12} G for most elements could easily be detected with L'Vov's apparatus).
Work was also done with strip or filament atomizers (7,
8). Commercial ETAA instruments became available in the
1970's. Most instruments have had rapid, sometimes
annual, design improvements as knowledge of ETAA has
grown. Most ETAA instruments used in the U.S. are made
by (alphabetical order) Instrumentation Laboratory,
Inc., Wilmington, MA., Perkin-Elmer Corp., Norwalk, CT.,
and Varian-Techtron, Palo Alto, CA.

     ETAA is now used for many ultratrace (subnanogram)
metals analyses in a wide variety of samples. Among
its advantages over the more traditional flame sources
are increased sensitivity, smaller sample size, and
minimized sample preparation. ETAA does require 1-2
minutes per sample, some operator skill, and the usual
precautions needed for contamination-free ultratrace
work.

     The fundamental differences between ET and flame AA
are in the methods of neutral atom vapor production and
containment. Flame AA requires the aspiration of liquid
or dissolved sample into the flame, where the droplets
are then desolvated, vaporized, atomized to produce the
neutral atomic state of the metal of interest, and transported through the observation beam of the spectrophotometer. The flame provides the energy and mechanisms for
atomization (which are rather complex and beyond the
scope of this discussion)(9). Observation time is
quite short (typically milliseconds, depending on the
flame gases and burner design)(10). At least 1 milli-
liter of sample is required, though less can sometimes
be used with peak detection (11). An equilibrium absorbance signal results, which increases as the nebulized sample enters the observation volume, maintains a

more-or-less steady state as the sample is continuously nebulized, and decays when the sample is removed. During the steady-state portion, the magnitude of the absorbance signal is proportional to the concentration of analyte atoms in the sample, subject to other considerations such as correct standardization and freedom from/compensation for interferences. Integrating this signal over time simply reduces the effect of random noise (instrument noise, fluctuations in nebulization, transport, etc.). Viscosity effects (different solvents, high percent solids) which affect the quantity of droplets per unit time, and an interesting collection of chemical and physical interferences, such as ionization, refractory compound formation, etc. can affect both accuracy and precision. For most elements the lower practical limit of flame AA is in PPM.

The atom generation process is quite different in ETAA. The furnace consists of a substrate (carbon, graphite, or metal) and an inert fill gas which may be flowing, evacuated, or pressurized. The liquid or solid sample is inserted onto the substrate (usually 1-100 µL or mg are sufficient) and dried if necessary. The optimum times and temperatures for drying, pyrolyzing or ashing, and atomizing the sample are selected, and automatically implemented by the programmable power supply. Fill gas program can be varied in some units. At atomization, the cloud of neutral gaseous metal atoms is generated by a variety of reactions, primarily reduction, and is permitted to diffuse into the observation volume or swept away by flowing fill gas. Observation time is much longer than with flame work (tenths to several seconds). Either peak height or peak area may be used to indicate the concentration of analyte. The effect of integration is to not only average out noise but also to ameliorate certain sample effects and some interferences (4,12,13). The actual reactions causing atomization (15), some interesting interferences (14, 17), competing reactions (16), fill gas role (16), and substrate effects (16) are all topics of current research.

Today ETAA is used for ultratrace metals analysis in drinking water, biological fluids, tissues, foods, sewage, petrochemicals, etc., as well as for forensic applications, commercial metallurgical studies, and the obvious use in basic research.

Since some of the energy for atomization is thermal,
the most sensitive elements are usually the more volatile metals. One recent advance in ETAA has been in the
increasing ability of commercial instruments to analyze
refractory elements such as V, Ba, Ti, and Mo. ETAA
problems with refractories include insufficient temperature, reactive fill gas, stable oxides, volatile molecular species in some samples, and reaction with certain
substrates. Some solid samples are used - recent applications include alloys (18,24), organics (19), foods
(22) and some biological samples (23). The direct analysis has some difficulties but minimized sample preparation often outweighs them.

Today analysis of elements such as chlorine by AA
is done indirectly, usually by precipitation with a
metal and subsequent analysis of the metal. This is a
laborious procedure and at ETAA levels solubility products
have some surprises for the analyst.

Current instrumentation can analyze two elements
simultaneously (20) - but since ETAA is slow, multielement analyses would be a distinct advantage. In the
last year, commercial instruments have refined their
power control methods to provide far better compliance
to the selected time/temperature program and improved
rates of heating. Since ETAA is a non-equilibrium process, this change has improved sensitivity, especially
for refractory elements.

Present ETAA theory describes some of the automiza-
tion processes for very limited situations, most especia-
ally those involving a carbon reduction of the metallic
oxide (21). Certain reactions during pyrolysis or ashing are also understood. Some factors affecting the
signal evolution peak height (P/H) vs. peak area (P/A),
were quantitated by L'Vov (4), but are not universally
applicable due to substrate geometry, material, temperature, and fill gas differences between the commercially
available ETAA units (16,25). There is still some discussion of which measurement, peak height or peak area,
should be used for ETAA!

Three areas in which ETAA is likely to develop in
the next five years are (A) increasing sensitivity of
analysis, (B) multielement capability (primarily due to
advances in spectrophotometers and the use of ETAA as a
source for non-AA spectrophotometers), and (C) application to a wider variety of samples. Advances in theore-

tical understanding will promote (A) and (C), aided by
the recent development of ETAA instruments with more accurate temperature control. This makes possible more
solids analyses, speciation and kinetics studies substrate interaction observations, and study of the reactions occuring during pyrolysis (or ashing) and atomization. In situ sample pretreatment will be more widely
used to permit cadmium, and other volatile elements. As
the prices of dye lasers become more in line with the
price of ETAA units, they will see greater applicability
to ETAA. CEWM-AA (continuous emission wavelength modula-
tion-AA)(25,26,27) and plasmas (29) offer multielement
potential, the first by true AA multielement, the second
by using the furnace as an atom generator for entry into
the plasma. Zeeman effect AA (28) has some advantages
for high-background samples, but probably will not impact
ETAA as much as the multielement capabilities of new instrumentation and automation. Already several automatic
sample entry systems are available for ETAA, though only
for liquids at this time. Microprocessor-controlled ETAA
units may become a reality within the next five years.

        Three aspects of our research are relevant to these
areas of possible ETAA advance. These are (1) fill gas
effects, both chemical and physical; (2) substrate effects; and (3) speciation.

(1) Fill Gas Effects - Chemical and Physical:

        L'Vov (4) considered and quantitated (for his system)
the P/H and P/A effects of changing the fill gas pressure and type - assuming that the only interaction with
the analyte atoms was physical. We have found his pre-
dictions correct when that assumption holds. One of our
studies employed rubidium, which does not react with nitrogen, in both both argon and nitrogen - we observed
precisely L'Vov's predicted signal decrease in nitrogen
(argon was the reference). Pressurization also had the
predicted enhancement of P/A and slight decrease in P/H.
However, for both barium and vanadium, the use of nitrogen produced far more than the theoretical decrease and
pressurization did not ameliorate this effect. Further
work with elements which form stable nitrides vs. those
which do not, led to the hypothesis of a gas phase, relatively slow (as compared to atomization and carbide
and oxide formation) reaction between the nitrogen and
the analyte. Aside from decreasing the sensitivity -
which is not always undesirable- these can sometimes

cause problems by enhancing other interference effects, especially interferences which retard the atomization rate. This does so because the signal (P/H, P/A) is a result of competing reactions - the fixed rate gas phase depletion by nitrogen and the variable atomization rate retardation. We observed this effect in barium analysis in gunshot residues. As predicted, precision in nitrogen is poorer in certain samples and P/A (which usually compensates for slight variations in atomization rate) did not improve precision in nitrogen, but did in argon fill gas.

(2) Substrate Effects:

Our studies on substrate effects were prompted by some puzzling atomization behavior. While investigating the properties of different furnace materials, the following anomalies were observed in P/H signals:
(A) with a tantalum substrate, in argon fill gas the barium atomization signal was rather poor and exhibited considerable memory, yet the addition of hydrogen - which usually decreases ETAA signals - dramatically improved the barium signal. A similar situation but without memory occurred for cobalt; here, nitrogen was tried. The cobalt signal improved after 5-6 firings in nitrogen. Silicon and aluminum were not very sensitive although hydrogen helped slightly, but vanadium was observable if argon was used as the fill gas.

(B) With a tungsten substrate, platinum was simply not visible no matter which fill gas was employed. Barium was very sensitive in argon but the signal decreased in hydrogen. The cobalt signal decreased in nitrogen. Neither exhibited memory. Again, silicon and aluminum were not very sensitive but vanadium was analyzable in argon.

(C) With a carbon substrate, barium was less sensitive and exhibited some memory; both hydrogen and nitrogen dramatically decreased the signal and nitrogen seemed to decrease the memory. Vanadium behaved similarly. Both vanadium and barium were less sensitive than on the tungsten substrate. With a careful choice of pyrolysis, both silicon and aluminum were extremely sensitive in argon, less so in nitrogen, and still less so in hydrogen. Platinum was observable without memory. Cobalt was much less sensitive than with tantalum in nitrogen, though more sensitive than with tantalum in argon.

(D) With a pyrolytic-graphite coated substrate (with

the planes of the "fishnet" oriented parallel to the
furnace longitudinal axis), both barium and vanadium
were much more sensitive in argon than with carbon and
exhibited no memory (nitrogen and hydrogen decreased
the signal). Barium was as sensitive as with the tungsten substrate in argon. Vanadium was slightly more
sensitive than with either tungsten or tantalum. Silicon and aluminum were less sensitive than with carbon
but far more sensitive than on the metal substrates.
Platinum was about 5 times more sensitive than on carbon, and cobalt was greatly improved.

Our subsequent substrate investigations provided
the following tentative explanations (see also the fill
gas studies above):
(A) Metal substrates can sometimes form intermetallic
solutions and/or compounds with the analyte, decreasing
the signal and sometimes causing memory. Due to its
larger lattice parameters and chemical behavior, tantalum seems to be worse than tungsten in this respect.
Nitrogen reacts with the tantalum surface and forms a
compound more stable than certain intermetallics, so
atomization is feasible - but it apparently requires
several atomization cycles to establish full coverage.
Hydrogen dissolves readily in tantalum and appears to
both block alloy formation and enhance atomization. We
hypothesize that most alloying, blocking of same, and
atomization phenomena on metal substrates occur at the
grain boundaries.

(B) Reactive carbon can form carbides with susceptible
elements, retarding atomization and causing memory - or
enhancing atomization. In the case of barium and vanadium, the carbides (or carbon compounds) hinder atomization. Fill gas has little effect on this; the apparent
nitrogen effect on memory appears to be depletion by
gas phase reaction. In the case of silicon and aluminum,
the carbon reaction can eliminate the stable oxides
(which on metals may tend to boil as molecules rather
than atomize). Fortunately the silicon and aluminum carbides can be quantitatively decomposed with excellent
analytical results.

(C) Pyrolytic graphite has a much lower gas permeability
than carbon, so it enhances sensitivity for some elements merely by retarding the escape of the atom cloud
through the furnace walls. It is also much less reactive, so doesn't readily form carbides - hence the
sensitivity increase for barium and vanadium and the decrease for aluminum and silicon.

(3) Speciation:

The predominant anion can affect atomization rate significantly, as well as determine the fraction of analyte atoms actually atomized. This often creates standardization problems (for a total analyte analysis, it is critical that all the analyte is in the same form at atomization). However, a skillful analyst can also use this to identify and quantitate metal species. Successful speciation usually requires (A) a case where the atomization rates are sufficiently different to distinguish the peak tops (computation can do the rest of the separation if the calculation is correctly set up), (B) correct choice of a time/temperature program, (C) slower atomization to enhance the peak separation, (D) very repeatable analysis conditions, and (E) correct standardization. Analytical use of speciation will be facilitated by automation of ETAA.

These three areas may impact the future of ETAA as more research is done. For example, there may be different optimum substrates for various analyses. Today most manufacturers choose one (to simplify instrument design). Automation could permit simple exchange of substrates if the ultimate in sensitivity and freedom from interferences is required for a specific element and/or sample. Changing substrates also offers great advantage in atomization research – and more theoretical understanding will help enhance sensitivity and applicability. Changing fill gas also affects both these areas. Multielement capabilities may involve tradeoffs in instrument conditions in which substrate and fill gas choice could prove important. The analytical uses of speciation are obvious and exciting in such a relatively inexpensive technique.

REFERENCES

1. King, A.S., Astrophysical J., Numerous Publications 1900-1940.

2. L'Vov, B.V., Spectrochimica Acta, 1961, 17: 761-770.

3. L'Vov, B.V., Spectrochimica Acta, 1969, 24B: 53.

4. L'Vov, B.V., Atomic Absorption Spectrochemical Analysis, Adam Hilger Ltd., London, U.K., 1970.

5. Massman, H., Spectrochimica Acta, 1968, 23B, 215-216.

6. Woodriff, R., et. al., Numerous Publications in Applied Spectroscopy, Anal. Chem., Spectrochim. Acta and Appl. Optics, 1968-1977.

7. West, T.S. and Williams, Y.K., Anal. Chim. Acta, 1969, 45: 27 and subsequent issues.

8. Donega, H.M., and Burgess, T.E., Anal. Chem., 1970, 42(13): 1521.

9. Christian, G.D. and Feldman, F.J., Atomic Absorption Spectroscopy, Wiley-Interscience, N.Y. 1970.

10. Price, W.J., Analytical Atomic Absorption Spectrometry, Heyden and Sons Ltd., London, U.K., 1972.

11. Hwang, J.Y., Unpublished Data, 1975.

12. Bancroft, M.F., Smith, S.B., Schleicher, R. and Coumas, J., Unpublished Data.

13. Schleicher, R.G., Hwang, J.Y., Bancroft, M.F., Emmel, R.H. and Kahn, H.L., Paper #43 at 1976 Pittsburgh Conf., Cleveland, OH.

14. Glass, E.D., Lichte, F.E., and Koirtyohann, S.R., Paper #43 at 1977 Pittsburgh Conference, Cleveland, OH.

15. Fuller, C.W., Analyst, 99: 1974, 739-744.

16. Bancroft, M.F. and Smith, S.B., Unpublished Data.

17. Leisz, D., Varnes, A. and Duff, R., Paper #174 at the 1976 Pittsburgh Conference, Cleveland, OH.

18. Marks, J.Y., Paper #44 at the 1977 Pittsburgh Conference, Cleveland, OH.

19. Sotera, J., Bancroft, M.F., and Hwang, J.Y., Paper #171 at the 1976 Pittsburgh Conference, Cleveland, OH.

20. Emmel, R.H., Stux, R. and Leighty, D., Paper #98 at the 1977 Pittsburgh Conference, Cleveland, OH.

21.   Stone, R.W. and Woodriff, R., Paper #66 at the 1977
      Pittsburgh Conference, Cleveland, OH.

22.   Kopur, J.K. and West, T.S., Anal. Chim. Acta, 1974,
      73: 180-184.

23.   Langmyhr, F.J. and Thomassen, Y., Z. Anal. Chem.,
      1973, 264: 122-127.

24.   Langmyhr, F.J. and Thomassen, Y. and Massoumi, A.,
      Anal. Chim. Acta, 1973, 67: 460-464.

25.   O'Haver, T.C., Turk, G. and Hornly, J.M., Paper
      #150 at the 1977 Pittsburgh Conference, Cleveland,
      OH.

26.   O'Haver, T.C., Zander, A.T. and Hornly, J.M.,
      Paper #152 at the 1977 Pittsburgh Conference,
      Cleveland, OH.

27.   O'Haver, T.C., Zander, A.T., and Hornly, J.M.,
      Paper #210 at the 1977 Pittsburgh Conference,
      Cleveland, OH.

28.   Koizumi, H. and Yasuda, K., Paper #208 at the 1977
      Pittsburgh Conference, Cleveland, OH.

29.   Koop, D.J., Silvester, M.D., and Van Loon, J.C.,
      paper #100 at the 1977 Pittsburgh Conference,
      Cleveland, OH.

Discussion

ROBILLARD - In your talk you mentioned dual channel ETAA and also CEWM-AA which was multielement analysis. In either of these two techniques do you sacrifice sensitivity or detection limit when compared to single element analysis?

BANCROFT - Sometimes, but not always. It depends on how much of a trade-off you have to make, in the time and temperature conditions chosen, for the analysis. It also depends on the optics of the instrument - poor resolution lowers sensitivity. Usually if you have an instrument which has adequate optical resolution and a very responsive detection system, the answer is no, because you simply optimize the time-temperature program of the ETAA unit for the most refractory element being done, and the more volatile elements merely atomize much faster than normal. Sometimes, however, in high background samples you must slightly sacrifice the sensitivity of the refractory element by choosing a more manageable background signal. A refractory element is typically best atomized by a very rapid step to a high temperature and this can result in an increased background signal, which may be a problem. Certain temperature control units can make a series of temperature steps, and then you can pick off each element at a temperature which is more suitable for its analysis. The important factors include instrument optical design, response times, atomizer design, the elements to be deter-mined simultaneously, and the matrix characteristics. I'd suggest contacting Dr. O'Haver about other factors in CEWM-AA.

ROBILLARD - Then what you're doing really is sequential analysis? Is that correct, rather than a simultaneous analysis of the various elements?

BANCROFT - You sometimes have the choice of doing either. With the dual channel ETAA method, you have the choice of doing two simultane-ously, or two slightly sequentially - depending on the atomizer characteristics. With the continuum emission wavelength modulation AA(CEWM-AA), which uses a powerful continuum source, the wavelength separation (resolution) is obtained by modulation of the signal and detection is effectively simultaneous so depends again on the temperature program of the atomizer. To a certain extent this depends on the elements that you have and the sample that you have and you can sometimes have the choice of either true simultaneous or slightly sequential analysis. The time span is seconds, though, not minutes.

ROBILLARD - Are there any elements that you cannot do together by the CEWM-AA method?

BANCROFT - Well, elements which have spectral lines which overlap (spectral interferences) or elements which alloy together and form a stable compound that won't easily atomize (chemical interference) are two problem categories. There are certain trace refractories in other refractories, as a common example of alloy formation - Co in Ta, for example.

SMITH - You might mention that the CEWM-AA method, since it uses a continuum source doesn't have the resolution of a single channel so it won't have quite the sensitivity of a single or dual channels which use a hollow-cathode source.

BANCROFT - Yes, I should have mentioned that the absolute sensitivity is improved by the use of line sources (hollow cathodes). Recent data with CEWM-AA as shown in comparison with Zeeman AA- shows that it is getting better, so there is some hope and sometimes this sensitivity loss can be traded off. Absolute sensitivity is still best with hollow cathode source ETAA.

ROBILLARD - This CEWM-AA and the instrumentation that is being developed to do this type of analysis, how expensive is it?

BANCROFT - At the moment (May 1977) it is only a university research tool. The last time I talked to Dr. O'Haver he was mumbling something about sixty or seventy thousand dollars. (This compares to around $20,000 for dual-channel ETAA.) Obviously this isn't competitive at the present time. However, in recent developments in microprocessors, computer programming, cost of memory and most everything else in the computer industry, the trend is for more computing power at drastically reduced costs. A significant cost in CEWM-AA is the mini-computer, disc, and memory. As an example, I'm watching a new small systems language, URTH, developed by K. Hardwick at the U. of Rochester, very carefully, for its impact in this area - this provides power and speed in minimum memory and minimizes the assembler programming required. For real-time spectroscopy those are important factors. In the next five years, if this trend continues, the cost of automating ETAA systems will become more in line with the cost of the ETAA unit itself ($\sim$$5,000).

ROBILLARD - I have one more question. Would you care to comment upon the comparison of the CEWM-AA method with inductively coupled plasma (ICP) spectroscopy? If you could only have one instrument, which do you think you would prefer?

BANCROFT - Well, at the moment, if I have to answer that question, I would have to say ICP but that's only because it's commercially available and CEWM-AA isn't. That factor aside, I think it depends on the analysis that I was attempting, because the ICP is an emission technique. There are many cases where emission techniques are equally sensitive and their inherent multi-element ability is useful. On the other hand, the CEWM-AA is a true atomic absorption technique and atomic absorption has the advantages of specificity and sensitivity over emission techniques in many cases. For ultra-trace work in small samples, I'd still pick conventional ETAA with a hollow cathode source.

SMITH - The plasma isn't quite as sensitive as the flame absorption or ETAA.

BANCROFT - Yes, to answer that question as ICP or CEWM-AA, then I would have to have a specific case of what analyses I was charged with doing. Both are inherently rapid multi-element analyses re-quiring automation, so sample volume could be important to offset initial equipment cost.

FILBY - At one time it looked like atomic fluorescence was much more suitable for a multi-element technique, because you could use a continuum source if you could get one strong enough. This could also be applied to flameless AA. Is anybody working on that as a multi-element technique?

BANCROFT - Several people have done work on atomic fluorescence (AF) and multi-element ETAA. One name that comes to mind immedi-ately is Dr. Winefordner at the University of Florida at Gaines-ville. However, it doesn't seem to have caught on as much as some of these other techniques. The range of applicable elements seems less than with either AA or AE. I have briefly experimented with atomic fluorescence and ETAA, and one thing I found very disappoint-ing was the unfortunate fluorescence quenching effects in ETAA, at least in matrices other than water. The interferences seemed to be, at least in my limited experimentation, more severe than with normal ETAA. AF also imposes some design constraints on the atomizer in order to get the 90° observation angle. Dr. Winefordner has had far better results.

TANKER TRAGEDIES: IDENTIFYING THE SOURCE AND DETERMINING THE

FATE OF PETROLEUM IN THE MARINE ENVIRONMENT

Chris W. Brown and Patricia F. Lynch

University of Rhode Island, Department of Chemistry

Kingston, Rhode Island 02881

ABSTRACT

Three major oil spills this winter in the New England coastal area have forced immediate action on the part of oceanographic,
chemical and biological researchers to determine the fate of the
oil in the environment and its effects on aquatic life. This paper discusses the infrared and gas chromatographic analyses of air,
water, sediment and tarball samples collected at one or more of
these spill sites. The chemical partitioning of these petroleum
samples in the environment is discussed and the possible source of
"mystery" tarballs is conjectured based on infrared analyses.

Introduction

The United States has long been dependent on the importation
of foreign oil for a great percentage of her supplies. To reduce
this dependence, major oil companies have successfully lobbied for
off-shore drilling along the New England Coastline in the area of
Georges Bank. This offshore area is one of the richest fishing
beds in the world. Hence, environmentalists, oil companies and
fishermen have found themselves locked on the horns of a dilemma.
If an event such as happened in Santa Barbara (or most recently
in the North Sea) were to repeat itself in Georges Bank as a result of well blowout, the New England fishing industry could be
severely crippled.

This situation has renewed and intensified scientific research concerning the environmental impact of an oil spill. It
has raised questions such as "Where does the oil go after it is

spilled?" "By what mechanisms does it get there?" "How much can
we expect to find in the air, water, sediments?" "How could this
affect marine animals?"

Three major oil spills this winter have forced immediate action on the part of oceanographic, chemical, and biological researchers to answer these questions and has provided them with
some very surprising answers.

In early December, 1976, the oil tanker, Argo Merchant, went
aground in the Nantucket Shoals and slowly spilled 7.5 million
gallons of a very viscous No. 6 fuel oil into the cold Atlantic
waters. At first the oil formed a coherent floating patch that
measured 30 miles wide and 100 miles long. This gradually broke
up into "pancakes" which sunk below the surface. On December 28,
three chemists from this laboratory left on the research vessel,
Endeavor, to collect water samples for analysis of hydrocarbon
levels in the area of the spill.

Experimental

Infrared spectra were measured on a Beckman 4260 Infrared
Spectrophotometer with a spectral slit width $<10cm^{-1}$ from 4000 to
2000 cm^{-1} and <4 cm^{-1} from 2000 to 600 cm^{-1}. Gas chromatograms
were measured on a Hewlett Packard 5710A using FID detection and
a 3m column of 3% SP2250 on 100/200 Suppeloport; the column was
temperature programmed between 90° and 300°C at 8°/min.

All solvents were reagent grade from Matheson, Coleman and
Bell, Inc. or Mallinckrodt, Inc., and all were redistilled in
glass prior to use.

Results and Discussion

A map of the area in which the Argo Merchant went aground
is shown in Figure 1. Sampling stations 1, 2 and 3 are marked,
and, at each site, 1 liter samples were collected (when possible)
at 1, 6 and 40 meters. At station 1, (Figure 2) which was southwest of the spill site, the largest hydrocarbon concentration was
200 ppb at 40 meters (1). At station 2, which was the closest to
the wreck site, the highest concentration of extractable organics
was found to be approximately 470 ppb. Station 3 was southeast
of the wreck, along the path of travel of the spill. Due to
weather conditions only a surface sample could be collected and
extractable organics in this sample were approximately 450 ppb.
Brown (2) et. al. found median levels of between 18 and 33 ppb
extractable organics in normal sea water samples. In some cases,
the concentrations in the waters near the site of the wreck represent a 50-100 fold increase over "normal" levels; however, they

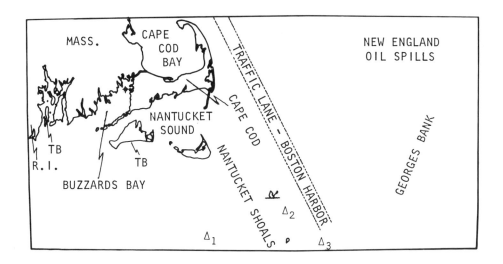

Figure 1. Map of New England Coastal and Georges Bank area.

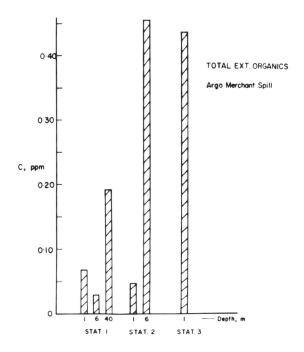

Figure 2. Histogram showing Quantitative Infrared results of
water samples collected on the first cruise to the site of the
Argo Merchant wreck.

still seem to be low considering the size of the spill. During
this 3-day sampling trip, no detectable amounts of oil were found
in sediment samples taken at the same sites and only one siting
of very small clumps of floating oil was made.

After obtaining these results, laboratory experiments were
begun in which 30.9 mg oil from the cargo hold of the Argo Merchant were "spilled" on the surface of 5 liters of Instant Ocean
(a solution of ocean salts in distilled water). Slow agitation
was provided by a magnetic stirrer and two 1 liter subsurface
samples were collected after 1 and 24 hours. These were analyzed
along with a control by infrared spectroscopy for extractable organics. At each sampling, one liter was analysed for total organics, whereas the other was passed through a Milipore Filter
(.45μ) and analyzed for dissolved organics. If the total oil
sample were incorporated into the 5 liters, a concentration of at
least 6,200 ppb extractable organics would be expected. However,
as is seen in Figure 3, for the unfiltered sample, the control
has 230 ppb organics, whereas the one hour oil/water sample contains only about 110 ppb as does the twenty four-hour sample. Instant Ocean was found to contain a high level of organic contamination, including polymers. One possibility for the decrease in
extractable organics in the oil/water sample is that when the oil,
which is highly water insoluble, is added to the water two liquid
phases are formed, i.e., an oil phase and a water phase. The organic contaminants in the Instant Ocean selectively adsorb onto
or dissolve into the organic layer; thereby, decreasing their concentration in the water layer. When one liter of the water layer
is then extracted, it contains less organic material than does the
original Instant Ocean control.

For the filtered sample, there is very little difference in
amount of dissolved organics between the control and the one hour
sample (70 ppb). There is a slight increase after twenty four
hours (20 ppb) but, from Figure 3, it can be seen that this oil
has a very low water solubility.

In the next two laboratory experiments a volatile crude,
South Louisiana, was used. Fifteen mg of the oil were added to
5.1 liters (3000 ppb) of Instant Ocean after two one-liter controls had been taken. Agitation was as before. Figure 4 again
shows the unfiltered control to contain more organics (240 ppb)
than the one hour sample (190 ppb). However, after twenty four
hours, more oil is "accomodated" by the water (290 ppb) than at
one hour. The filtered samples indicate that very little of the
oil is truly soluble even after twenty-four hours, since the control and two samples contain about the same amount of organics
(80-90 ppb).

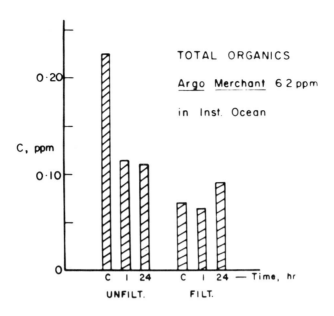

Figure 3. Histogram showing quantitative infrared results of laboratory experiments using cargo oil from Argo Merchant and Instant Ocean.

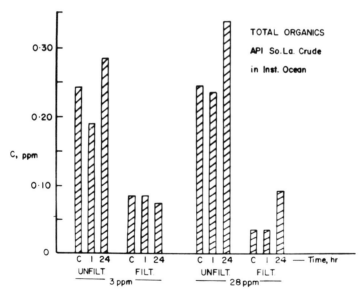

Figure 4. Histogram showing quantitative infrared results of laboratory experiments using A.P.I. S. Louisiana Crude and Instant Ocean.

Using the same oil, another experiment was begun in which
120 mg of oil were added to 4.2 liters of Instant Ocean (28000
ppb) after control samples were taken. Figure 4 reveals that
the level of organics in this control is the same as in the
first experiment (i.e., 240 ppb). However, the one-hour sample
does not show much of a decrease in organic levels compared to
the control. This follows, since in this experiment the Instant
Ocean was "supersaturated" with oil. After twenty-four hours the
unfiltered sample contains 330 ppb organics.

The filtered samples show that very little of this organic
matter is truly dissolved after one hour. There is a slight increase in dissolved organics (40 ppb) after twenty-four hours.
Furthermore, the amount of organics in the filtered samples in
this experiment as well as the previous one are 80-90 ppb, suggesting that a saturation point has been reached.

Quantitative results from these three laboratory experiments,
as well as from the Argo Merchant spill samples, underscore the
fact that oil is highly insoluble or immiscible in water. This
being true, where does oil go after a spill, if very little of it
dissolves in or is accomodated by the water?

The answer to this is very complicated because it depends on
a number of variables, e.g., location of spill, oil type, sea
state, water and air temperature, etc.

In January 1977, a second major spill occurred in Buzzards
Bay, Massachusetts. A barge carrying 300,000 gallons of No. 2
fuel ran into heavy ice in the canal and sprang a leak which deposited the cargo into the congested canal area. Because of the
long period of severe cold weather, it was thought that little of
the oil would penetrate below the hundreds of feet of ice and
snow and pollute the bay. The oil-coated ice was plowed up and
hauled away in trucks in an effort to clean up the spill. A few
days later, we collected water samples at the spill site and returned them to the lab for analysis.

Figure 5 shows the infrared spectra (1700-650 cm^{-1}) of the
oil extracts from the ice and the water sample as well as the oil
from the barge cargo. The cargo spectrum, typical of light fuel
oils, has many sharp absorption bands between 900 and 650 cm^{-1}.
The spectrum of the oil extracted from the ice is almost identical
to the cargo spectrum, suggesting little chemical change or weathering of the ice sample. The spectrum of the oil extracted from
the water below the ice layer is very similar to the other two,
however, some of the bands have decreased in intensity because of
a loss of the more soluble components into the water. These results show the high level of oil accomodated in the water just

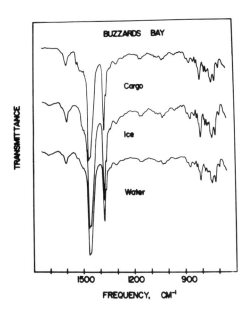

Figure 5. Infrared spectra (1700-650 cm^{-1}) of the No. 2 fuel oil spilled in Buzzards Bay, the spilled oil extracted from the ice, and the oil extracted from the water beneath the ice.

below the ice which could pose an environmental hazard to the aquatic life in the bay.

In addition to these samples, a water sample was collected at about 1-2 meters depth, as well as a sediment sample taken directly below the water sample. Figure 6 shows the infrared spectra (1750-650 cm^{-1}) of the oil extracted from these two samples, as well as the spectrum of the oil extracted from the surface water.

Comparison of the surface sample with the column sample shows the relative increase of the bands at 810 and 870 cm^{-1} (due to aromatic hydrocarbons) in the water column. Comparison of the surface spectrum to the sediment spectrum shows a definite relative increase in the 720 cm^{-1} band (due to paraffins) in the sediment sample.

The trend in partitioning in this spill seems to be that the aromatics, being more water "soluble", concentrate in the water column whereas the paraffins, which are relatively "insoluble", pass through to the sediments.

At the spill site, the air was saturated with an oil odor.

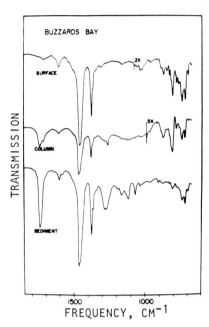

Figure 6. Infrared spectra (1850-650 cm⁻¹) of the oil extracted
from surface water, the water column (1-2 meters), and the sediments at the site of the Buzzards Bay oil spill.

Using a portable pump, we collected air samples on charcoal tubes
and returned the tubes to the lab for analysis. The organics were
extracted from the charcoal with CS_2 and analyzed by G.C.   In
addition, a laboratory apparatus was constructed to simulate a
spill site and to allow for the collection of organic vapors over
the spill. A diagram of the laboratory air sampling device is
shown in Figure 7. Seawater is placed in the inner glass cup and
the oil is "spilled" on top of this. A charcoal prefilter is
placed over the top of the apparatus and air is pumped from the
atmosphere, through the prefilter, around the spill area and
through the charcoal cartridges. Air flow is cut off to the cartridges by the stopcock valves above the cartridges.

     Using this apparatus, Buzzard's Bay cargo was spilled onto
Instant Ocean and air samples were collected in the laboratory.
These were also analyzed by G. C. Figure 8 shows the chromatogram
of the field sample and that of the laboratory sample. The basic
fingerprint due to the oil sample (shown in the laboratory chromatogram) can be seen in the field sample. The field sample also
has extraneous peaks from other air borne organics.

     The partitioning of the oil between the air and the water is

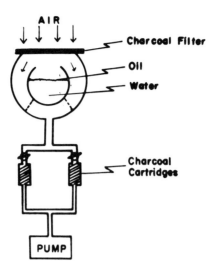

Figure 7. Schematic of the air sampling device used to collect hydrocarbons into charcoal tubes.

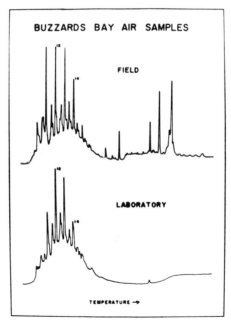

Figure 8. Gas chromatograms of oil vapors collected on charcoal tubes at the site of the Buzzard's Bay spill and oil vapors collected in the lab using the air sampling device and the cargo oil form the grounded barge.

shown in Figure 9. The bottom chromatogram is of the total oil
sample. Its components range from a little below C_{10} to a little
above C_{22}. The chromatogram of the water column sample contains
oil whose components range between C_{14} and C_{22}. The volatile
fraction lost to the air contains components lower than C_{15}.

This spill enabled us to trace the course of travel of oil
components into the air, on the ice, at the water surface, in the
water column and into the sediments. However, because of the
high viscosity of the No. 6 oil from the Argo Merchant, we could
not monitor that spill with the same success. A third tragedy has
provided us with a clue as to the ultimate fate of this type of
heavier oil in a spill situation.

A few days after the Argo Merchant ran aground, the tanker,
Grand Zenith, was lost somewhere in the Atlantic with a cargo of
8 million gallons of No. 6 fuel oil. The tanker has never been
found and is presumed sunken. But what of its cargo?............

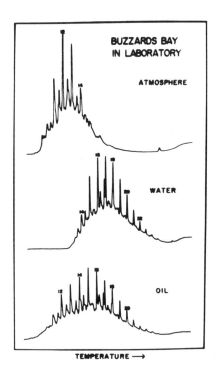

Figure 9. Gas chromatograms of the volatile (atmosphere), and
soluble (water) fractions of the spilled oil as well as the total
oil sample from the barge.

The top infrared spectrum in Figure 10 is the fingerprint of
the cargo oil from the Argo Merchant. The middle spectrum is that
of the Grand Zenith Cargo oil. The major spectral differences between the two samples are the relative intensities of the aromatic
bands (740, 810, and 870 cm^{-1}) and the ratio of the paraffin (720
and 725 cm^{-1}) to the aromatic bands.

In mid-February, large tarballs washed up on Martha's Vineyard and samples of these were analyzed by infrared. The bottom
spectrum is that of one of the tarballs. Visual comparison of
this spectrum to the two cargo spectra show that the cargo from
the Grand Zenith is a much closer match than the Argo cargo. The
only difference between the Zenith spectrum and tarball spectrum
are the paraffin bands at 720 and 725 cm^{-1}. Previous analysis
of various tarball samples have shown that this increase in the
paraffin band intensity is characteristic of tarball formation and
weathering.

In mid-February, tarballs were also washing up on the beaches
of Jamestown, Rhode Island. A ten pound tarball was collected and
an infrared spectrum was measured of a central core sample from

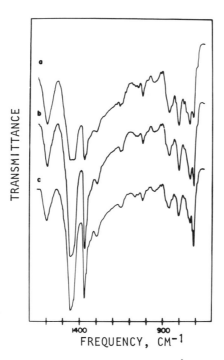

FREQUENCY, CM^{-1}

Figure 10. Infrared spectra (1650-650 cm^{-1}) of cargo from a. Argo
Merchant, b. Grand Zenith, and c. a tarball washed ashore at
Martha's Vineyard, Massachusetts.

this tarball. In Figure 11 the top spectrum is that of the tarball from Martha's Vineyard while the bottom spectrum is the
fingerprint of the Jamestown tarball. The two spectra are virtual
overlays.

      Considering the similarity in fingerprints of the Jamestown,
Martha's Vineyard and Grand Zenith samples, it was decided to artificially weather the Grand Zenith cargo sample on Narragansett
Bay. By "spilling" the cargo into a fiberglass tank containing
circulating seawater and by periodically taking samples of the oil
for analysis, the changes in the infrared fingerprint during the
weathering process could be monitored. By weathering the cargo it
was hoped that the fingerprint of the cargo would more closely
match that of the other tarballs. During the weathering experiment,
tarballs were washing up on beaches of Cape Cod. Some of these
were also analyzed by infrared. The top spectrum of Figure 12 is
the fingerprint region of a tarball sample at Nanset Beach, Cape
Cod, Mass. The bottom spectrum is a sample of Grand Zenith cargo
weathered 504 hours in seawater. The special feature of these two
spectra is that the relative paraffin band intensities are almost
identical in the two samples. The increased intensity of the band
at 1020 cm^{-1} in the artificially weathered sample is due to the
fact that, during the time that the cargo sample was weathered, the
temperature of the Bay water was higher than that to which the tarball samples would have been exposed. Higher water temperatures

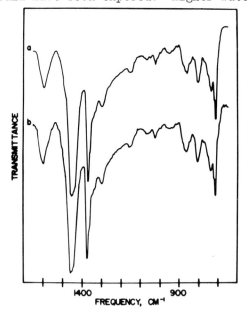

Figure 11. Infrared spectra (1650-650^{-1}) of tarballs washed ashore
at a. Martha's Vineyard, Massachusetts, and b. Jamestown,
Rhode Island.

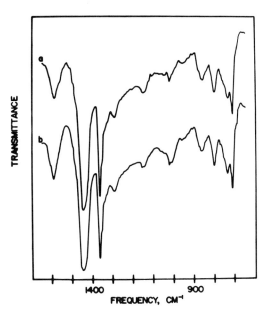

Figure 12. Infrared spectra (1650-650 cm^{-1}) of <u>a</u>. a tarball washed ashore at Nanset Beach, Massachusetts, and <u>b</u>. cargo from the Grand Zenith that had been artificially weathered on Narragansett Bay for 504 hours.

promote more rapid sample oxidation, causing an increase in the fingerprint bands at 1020 and 1700-1725 cm.

   These spectral comparisons furnish strong evidence that the Grand Zenith cargo could be the source of most of the tarballs collected at Jamestown, Martha's Vineyard and Cape Cod. It seems very unlikely that the Argo Merchant is the source. However, since oil tankers routinely travel this route, and since the cargo of these tankers have not been similarly analyzed, no definitive conclusion can be reached concerning the source of these tarballs. It must also be taken into consideration that the sample of oil labelled "Grand Zenith Cargo", was, in fact, taken from the pipelines through which the oil is passed on loading the tanker. Therefore, the oil from the cargo hold, if it mixed with any oil left from a previous shipment, may have a slightly different infrared fingerprint than the same oil before it was loaded into the tanker.

   When more is known about the unique mechanism of tarball formation, laboratory experiments could be more accurately modeled after real environmental conditions. In this way, tarballs could be artificially formed in the laboratory just as we can now artificially weather oil spills.

From these three tragedies, we have gained some insight into the problems of tracing the fate of an oil spill in time. Partitioning of the components of the oil takes place between the surface slick and the air, the water column, and the sediments. The amount of partitioning seems to depend on environmental conditions at the time of the spill and on the type of oil spilled. The surface oil slick will eventually dissipate. The residual components of a light oil may become incorporated in bottom sediments. The residual components of a heavier oil may also sink to the sediments or they may incorporate sediment, shells, sand, etc. in them to form tarballs. The types of components that partition into the air, the water, and sediments become important when related with their relative toxicity to the plant and animal kingdom. And only when we find out what components go where can we accurately study their impact on the environment.

Acknowledgement

The authors wish to express their appreciation to the Environmental Control Division, the Energy Research and Development Administration (contract no. E(11-1)-4047) for support of this research. We are also indebted to M. Ahmadjian, M. M. Brady, and F. E. Franklin for collecting samples and performing analyses.

References

1. Brown, R. A., Elliott, J. J., and Searl, T. D.: Measurement and characterization of nonvolatile hydrocarbons in ocean water. AID. 4BA. 74 (1974).

2. Gruenfeld, M.: Extraction of dispersed oils from water for quantitative analysis by infrared spectrophotometry. Environ. Sci. Tech. 7 (1973) 636.

WATER POLLUTION STUDIES USING RAMAN SPECTROSCOPY

Laurent Van Haverbeke* and Chris W. Brown

University of Rhode Island, Department of Chemistry

Kingston, Rhode Island 02881

ABSTRACT

Methods for adapting laser Raman spectroscopy for monitoring water pollution have been developed and tested. Both conventional and resonance Raman spectra have been measured; the latter effect lowers the detection limits by $\sim 10^3$. In general, Raman spectroscopy has the following advantages in pollution studies: (i) water is a very weak Raman scatterer, (ii) each chemical has a characteristic Raman fingerprint, (iii) quantitative measurements can be made easily, (iv) remote and flow-through detection systems are feasible. These advantages and the difficulties with the method are discussed herein.

I. Introduction

The increase of pollution by hazardous chemicals and man's awareness of their possible dangers have created a great need for suitable techniques for controlling and monitoring air and water pollution. In the last decade, a number of known analytical techniques have been adapted for pollution analysis; Raman spectroscopy is not an exception in this trend. The special characteristics of this technique suggest that it will be a very useful tool for pollution detection, especially in water pollution studies.

Raman spectroscopy is part of the group of optical spectroscopic techniques, which also includes visible and uv spectroscopy, infrared spectroscopy, fluorescence spectroscopy, etc. (6,7) It is based on light scattering. If monochromatic light (usually in the visible region) is sent through a sample, which contains one or more kinds of molecules in vibrational and rotational states according to the Boltzmann-distribution, most of the incident light passes through the sample. A small portion of the light, however, is scattered spherically. This scattering may happen in two different ways; elastic and inelastic. If the light is scattered elastically, the frequency of the scattered light is the same as the one of the incident light. This phenomenon is called Rayleigh scattering. The light is scattered inelastically when a molecule goes from its original vibrational state to a lower or higher one; the frequency of the scattered light is accordingly higher or lower. This phenomenon is called the Raman effect. The combination of these two phenomena gives rise to a frequency spectrum of the scattered light, dominated by an intense central band at a frequency equal to the one of the incident light, plus a number of smaller bands on both sides of this central band. The lower frequency lines, called Stokes lines, appear at the same distance from the central band (the Rayleigh line) as the higher frequency lines, which are anti-Stokes lines. The intensity of the former are, however, much greater, and usually only this part of the spectrum is measured.

The intensities of the different Raman lines in the spectrum of a particular sample depend on a large number of intrinsic and external factors. First, they depend on the probability of going from one particular vibrational state to another. Secondly, they depend on the polarizability characteristics of the molecule and of the individual transitions. These factors are inherent to the sample being studied and can only be varied under special conditions, which will be discussed later.

Externally controllable parameters include, at first instance, the intensity and the frequency of the incident light. It is easy to see that the Raman band intensity is linear with the intensity of the incident light. Furthermore, it has been proven that the intensity of a Raman spectrum is proportional to the fourth power of the frequency of the incident light, which accounts for stronger spectra when going from red to violet light. It is obvious that the efficiency of the detection equipment also affects the observed Raman intensities.

A special situation occurs when the frequency of the incident light is the same as the maximum of an electronic absorption band or falls within the absorption band envelope. The Raman bands corresponding to those vibrations that contribute to the absorption band are enhanced by several orders of magnitude in comparison with conventional Raman spectroscopy. This phenomenon is called resonance Raman spectroscopy (3). Since the contribution to the absorption band is not the same for every vibration, the relative band intensities may differ from those in the conventional spectrum.

A schematic diagram of the instrumentation to obtain Raman spectra is given in Figure 1. Since the Raman bands are very weak, an intense monochromatic light source is needed. Lasers are the most used sources for incident light. Currently, He-Ne lasers and Kr lasers are used for the red part of the visible spectrum, Ar^+ lasers for the rest of the visible spectrum and N_2 lasers for the near uv part of the spectrum. A variable frequency source can be obtained by using these in conjunction with dye lasers.

The conventional way of obtaining Raman spectra is with a 90° configuration. In this case, the scattered light is collected perpendicular to the incident light. The spectrum is analysed through a monochromator and detected by a photomultiplier. Signal amplification is usually done by means of photon counting electronics, and then displayed on a recorder or transmitted to a data handling unit.

Raman spectroscopy has a number of advantages and disadvan-

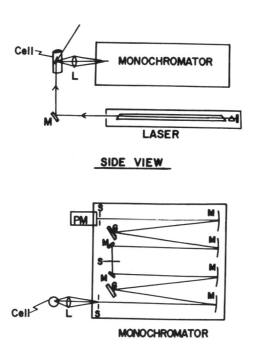

SIDE VIEW

TOP VIEW

Figure 1: Schematic diagram of a Raman spectrometer.

tages in comparison with other techniques used in pollution analysis. By far, the major advantage is the fact that the presence
of water does not significantly disturb the sensitivity of the
effect. Water, even in large quantities, shows only a very weak
Raman spectrum. This implies that solutes can be detected in
water, even at low concentrations, without previous extractions
that are necessary with most other techniques. Furthermore,
because of the complexity of the spectrum, it can be used for
identification purposes better than by other techniques.

One of the major disadvantages of Raman spectroscopy is the
weakness of the Raman signal. However, this can be overcome by
using more intense incident light, by using resonance Raman
spectroscopy or by computer data handling techniques. Another
disadvantage is the occurrence of fluorescence by most natural
waters and by certain hazardous chemicals. Studies are now underway to deal with these problems.

II. Detection Limits For Water Soluble Compounds

The application of lasers in Raman spectroscopy during the
middle sixties made it possible to obtain Raman spectra from relatively low concentrations. However, the first efforts to apply
Raman spectroscopy to pollution studies were not made until the
early seventies. The first studies using resonance Raman spectroscopy were performed during the past two years.

II.1 Conventional Raman Spectroscopy

In 1970, Bradley and Frenzel (4) reported a preliminary study
on the use of laser Raman spectroscopy for detection and identification of molecular pollutants in water. They reported the detection of benzene in water at concentrations around 50 ppm. In
1972, Baldwin and Brown (2) examined the detection level for some
inorganic anions in water. They could obtain minimal detectable
concentrations between 25 and 75 ppm, depending on the type of
anion. Recently, using more advanced equipment, Cunningham, Goldberg and Weiner (5) were able to go down to 4 - 40 ppm for a
series of ionic and molecular species in water.

All these measurements were done in pure water, and had the
advantage of ideal conditions. In their study on anions, Baldwin
and Brown (2) also measured the spectrum of sea water and were
able to demonstrate the presence of sulfate ions at a concentration around 2400 ppm as is shown in Figure 2. A general conclusion
that could be drawn from these experiments is that under ideal
conditions, the detection limits were hardly low enough to be of
practical use. Using the technique in real circumstances would
even be less favorable.

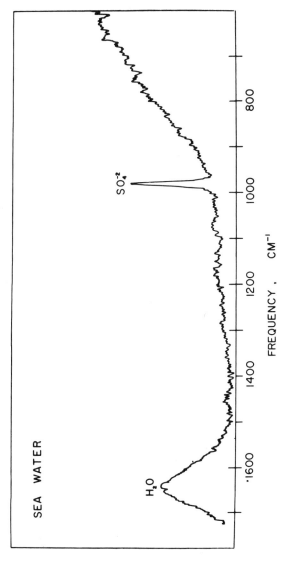

Figure 2: Raman spectrum of sea water showing a strong SO_4^{--} band at 980 cm^{-1}

In order to lower the detection level of pollutants in water
when using Raman spectroscopy, it is necessary to increase the
spectrum intensity without increasing the one of water which occurs
as a background. Increasing the laser intensity only partially
solves the problem. An interesting way of increasing the solute
spectrum without increasing the one of water is by taking advantage
of the resonance Raman effect.

II.2 Resonance Raman Spectroscopy

In a recent study, Van Haverbeke, Lynch and Brown (11) have
examined the applicability of resonance Raman spectroscopy in
pollution measurements. Because of the limitations of their equipment (visible laser and optics), they were restricted to compounds
that absorb in the visible spectrum. Their study lead to detection
limits of 30 - 55 ppb for industrial fabric dyes, which is approximately three orders of magnitude lower than those of conventional
Raman spectroscopy. Their comparison with uv - visible spectrophotometry shows the same sensitivity for both techniques (see
Table I). Moreover, they were able to obtain spectra suitable
for identification purposes at concentrations between 75 and 175
ppb. Spectra of the four dyes they examined at concentrations
close to the latter are shown in Figure 3. They also tested the
method under actual conditions using artificially "polluted" river
water. A spectrum at a concentration of 300 ppb clearly showed

TABLE I

Minimal Detection and Identification Levels.*

| Dye Name | Raman | | Visible |
	Detection	Identification	Spectra
Superlitefast Rubine	35	75	66
Procion Red	30	140	30
Lyrazol Fast Red	45	160	40
Direct Red 83	55	175	50

* All values in parts per billion (ppb)

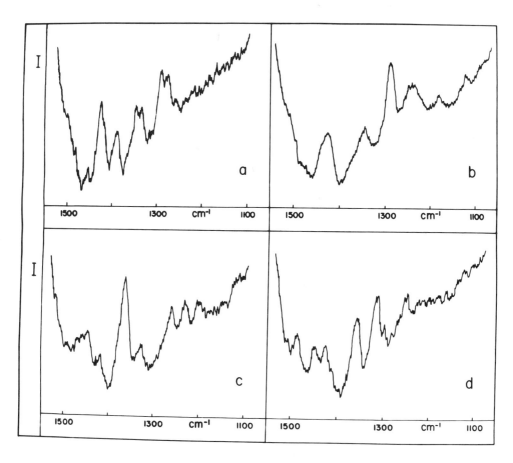

Figure 3: Resonance Raman spectra of a) Superlitefast Rubine (88 ppb), b) Procion Red (192 ppb), c) Lyrazol Fast Red (210 ppb), d) Direct Red 83 (190 ppb).

the presence and revealed the identity of the dye as is shown in
Figure 4. A similar study on pesticides gave analogous results (8).

II.3 Limiting Factors

There are two major factors that limit the use of this technique in practical applications. These are the spectrum of the
solvent, water, and the presence of fluorescence.

Although water is a very weak Raman scatterer, the presence
of large amounts of it together with very small amounts of the
compound to be investigated poses some problems. Indeed, the water
spectrum appears as a very huge background above which the sample

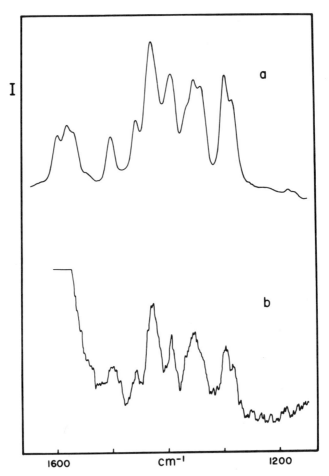

Figure 4: Resonance Raman spectra of: Superlitefast Rubine: a)
          in distilled water (24.30 ppm) b) in riverwater (300 ppb)

spectrum has to be observed. However, using detection electronics that have signal offset and expansion capabilities, one can overcome these problems.

The most important limiting factor is the occurrence of fluorescence. The sample is radiated by the incident source having a certain frequency. If, by some mechanism, the sample is able to convert part of the energy to internal effects other than vibrations and rotations and emit the rest of it as light of a frequency other than the incident light, fluorescence occurs. Experimentally, fluorescence causes a huge, broad band in the spectrum and the occurrence of a lower signal to noise ratio.

Many compounds may exhibit this phenomenon. It may be due to the compound studied, or it may be due to the environment of the compound, i.e. the nautral water. The latter situation is especially important, because it has been observed that a great number of natural organic compounds exhibit fluorescence. Since in natural waters organic compounds are present in the form of aquatic life, most natural water samples show fluorescence.

Presently, studies are underway to eliminate the fluorescence problem. These include instrumentation with pulsed lasers and gating electronics, rapid scanning techniques based on television camera type detectors, and data-handling by means of automatization equipment.

III.   Qualitative And Quantitative Properties

In addition to the detection of water pollutants, Raman spectroscopy, both conventional and resonance, can be used to identify pollutants and to determine their amounts.

III.1 Identification Capabilities

As is explained above (Section I) the origin of Raman spectra is in vibrational rotational modes of the compound studied. It is easy to see that when the vibrational characteristics of two molecules differ greatly, the Raman spectra will also be very different. These vibrational characteristics depend on the kind and number of functional groups in the molecule. Thus, if two compounds have totally different groups, their spectra will differ completely.

Furthermore, even if functional groups are the same, summetry properties and distribution of the electronic charges cause differences in the positions and the intensities of the Raman bands. In addition to this, when using resonance Raman spectroscopy, the contribution of the different vibrations to the electronic absorption band may even enhance the differences in band intensities.

The overall result is that each compound has a spectrum or
even a part of a Raman spectrum that is unique and different from
that of any other compound. By analogy with infrared spectroscopy,
where the same phenomenon occurs, we can speak about a "fingerprint" which can be used for identification purposes. The spectra
of four dyes at low concentrations are shown in Figure 3. Although
the composition of these dyes is not very different (they all are
poly-aromatic diazo dyes), the differences in their spectra are
great enough to differentiate one from another (8,11).

III.2 Quantitative Measurements

Raman spectroscopy can also be used for quantitative measurements. However, some very important aspects are to be considered
here.

Although in theory not impossible, it is very difficult to
perform absolute measurements of Raman band heights or intensities.
The main reason for this is the uncertainty regarding the efficiency
of the spectrum analyser (monochromator) and the detection system,
and also the reproducibility of the optical sample arrangement.
Therefore, it has become common use to measure the relative height
or intensity of the compounds versus an internal standard.

The next problem is the choice of the internal standard. One
possibility is very obvious. Since we always have the same solvent
(water) present in more or less the same amount, we can easily use
one of the two bands of water (around 1600 and 3500 cm^{-1}) as internal reference band. The disadvantage of this choice is the fact
that both bands are extremely broad, which tends to introduce errors
in their height and area calculation.

On the other hand, we could add an internal standard to the
solution by adding a foreign compound. Preferably, this standard
should have Raman bands which would not interfere with Raman bands
of the pollutant. This means that probably no universal standard
can be found, and that the selection of it depends on knowledge of
the pollutants present. The band shape poses no problem, if an
adequate standard is chosen. However, it is virtually impossible
to use an added standard for remote detection systems or flow
through systems, as will be discussed in the next section. Overall,
the choice of the standard will be indicated by the kind of measurement that is being performed.

Both for sample and reference bands it is preferrable to use
integrated band intensities instead of the band heights, since the
height is much more sensible to differences in monochromator slit
variations than the integrated intensity. Integrated intensity
calculations can be done by several methods with varying accuracy
(9). We have been most successful with those based on band contour

analysis using asymmetrical mixed Gauss-Lorentz profiles in combination with a minicomputer (10).

When using conventional Raman spectroscopy, one obtains a
linear relationship between the relative intensity and the
concentration. Indeed, all factors affecting the uncertainties in
the absolute intensity of sample and reference bands vanish when
calculating the intensity ratio.

A slightly different situation occurs when using resonance
Raman spectroscopy. In this case, we have to account for the
absorption of light by the sample. The amount of light reaching
the point of scattering is the same for both sample and reference
(H_2O) molecules. However, since the frequencies of the Raman bands
of sample and reference are different, the effect of absorption
will be different on each, and their intensity ratio is expressed
by the following equation:

$$\frac{I_{sample}}{I_{ref}} = (Const)\ c_{sample}\ exp(-(\varepsilon_s - \varepsilon_r)\ \ell\ c_{sample}),$$

where ε_s and ε_r are the molar absorptivities of the sample at
the frequencies of the sample band and reference band, and ℓ is
the optical path length between the scattering point and the end
of the sample cell. As can be seen, an exponential deviation from
a straight line is obtained (12).

This relationship can still be used for quantitative measurements by constructing a calibration curve. However, care must be
taken to do all measurements in the same optical arrangement, so
that ℓ remains constant. This requires rigid sample holders and
perfectly matched cells. An example of such a calibration curve
for a dye in the region 0 - 6 ppm is shown in Figure 5.

IV. Continuous Monitoring Systems

Since Raman spectroscopy measurements may be performed without
extraction, it is possible to make remote and continuous measurements.

IV.1 Remote Sensing

A remote sensing system has been proposed by Ahmadjian and
Brown in 1973 (1). Since it is impossible to use a 90° system for
remote sensing, they used the back-scattering principle (collection
of light at 180°). A schematic diagram of their arrangement is
shown in Figure 6. The laser beam and scattered light travel along
the same route. However, due to the large difference in diameter
between the laser beam (1mm) and the scattered light beam (13 cm),

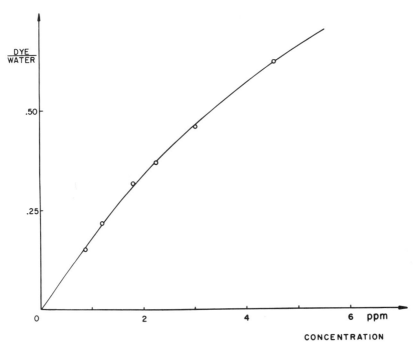

Figure 5:   Dye to water band intensity ratio versus concentration
            for Superlitefast Rubine.

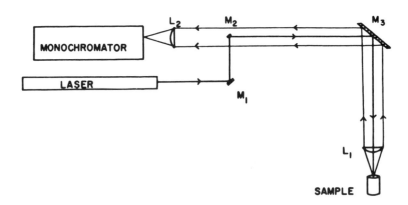

Figure 6:   Schematic diagram of the optical arrangement for remote
            sensing.

the reflection of the laser beam on the sample surface can be removed using a small mirror.

The authors tested their system by means of conventional Raman spectroscopy on nitrate ions. They obtained a detection limit around 150 ppm. This means that the loss of sensitivity amounted to 1/6 of that obtained with the conventional setup.

The method is particularly interesting because there is virtually no limitation to the distance between the instrument and the sample position. Moreover, by using a shielded lightpath, one can avoid interferences from daylight and other light sources.

IV.2 Flow Through System

Another approach to continuous monitoring of natural waters by means of Raman spectroscopy was proposed by Van Haverbeke and Brown (12). They mounted a flow cell in the standard illumination chamber of the spectrometer. By means of a continuous pumping system and tubing, water is drawn from the source to be monitored and passed through the cell.

Comparison with classical Raman measurements showed that the background is enhanced and the signal to noise ratio lowered by approximately 30%. This is caused by the flow of the solution through the cell. Experiments have shown that a cell with square cross-section gives the best results. The sensitivity is, consequently, 30% less than for conventional measurements.

The method was tested for response when gradual concentration changes occur. This was done by measuring the spectra at certain intervals when a time dependent solution was flown through the cell. The results are given in Figure 7. As can be seen, the agreement between the actual concentration (solid line) and the experimental determinations (circles) is very good, even in the low concentration range where the Figure is 10x expanded in the concentration axis.

The method was also tested for its response on sudden concentration changes. Three questions were asked here: a) How much time goes between the actual change and the moment it is noticed on the instrument? b) What is the minimum duration of such a sudden change to be able to determine its concentration? c) Can the amount released during the total duration be determined?

To determine this, water was passed through the cell. At a certain moment and for a certain time, a certain concentration was injected, after which again pure water was flown through. The reaction of the instrument at a particular frequency was recorded.

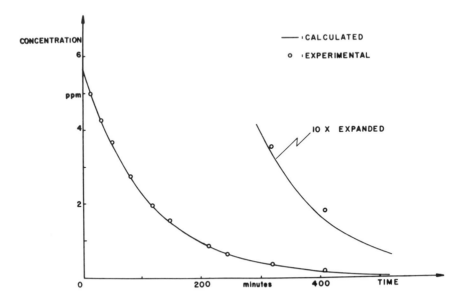

Figure 7:   Response of a flow through system on gradual concentra-
             tion changes.

The results are shown in Figure 8. It can be noticed that the instrument reacts approximately 1.25 minutes after the injection and that the injection must at least last 1.5 minutes to be able to record the actual concentration. It must be mentioned that these values depend entirely upon external conditions such as flow rate and distance between instrument and sampling spot.

To determine the capability to detect the amount of pollutant, the area beneath each of the curves was determined. Since all injections were done with the same concentration, the total amount is relative to the duration of the injection. A plot of area versus duration is given in Figure 9. Since these measurements were done with resonance Raman spectroscopy, a linear relationship could not be observed. However, the curve can be used as a calibration curve.

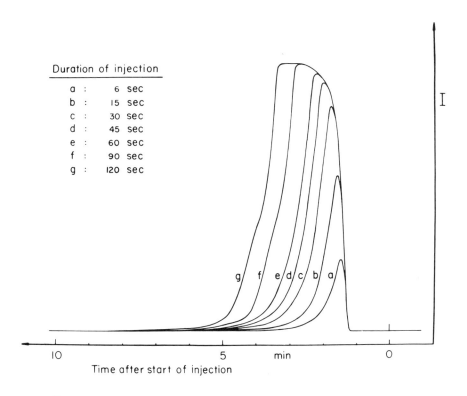

Figure 8: Response of the flow through system on sudden concentration changes.

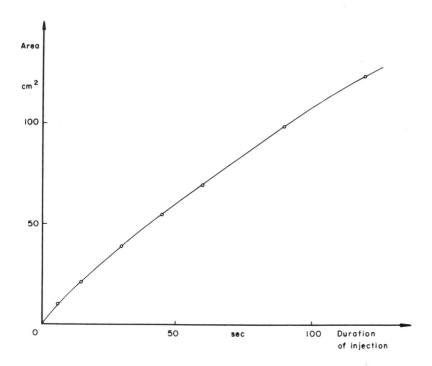

Figure 9:   Flow through system: Area under the recorded curve
            versus the duration of the injection.

V. Conclusion

From the foregoing, it can be concluded that Raman spectroscopy may be a very useful analytical tool in pollution monitoring.
In order to obtain good sensitivity, the use of resonance Raman
spectroscopy is desirable and with use of uv lasers a larger
number of compounds can be analysed by this technique.

Qualitative as well as quantitative measurements can be made
using the Raman technique. Above all, the most important feature
is the fact that Raman spectroscopy can be used directly on the
aqueous solution, avoiding complicated extraction procedures that
may modify the results of the method. Because of this feature, it
can be used as a continuous monitoring technique without great
complications.

Acknowledgement

This work was supported by the National Sea Grant Program
(National Oceanic and Atmospheric Administration), and by the

Office of Water Resources Research (Department of the Interior).
One of us (L.V.H.) gratefully acknowledges financial support from
the State University Center of Antwerp and a NATO Research Fellowship.

References

1. Ahmadjian, M. and Brown, C. W.: "Feasibility of remote detec-
    tion of water pollutants and oil slicks by laser-excited
    Raman spectroscopy." Env. Sci. Technol. 7, 452 (1973)

2. Baldwin, S. F. and Brown, C. W.: "Detection of ionic water
    pollutants by laser excited Raman spectroscopy." Water Res.
    6, 1601 (1972)

3. Bernstein, H. J.: "Resonance Raman spectra." Advan. Raman
    Spectroscopy (1972), 305

4. Bradley, E. B. and Frenzel, C. A.: "On the exploitation of
    laser Raman spectroscopy for determination and identification
    of molecular water pollutants." Water Res. 4, 125 (1970)

5. Cunningham, K. M., Goldberg, M. C. and Weiner, E. R.: "Inves-
    tigation of detection limits for solutes in water measured by
    laser Raman spectrometry." Anal. Chem. 49, 70 (1977)

6. Gilson, T. R. and Hendra, P. H.: "Laser Raman spectroscopy."
    Wiley Interscience, New York   (1970)

7. Herzberg, G.: "Molecular spectra and molecular structure.
    II infrared and Raman spectra of Polyatomic molecules." Van
    Nostrand, Princeton, New Jersey   (1960)

8. Thibeau, R. J., Van Haverbeke, L. and Brown, C. W.: "Detec-
    tion of water pollutants by laser excited Raman spectroscopy.
    II. pesticides and fungicides." Appl. Spectroscopy submitted
    for publication.

9. Van Haverbeke, L. and Desseyn, H. O.: "De geintegreerde
    intensiteit van absorptiebanden in infrarood spectrometrie.
    Deel I: bepalingsmethoden." Ind. Chim. Belge 39, 142 (1974)

10. Van Haverbeke, L., Brown, C. W. and Herman, M. A.: "Infrared
    intensity measurements by band contour analysis." Appl.
    Spectrosc. submitted for publication.

11. Van Haverbeke, L., Lynch, P. F. and Brown, C. W.: "Detection of water pollutants by laser excited resonance Raman spectroscopy. I. Detection and identification limits for industrial dyes." Anal. Chem. submitted for publication.

12. Van Haverbeke, L. and Brown, C. W.: "Detection of water pollutants by laser excited resonance Raman spectroscopy. III. Flow through system." Anal. Chem. submitted for publication.

*First Assistant
Laboratory for Inorganic Chemistry
State University Center of Antwerp
Groenenborgerlaan, 171, B2020 Antwerp (Belgium)

Discussion

GEISS - I have a question for Dr. Haverbeke. Do you use dye lasers in the visible?

HAVERBEKE - No, we haven't. We have not used them and I'll tell you why. If you use Raman spectroscopy you need all the power you can get to excite your sample. From my experience if you are going up to a dye laser you're going to lose 95% of your actual power and that's too much loss to bear. It is better in such a case to choose a line of your argon laser which is in the vicinity of your absorption band, and you can use the full power of the laser.

GEISS - I have another question. What about disturbance from fluorescence, if you are using practical samples?

HAVERBEKE - That is one of the major problems. I must say that at the moment studies are on the way to deal with fluorescence. Certainly, if you have natural waters you have organic compounds that do fluoresce.

GEISS - We have learned that Raman spectroscopy in air pollution completely failed, first of all of course because the sensitivity is too low, and second if it were high enough it would be cancelled out by fluorescence.

HAVERBEKE - That's right. Well fluorescence is a major problem, but I know for the moment there are a number of studies underway to deal with fluorescence and I have seen some preliminary results, and they really are fantastic. They can diminish fluorescence from one tenth to one twentieth of the normal level.

GEISS - This means taking the higher frequency.

HAVERBEKE - No, they mostly work with pulse lasers and grating optics.

HEINRICH - In your telescope type instrument what was the distance that you could cover?

HAVERBEKE - Approximately five to six feet long and three feet wide, I think.

HEINRICH - I mean the distance when you had it in parallel - the primary being the secondary thing. How far --

HAVERBEKE - It has been done for twenty-one feet, There is actually no limitation. Well, you're going to lose some sensitivity, but theoretically there is no limitation to the distance between the instrument and the sampling spot.

HEINRICH - So you could conceivably look at a smokestack.

HAVERBEKE - You could, and I think there have been experiments on that. They have been partly successful and partly failures.

BEAMAN - How good is your technique compared with using X-ray fluorescence for trace elements as a tag in the oils?

BROWN - To be perfectly honest I don't know anybody who has used X-ray fluorescence of trace elements. The Coast Guard is looking for an instrument they can use in the field and it should be relatively inexpensive. What they would like to do is have instrumentation in each port captain's office and there are about 64 port captain's offices.

HEINRICH - Dr. Brown, given the percentages of probability you'll get, how would the analysis like that stand up in court?

BROWN - I was trying to avoid that question. They have never really gone to court. Generally, the Coast Guard can settle all these cases outside of court. If they have 99% probability from infrared and if they can get a 99% probability from another technique, they don't have to go to court. The Coast Guard sets the fines themselves, anywhere up to $5,000. Now, if the spiller wants to contest it, then it could go to court; however, they generally negotiate the fine.

FELDMAN - Well, for a little fine like that, there is no reason for them to bother.

BROWN - Yes, that's right. It is cheaper to pay the $5,000 when it costs $100,000 to pick it up.

GEISS - I suggest that if they put the captain in prison it would be more effective.

FELDMAN - Well, that's true for all crime.

ETZ - I'm wondering about the analysis of a water sample by your technique. To what extent would you have to remove suspended particulates, fine particles, in order to --

HAVERBEKE - You have to.

ETZ - What will happen when you do this?

HAVERBEKE - In such a case the only thing you can do is filter your sample and treat the particulates and the solution separately, because with particles in your solution you get a reflection in Raman spectroscopy. Certainly that will enhance the background of the spectrum, so you have to remove them.

GEISS - Do you use the remote sensing arrangement for ranging as well?

HAVERBEKE - No.

SAWICKI - Where did you get the samples. Do you know?

BROWN - No, I don't. The Coast Guard brings these samples in to us. Sometimes we'll know the general locality, but many times we don't know exactly.

SESSION III:

METHODS FOR FIELD USE

PASSIVE SAMPLING OF AMBIENT AND WORK PLACE

ATMOSPHERES BY MEANS OF GAS PERMEATION

Philip W. West

Environmental Sciences Institute
Department of Chemistry
Louisiana State University
Baton Rouge, Louisiana 70803

ABSTRACT

Sampling is often the most critical and most difficult step
in the analytical process. Passive sampling of atmospheres holds
great promise in the study of ambient air and work place atmospheres.
The permeation of gases through membranes has been introduced as a
means for passive sample collection that provides both accuracy
and convenience. The permeation technique has the advantage that
it is relatively free from disturbing effects such as variations in
temperature, humidity, movement and the presence of copollutants.
Because no power is required the sampling units are small, light,
convenient and essentially indestructible. Thus, permeation
devices offer the ideal approach for personal monitors and fills
a unique need for industrial hygiene studies.

Introduction

The study of air quality generally involves one or more
sampling approaches and the validity of the sampling will determine to a large extent the validity of the analytical results
obtained. Stack sampling provides information regarding the
emissions from point sources and thus provides a means for predicting the impact of emissions on the receiving environment.
Ambient air sampling is the accepted method for determining air
quality in general. The exposure of workers in potentially
hazardous environments presents a special problem. Personal
monitors carried by the individual workers are dictated in many
instances and these may be supplemented or substituted by area

monitors that provide information on conditions within a limited
and confined space area.

Passive sampling has limited application in studies involving stacks or other point sources. For ambient air studies and
area and personal monitoring, however, passive sampling often
has considerable appeal, and in the case of personal monitoring,
there appears to be nothing known at this time that offers the
many advantages provided by the permeation technique.

Passive Methods for Sampling

The analytical process is initiated by some need. Based on
need, samples are usually collected, although the ideal approach
would be an _in situ_ method of measurement that would determine the
amount or concentration of material of interest by means of some
direct sensing device. Alternatively, certain effects measurements can be made without sample collection, as for example the
use of lead peroxide candles or sulfation plates in determining
relative amounts of sulfur dioxide in atmospheres, the use of
corrosion coupons to define relative corrosion rates, or the
study of defined physiologic parameters as a measure of health
effects associated with certain activities and exposures. In
general, effects measurements do not quantify in terms of what is
present and this detracts from potential uses in controlling
causative agents. Because the collection of samples is generally
necessary it is well to consider how samples can be taken and
processed. Where possible, the analyst should have an input into
determining the location, manner, duration and frequency of
sampling. The analyst should also determine how the collected
material is to be stablized and processed, and finally, which
measurement techniques are best suited for the samples collected.

The study of air quality introduces a number of analytical
problems that are the result of the complexity and great dilution
of most atmospheric systems. Both particulate and gaseous
materials are present in the air, but the present discussion will
be limited to gases and will emphasize the use of permeation as a
means of passive sampling as required for either industrial
hygiene or ambient air studies.

Passive sampling includes effects measurements, the use of
evacuated bottles, diffusion systems and permeation devices. As
mentioned previously, effects measurements cannot be related
satisfactorily to actual atmospheric concentrations. Evacuated
bottles or flasks are often used for collecting grab samples,
but they may also be used as a passive device for collecting
integrated samples over a predetermined time period. For personal

monitoring the sampling rate can be determined by means of a
critical orifice and the sample volume determined from the flow-
rate and the time involved. Diffusion has been used in sample
collection as a means of minimizing the various parameters that
affect sampling rates. The diffusion technique was first introduced as a means of preparing standard gas atmospheres (3) and it
has been used quite successfully for such purposes (1, 7) where
conditions can be kept reasonable constant. In preparing standard
gas mixtures a constant concentration of reference material in
liquid form is allowed to diffuse into a dynamic dilute system.
The conversion of a concentrated reference to a dilute mixture
can be established with considerable reliability. The reverse
process applied in sampling involves the collection of material
from a dilute system with subsequent concentration, stabilization
and ultimate measurement of the collected material. In our
experience, temperature and humidity effects which are difficult
to control in the field detract seriously from what would otherwise
be a convenient means for sample collection. Diffusion also lacks
flexibility because it is not discriminatory, and this together
with the variations induced by changing conditions detract from
the ultimate utilization of this process.

The permeation approach for sample collection was introduced
by Reiszner and West in 1973 (6). Ambient air was sampled for
sulfur dioxide by means of a simple cell, containing a solution
of tetrachloromercurate(II), stoppered at the top and enclosed on
the bottom by a permeable membrane of a silicone polymer (See
Figure 1). The device was placed in the atmosphere to be sampled
whereupon sulfur dioxide in the atmosphere permeated the membrane
and reacted with the tetrachloromercurate(II) and was thus captured
and stabilized as dichlorosulfitomercurate(II). The resultant
complex was then determined by the West-Gaeke spectrophotometric
procedure. The amount cf SO_2 found could be related to its
concentration in the atmosphere being sampled through the use of
the predetermined permeation constant. The results provided a
time-weighted average for the SO_2 concentration existing in the
atmosphere under study for the exposure time. The first study
demonstrated the utility of permeation for ambient air studies and
introduced the possibility of permeation as a means for sample
collection for personal monitoring.

Permeation of a given gas through a given membrane is a
function of the concentration of that gas. Under specified
conditions the amount of gas passing through the membrane can be
established from the predetermined permeation constant. In a
physical sense, permeation can be compared with dissolution.
Thus, a gas molecule coming in contact with a liquid dissolves and
permeates into that liquid without the benefit of stirring. For
the specific case of sampling gas molecules, the sample atmosphere

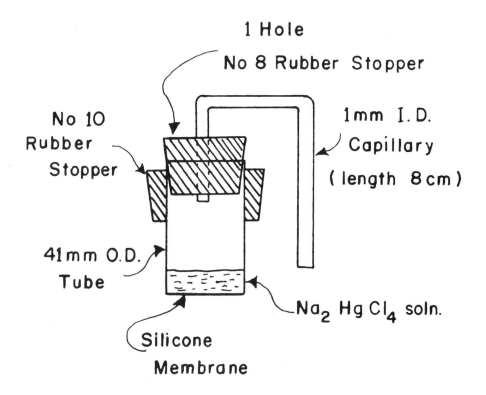

Figure 1. Permeation device for ambient air monitoring.

contacts the polymer surface. The gas molecules dissolve and permeate into the polymeric barrier. If a collecting agent is on the
other side of the barrier the permeating gases are removed from the
membrane, thus preventing the establishment of an equilibrium. Of
course, if there is no collecting action established to remove the
permeated molecules the membrane simply becomes saturated with the
molecules in question and an equilibrium is established between the
atmospheric gas and the gas that is dissolved in the membrane.

Because permeation follows a fixed law, the behavior of a
given gas for a given membrane can be summarized by the following
equation:

$$k = \frac{Ct}{w}$$

where k = the permeation constant, C = concentration of the gas of
interest in ppm, t = time of exposure and w = amount of gas adsorbed,
μg.

The permeation constant for each individual gas can be
determined by exposing the selected membrane to a known concentration of the gas in air (See Figure 2). The resultant constant
obtained by calibration can then be used for determining the average
concentration of the gas of interest in unknown atmospheres through
means of the equation:

$$C = \frac{wk}{t}$$

where C = the time-weighted average for the gas concentration, ppm.

Physical laws, such as those governing permeation apply under
specified conditions, and it is important that any varability in
conditions be kept at a minimum or corrective steps be taken to avoid
the variables or compensate for them. The sampling of gases is
generally affected by variations in humidity, temperature and the
presence of other gases. For permeation it was quickly established
that variation in humidity had only a negligible effect. Changes in
temperature, however, altered dramatically the rate of permeation
for the gases studied when membranes such as rubber and cellophane
were used. Fortunately, silicone membranes such as single-backed
dimethyl silicone (General Electric Company, One River Road,
Schenectady, NY 12305) exhibited little temperature effect. Also,
permeation of a given gas was found to induce negligible error due
to the presence of other gaseous species, movement of the air being
sampled, or movement of the sampler.

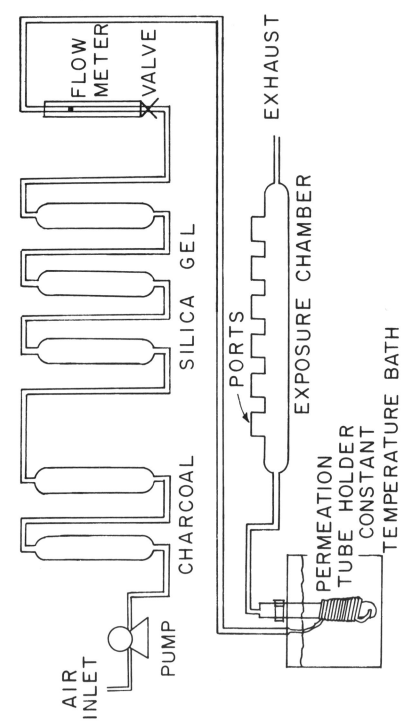

Figure 2. Exposure chamber for calibration.

Applications

Both ambient air and work place studies can be made using passive sampling techniques. Admittedly, most ambient air studies are made using dynamic samplers. Dynamic samplers are often an integral part of a monitoring system in which samples are collected, processed and measurements made and continuously recorded. For many monitors the results obtained are digitized and automatically transmitted by means of telemetry to a central laboratory for final processing. In general, it is difficult to fault automatic monitors but there are certain drawbacks such as complexity, cost, maintenance problems and the requirement of power that understandably are associated with such sophistication. Passive sampling by means of the permeation technique costs only a small fraction as much as conventional sampling and monitoring installations. Furthermore, the permeation devices do not wear out and, because there are no moving parts, they are essentially maintenance free. Because they are so very small, light-weight and require no power they can be installed almost anywhere and left unattended for prolonged periods of time.

As implied above, both short-term as well as long-term sampling may be of interest in determining ambient air quality. Passive methods of sampling offer distinct advantages for long-term sampling and has some distinct advantages for even short-term studies. In considering the relative merits of passive sampling and particularly the use of the permeation technique, it is well to consider the need for air quality studies.

For the moment, discussion of work place atmospheres will be postponed. For ambient air studies, however, the need for evaluation is based on the general interest man has regarding his environment. The impact of atmospheric contaminants on weather and climate modifications, corrosivity and aesthetic quality is not within the scope of the present discussion. Toxicity, however, is to be considered. If toxins are to be considered so should nutrients. This bit of prejudice can be excused possibly if one recalls that there is usually an optimum range over which contaminants may be desirable and only when the range is surpassed does the contaminant become a pollutant causing toxic or other undesirable manifestations. It might be added that from the standpoint of toxicity that there is justification in considering the welfare of both plant and animal species.

Toxins may be classified as being acute, cumulative or additive. Most attention has been directed toward acute toxins because of their dramatic effects. This may be a misdirected emphasis because often if lethal concentrations are not experienced, little if any harm results. For example, it appears that professors who taught the hydrogen sulfide qualitative analysis scheme for their professional lifetimes were usually healthy and lived to ripe old ages in spite of the fact that they were almost continuously

being exposed to unpleasant concentrations of this acute toxin.
Likewise, people living in Rotorua, New Zealand, have a normal
life expectancy in spite of continual exposure to relatively high
concentrations of hydrogen sulfide. Somewhat similar observations
can be made regarding exposures to chlorine, sulfur dioxide,
formaldehyde and carbon monoxide. The latter example is interesting because the cigarette smoker draws in a lethal concentration
of carbon monoxide at each puff, but because he has an intermittent
exposure he is not killed nor does he exhibit much evidence of
lasting injury. (Admittedly one might be tapped on the head with
a hammer intermittently for years without being killed but probably
there would be little benefit derived!)

The preceding discussion serves as an introduction to a
brief consideration of cumulative and additive toxins. These
toxins are insidious rather than dramatic and their significance
can hardly be determined by short-term monitoring. If there are
cumlative or additive effects associated with exposures to gaseous
materials, then time-weighted average concentrations obtained over
prolonged periods of time may provide vital information. Permeation
devices could be especially valuable for such long-term studies.

In concluding the discussion of long-term sampling, attention
is directed to possible beneficial aspects of contaminated air masses.
The direct benefits may be limited to vegetation where consideration
should be given to farm crops or vegetation in general which derive
much of their nourishment from the air. In addition to carbon
dioxide, plants are benefited by sulfur compounds and oxides of
nitrogen. In addition to the gaseous nutrients, trace elements in
dust and dissolved nutrients in precipitation bring life to the
plant world. Obviously, animals benefit also through secondary
contributions and in some cases, possibly through primary contributions. The adverse effects of air pollution on plant life is
well documented and it is obvious that exposures to sulfur dioxide,
ozone, and hydrofluoric acid should be monitored. If it is
important to monitor gaseous contaminants and pollutants in the air,
long-range sampling by means of permeation devices should prove to
be an invaluable tool.

Methods for short-term sampling for sulfur dioxide, nitrogen
dioxide and carbon monoxide have been investigated. The permeation
method for sulfur dioxide is valid for sampling periods ranging
from less than an hour to a week or more. Field studies have shown
that the data obtained from the permeation device compares most
favorably with results obtained with the conventional West-Gaeke
procedure or with a coulometric type continuous monitor (Table I).
For sulfur dioxide monitoring over periods of two weeks to three
months, or even longer, a modified permeation technique is under
study. Because, only gases permeate there is no danger of collecting sulfuric acid aerosol or sulfates.

TABLE I. Field studies of SO_2 sampling methods.

Date	West-Gaeke	Coulometric	Permeation Device
6/15/72	<2	<25	<5
6/17/72	<2	<25	<5
6/20/72	--	144	156
6/24/72	--	<25	23
6/25/72	--	<25	17
8/08/72*	158	--	156
8/09/72**	47	--	42
8/12/72	128	--	132
8/18/72	14	--	6
8/20/72	145	--	120
8/21/72	99	--	93
8/22/72	382	--	386

Results shown in $\mu g/m^3$

*Seven days exposure

**Two days exposure

Therefore, an approach based on sample permeation with conversion of the permeated SO_2 to sulfuric acid holds promise. By using a catalyst such as a manganese salt, rapid conversion to sulfate can be achieved and this in turn, because of the simplicity of the system, can be determined by either the barium chloranilate procedure or by ultimate conversion to perimidylammonium sulfate with thermal decomposition and final spectrophotometric or flame photo metric measurement.

An unpublished study has been made which indicates that nitrogen dioxide permeates readily. A monitor has been developed which weighs only 50 grams and functions by means of the permeation of NO_2 through a silicone membrane with the permeated gas being absorbed in an alkaline thymol solution. The resultant nitrite reaction product is determined colorimetrically. The method is sensitive to ambient air concentrations of nitrogen dioxide and exhibits essentially linear response to this pollutant up to the OSHA limit of 10,000 $\mu g/m^3$. Field evaluation indicates that the monitor yields high results in ambient air when compared with the accepted EPA-TGS-ANSA method. Studies are projected to determine whether the permeation method or the EPA method is the more correct.

A permeation method for monitoring carbon monoxide has been described (2) and an ongoing investigation indicates that hydrogen fluoride can be monitored by means of a permeation device. It is hoped that the HF monitor will have sufficient sensitivity that it can be used for either short-range or long-range monitoring. The

SO_2, NO_2 and HF monitors, if validated, could prove invaluable in evaluating the toxic/nutritional impact of these gases on vegetation.

Industrial hygiene studies may be based on the physiologic response of individuals to known stresses or analyses can be performed on feces, urine or blood to determine the presence and concentration of suspected toxins. In such approaches the individual is used as the sampling mechanism. Area monitoring has been used within the work place to establish the average concentration of hazardous material to which individual workers might be exposed. The ideal approach for establishing individual exposures is the use of personal monitors. In the past, personal monitors have been relatively troublesome and cumbersome. Personal monitors that have been available in the past have sometimes been more hazardous than the material that they monitored. They were relatively expensive and undependable. It has been estimated that the development of individual personal monitors would require a million dollars or more for each substance for which monitors should be developed. The introduction of passive sampling by means of the permeation technique introduces exciting new possibilities.

A personal monitor for vinyl chloride has been developed which utilizes the permeation technique for sampling (4). The permeated vinyl chloride is trapped on activated charcoal which is removed for subsequent determination by gas chromatography. The monitor is about the size of a standard film badge, weighs less than 35 grams and requires no source of power (See Figure 3). The monitor is insensitive to temperature and humidity effects and is free from significant interferences. The monitor is convenient to use and the analytic finish is essentially the same as that currently in use for vinyl chloride determinations. The monitor is ideally suited for monitoring as required by OSHA regulations because it provides data based on time-weighted average exposures with no data reduction steps required. The new vinyl chloride monitor has been field tested extensively in Great Britian, Canada and the United States. The field testing was conducted by impartial investigators who cooperated in establishing the validity of the approach. In a wide variety of test environments all results reported were shown to completely substantiate the reliability and convenience of the permeation type passive sampling monitors (Table II).

Conclusions

Permeation has been proposed as a new approach for the passive collection of gaseous samples for short and long-range ambient air monitoring and for personal monitoring for industrial hygiene studies. The rationale of the new approach is an extension

TABLE II. Field data for vinyl chloride.

Pump with Carbon tube	A.S.T.M.	Permeation badge
0.74	0.89	1.21
1.00	1.00	1.19
0.13	0.14	0.18
0.13	0.12	0.19
1.4	1.0	2.7
0.24	0.17	0.15
0.12	0.14	0.10

Results shown in ppm.

of the classic work of O'Keeffe and Ortman (5) who introduced the permeation principle for the preparation of primary standard gas mixtures. Preparation of standard gas mixtures involves the conversion of a concentrated (liquid) material to a dilute mixture of gases having an accurately known composition. In the reverse process a constituent present in a dilute atmosphere is concentrated by the permeation process using a suitable collecting and stabilizing agent with quantification of the process being derived from the exact nature of the permeation process. External influences which might otherwise invalidate the use of permeation for the collection of samples have been obviated with the result that a convenient, flexible passive technique is now available for atmospheric monitoring. The inherent reliability of permeation-type monitors is spectacular when compared with other sampling and monitoring devices and the simplicity of such monitors should insure widespread use where cost is a consideration. The accuracy of permeation is probably as good or better than most other means of sampling. Safety is often a consideration and permeation devices have the advantage that no motors or moving parts are involved and thus there are no sparks or hot surfaces possible to cause explosions or fires. The fact that personal monitors can be made that are no larger than a radiation dosimeter is also an advance from the safety standpoint.

The future for permeation badges should include their use in personal monitoring for gases such as vinylidene chloride, carbon tetrachloride, trichloroethylene, ethylenedichloride, methyl chloride, benzene and toluene. Inorganic gases such as arsine, phosphine, hydrogen cyanide, hydrogen fluoride, chlorine and bromine should be added to the list of toxins that require personal monitoring and which lend themselves to the permeation approach. The future for permeation devices for ambient air monitoring is likewise bright, especially for long-range studies and for studies of pristine areas.

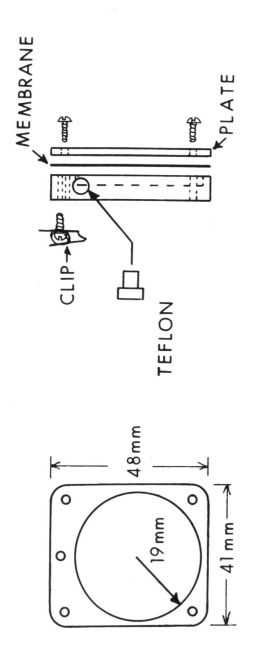

Figure 3. Design of permeation device for personal monitoring.

REFERENCES

1 ALTSHULLER, A.P. and COHEN, I.R.:   Anal. Chem. 32 (1960) 802.

2 BELL, D.R., REISZNER, K.D. and WEST, P.W.:   Anal. Chim. Acta
   77 (1975) 245.

3 McKELVEY, J.W. and HOELSCHER, H.E.:   Anal. Chem. 29 (1959) 123.

4 NEIMS, L.H., REISZNER, K.D. and WEST, P.W.:   Anal. Chem.,
   In press.

5 O'KEEFFE, A.E. and ORTMAN, G.C.:   Anal. Chem. 38 (1966) 760.

6 REISZNER, K.D. and WEST, P.W.:   Environ. Sci. Technol. 7 (1973)
   526.

7 SALTZMAN, B.E. and GILBERT, N.:   Am. Ind. Hyg. Ass., J. 20
   (1959) 379.

Discussion

HINKLEY - You mentioned a time span of from one hour to twenty-four hours for that permeation device. If someone were visiting a chemical factory and you wanted to insure that he was protected you might give him a device like that. If he went around for about one hour, would that limit the sensitivity that the device would measure? That is, would you be able to still get down to five parts per billion?

WEST - Oh yes, no problem on that provided a good gas chromatograph is available. This has been done by one of the companies in the Baton Rouge area.

NEWBURY - In our discussion of the problem of interferences. I think there were one or two chemicals near the bottom of the slide which say at the three ppm level were showing 100% interference.

WEST - That was only with thermal decomposition. Those were materials that, when one transferred the sample thermally, broke down to give the same response as vinyl chloride. With elution of the sample by means of carbon disulfide no interference was found. In the plants throughout the world where field tests were run no adverse criticisms have been raised by anyone who has used these devices even in complex industrial areas.

ETZ - What is the reading that you get when you walk around with two of these devices where one does not use the silicone membrane and capsulates the sample simply on a charcoal bed?

WEST - You mean with a Sipin pump?

ETZ - No, the same device with just the charcoal bed. You would be absorbing the gas too. Correct?

WEST - Yes, but in that case there is no way to relate the amount collected with concentration. The story really goes back to my being upset with the lead peroxide candle. This was an effects method basically. One would set out lead peroxide candles around a town or a city or whatever area under study and find the relative insult at any point as compared to any other location. The lead peroxide candles simply told one there's more SO_2 here than there. It didn't tell what the concentrations were but instead indicated qualitatively the relative insults from place to place. Almost all regulations that control exposures and health hazards are related to concentrations. As far as I know, this applies throughout the world; control agencies want to know what the average concentration is for an exposure period. The permeation badge does this, because it provides a means for determining the relationship between the amount of material collected and the amount of sample.

ETZ - I understand you and I appreciate the significance of sampling in the breathing zone of the worker.

WEST - Although you can't put these in your nose, you can put them
on a hardhat which is mighty close, or you can put them on the
jacket collar and you're getting mighty close, but you have a very
good point. Our friends at Ethyl for example found out that you
have a difference if you collect a sample at head level and compare
it with one collected at waist level. They ended up sample collecting with a Sipin pump with the tube beside the neck on one side
and the permeation badge on the other and obtained good correlations. Our friends in England did all sorts of studies in exposure
chambers and in four different plants, and they gave the monitor a
thorough workout and gave us all the statistical studies. The results provide complete substantiation for the permeation approach.

LODGE - I think perhaps the point here that hasn't been made quite
clear is that if you consider your charcoal as a capacitance, without the membrane that capacitance is in series with an extremely
variable resistor which will change with the air flow to it. A
surface layer that changes in thickness all over the map, whereas
the membrane constitutes a large fixed series resistance compared
to which the other changes owing to ventilation and so on are trivial. So that really is the difference between this device and
the previous passive samplers which were just hanging out in the
air naked.

ETZ - Does this gas exist other than as a monomer vinyl chloride
in the gas phase or does it dimerize or form other species than
the monomer?

WEST - I'm not qualified to comment on that. I don't know. All I
know is that the measurement is almost certainly made on the monomer and I do not believe the polymers are of health significance.
The monomer is what the regulations refer to and this is what the
measurement is made of. I suppose there would be some dimerization
any time you have an atmosphere on vinyl chloride, depending on
the concentration and what else is present, but what we're interested in is the monomer and that's what we're measuring.

FILBY - The fact that you have very little temperature variation
here is due to the fact that this is a solution phenomenon across
the membrane as opposed to a gaseous diffusion.

WEST - Yes, that's right.

FILBY - So this is also less sensitive to any changes in pressures,
is that correct?

WEST - We have no significant pressure effect. Of course the studies have always been made at ground level so I don't know what the effect might be if you were to use this at high altitude. You might have to calibrate the membrane if you were in Denver; it might show a different value for the permeation constant than for the membrane calibrated in Baton Rouge. I can't really comment on that because we didn't study the effect of pressure. However, I think that there is no question that any effect is well within the 50% limit set by OSHA.

STIEHL - What kind of variability is associated with the calibration? Do you calibrate each sheet of silicone membrane or does each device have to be calibrated?

WEST - We calibrated each and every device. Some of our industrial friends did make a study in which they obtained sheets that are about 36 inches square which they calibrated. They took samples out of four corners and out of the middle and found a variation of about 15%. The General Electric membranes are still pretty much batch made and thus are not completely uniform. There are other manufacturers such as Dow Corning who make a silicone membrane which is thicker. We had to use the G. E. membranes for vinyl chloride, but for benzene we would probably go to Dow Corning because benzene and toluene permeate at a much higher rate than vinyl chloride, so we can use a thicker membrane. The most variability that occurs or has been experienced so far is less than 20%, and of course that means according to the regulations of OSHA you're well within the 50%. So for industrial studies you might have a whole plant where you can have the secretaries wear the permeation devices now. Considering large volumes sales, it could be that as long as you're within 20 to 50% which is all the assurance you need, one would have to run only one calibration on each sheet. That would cut down costs.

TORIBARA - I've got a couple of questions. Recently I went to a meeting here at a center that studies mercury toxicity. The discussion concerned a plant in Almaden, Spain where they mine mercury, and I guess the mercury exposure is quite a problem there. They were talking about using some kind of a passive monitor, a personal monitor for those working. Is there something comparable to yours on the market?

WEST - Yes, that's manufactured by Minnesota Mining and Manufacturing, and it's a very clever approach in which they have a gold foil which the mercury vapor amalgamates. The badge is sent to 3-M after exposure and they measure the amount of collected mercury. They do this by measuring the change in resistance of the metal foil that results from the amalgamation. The problem there is that chlorine interferes, and for most places where you are going

to be studying mercury you're going to have a good bit of chlorine
in the atmosphere. In the mines in Spain, unless the rains on the
plains are chlorine-containing rains, they will be all right.
Chlorine is a problem that as far as I know has not been licked.
I don't know the extent of that interference but it is apparently
substantial. It's a very clever approach and hopefully the interference problem will be licked.

TORIBARA - I think they're just going to try this now. But I don't
know if those Spanish workers know that there is gold foil or
something in there, the instrument would not last very long. I
was going to ask one other question. In most of these plants is
there some kind of an alarm system, if you had a sudden release of
a large amount of vinyl chloride accidentally, so that the workers
could take cover?

WEST - Yes, a good many plants have tried to go that way. They
have area monitors or alarm systems. One of the problems in this
whole approach of personal monitoring has been that the workers at
every chance they get shut off the motor because it makes noise,
or they get around to some part of the plant where they will slip
out of the monitor. They just don't want to wear them. You can
imagine what it would be like on a day, say 85 or 90 degrees in
the shade, walking around with something like that strapped on you.
That alone is enough to discourage one, so they have a lot of problems like that. Incidentally, these permeation monitors are now
on the market. They are being sold, and I guess they will probably
be sold internationally soon. There is a great deal of interest
in manufacturing companies in Japan, England and Holland, as well
as in companies in this country. It is, I think, a very encouraging approach to a very difficult problem.

THE CONTINUOUS MEASUREMENT OF SULFUR-CONTAINING AEROSOLS BY FLAME PHOTOMETRY: A LABORATORY STUDY

James J. Huntzicker and Robert S. Hoffman

Oregon Graduate Center
19600 N. W. Walker Road
Beaverton, Oregon 97005

ABSTRACT

Aerosols of H_2SO_4, $(NH_4)_2SO_4$, and Na_2SO_4 have been continuously measured in the laboratory with a sulfur-specific flame photometer. Separation of the aerosol sulfur from gaseous sulfur (SO_2 and H_2S) was achieved by a diffusion stripper upstream of the flame photometer. The stripper is a tube coated with a substance which adsorbs SO_2 and H_2S. These gases diffuse rapidly to the wall where they are scavenged, but particles with much lower diffusion coefficients are efficiently transmitted. The flame photometer responds to $(NH_4)_2SO_4$ as $[S]^{1.9}$ and to H_2SO_4 as $[S]^{1.6}$ where $[S]$ is the sulfur concentration. Above 2 $\mu g(S)/m^3$ the response to $(NH_4)_2SO_4$ exceeded the H_2SO_4 response. An experiment involving signal averaging of the flame photometer output produced an order of magnitude improvement in the signal to noise ratio and indicated that a limit of detection of 0.5 $\mu g/m^3$ of sulfur (i.e., 1.5 $\mu g/m^3$ of $SO_4^=$) or lower can be expected. When a heater was inserted upstream of the diffusion stripper, the response to H_2SO_4 decreased to zero between 70 and 135°C and to $(NH_4)_2SO_4$ between 140 and 220°C. These phenomena result from the conversion of the aerosol sulfur to H_2SO_4, SO_3, and SO_2 gases which are scavenged at the walls of the heater and the diffusion stripper. Both the characteristic H_2SO_4 and $(NH_4)_2SO_4$ thermal profiles were observed in a $H_2SO_4/(NH_4)_2SO_4$ aerosol more acidic than NH_4HSO_4.

Introduction

In recent years the problem of sulfur-containing aerosols –
and in particular sulfuric acid aerosol – has received considerable
attention. This has resulted primarily because of evidence suggesting that sulfate aerosols are respiratory irritants (9,10).
Of the common sulfate aerosols toxicological studies have shown
that sulfuric acid is the most potent (1). Sulfate aerosol has
also been shown to play an important role in visibility reduction (24,26). Despite the acknowledged importance of sulfur-
containing aerosol there is no currently available instrument for
the measurement of particulate sulfur on a continuous basis. We
report here preliminary results on an instrument designed both for
the continuous monitoring of total aerosol sulfur and for distinguishing between different sulfur compounds in the aerosol.

The development of methods for the continuous measurement of
particulate sulfur is relatively recent. Most methods currently
in use involve the collection of the aerosol by filtration or
impaction followed by subsequent chemical analysis. The analytical
methods can be segregated into two major categories: those which
involve wet chemistry and those which do not. The wet chemical
methods usually consist of an extraction step followed by colorimetric, turbidimetric or nephelometric determination. Methods
which do not require a wet extraction include X-ray fluorescence
analysis, photoelectron spectroscopy (ESCA), and the flash
volatilization-flame photometric technique of Roberts and
Friedlander (21). Determination of the chemical nature of filter-
collected sulfur has been achieved by several methods – the most
successful being the solvent extraction method of Leahy et al (18)
and photoelectron spectroscopy (4). The non-continuous methods
suffer from two primary difficulties: they are time consuming and
subject to the possibility of artifact sulfate formation from the
conversion of SO_2 to $SO_4^=$ on the filter (20). An excellent review
of these methods has been given by Tanner and Newman (23).

Recently several continuous or quasi-continuous methods for
aerosol sulfur involving the use of a flame photometric detector
specific to sulfur have been reported. Mudgett et al (19)
described an instrument which first collects aerosol on a Fluoro-
pore filter and then measures any H_2SO_4 on the filter by vaporizing
the H_2SO_4 with warm, dry air. The vaporized H_2SO_4 is transported
into the flame photometer and measured. The instrument is made
quasi-continuous by automating the filtration and vaporization
steps and operating in a cyclical fashion. A second set of new
methods is based on the work of Crider (5) who demonstrated that
the sulfur-specific flame photometer responds directly to sulfuric
acid aerosol. Huntzicker et al (11,12,13) used a diffusion tube
stripper in conjunction with a flame photometric detector for the

continuous and selective measurement of aerosol sulfur. The stripper is simply a tube coated with a substance which adsorbs SO_2 and H_2S. These gases diffuse rapidly to the tube wall where they are stripped from the air stream, but particles which have much lower diffusion coefficients are efficiently transported into the flame photometer. A similar approach has been used by Durham et al (7). Kittelson et al (16) reported a method which measures both gaseous and aerosol sulfur. The aerosol is removed from the air stream by an electrostatic precipitator whose corona voltage is chopped at 0.2 Hz. The resultant AC component of the flame photometer output corresponds to the aerosol sulfur and is measured by a lock-in amplifier. The DC component is a measure of the gaseous sulfur concentration.

Huntzicker et al (11,12,13) and Kittelson et al (16) have shown that H_2SO_4 can be distinguished from $NH_4HSO_4/(NH_4)_2SO_4$ by heating the aerosol upstream of the diffusion stripper or the electrostatic precipitator. The sulfuric acid vaporizes at about 100°C whereas the ammonium sulfates decompose above about 150°C. The resultant vapors are scavenged at the heater wall or in the diffusion stripper producing a decrease in the flame photometer signal. The temperature range over which this decrease occurs is characteristic of the sulfur compound being sampled. A similar approach has been employed by Husar (15) who used an integrating nephelometer as the detector. A nephelometric approach for the separation of acid (H_2SO_4, NH_4HSO_4) and neutral (($NH_4)_2SO_4$) sulfates has also been used by Charlson and co-workers (2,3,25). In this method the light scattering coefficient, b_{scat}, is measured as a function of relative humidity which can be varied within the instrument. The presence of $(NH_4)_2SO_4$ is indicated by a sharp rise in b_{scat} at a relative humidity of 80%, which corresponds to the deliquescence point of $(NH_4)_2SO_4$. Acid sulfates are detected by the addition of NH_3 and the subsequent observation of the $(NH_4)_2SO_4$ deliquescence.

The discussion which follows is a summary of our previous results (11,12,13) supplemented by our most recently obtained data. We present first, however, a brief discussion of the principles underlying the flame photometric analysis of sulfur. In a fuel-rich hydrogen-air flame the presence of sulfur produces a characteristic emission spectrum in the 320 to 440 nm wavelength range. This emission results from the excited state molecule S_2^*. Because the formation of S_2^* requires the presence of two sulfur-containing molecules, the intensity of emitted light is theoretically proportional to the square of the sulfur concentration in the air stream. It has been shown, however, that the exact power law dependence is generally less than two and depends on the flame condition (8) and the particular sulfur compound being measured (6,22). Specificity for sulfur is obtained by isolating one of the S_2^* emission lines

(usually at 394 nm) with an interference filter. Directly above
the burner tip temperatures in excess of 1400°C have been measured
in the inner core of the flame. A few millimeters higher, however,
the temperature is between 400 and 500°C, and it is in this region
that the S_2^* is formed. Sulfur specific flame photometers have
been commercially available for a number of years and have been
used for ambient air analysis and as gas chromatographic detectors.

Experimental and Results

The experimental set-up used in these experiments is shown
in Figure 1. Polydisperse aerosol is generated by nebulizing
aqueous sulfate solutions in the 10^{-4}-10^{-3} M concentration range.
Excess charge is removed from the aerosol in the ^{85}Kr charge
neutralizer, and equilibration with the relative humidity (~50%)
of the dilution air is achieved in the 12 liter aging flask. The
aerosol stream is sampled by the flame photometer through the
diffusion stripper which removes essentially all of the H_2S and
SO_2. The transmission efficiency for particles is a function of
particle size. At 0.01 μm diameter the efficiency is 85% and
increases rapidly to near 100% as the particle diameter increases.
The stripper is simply a glass tube containing a cylinder of Whatman filter paper impregnated with $Pb(CH_3CO_2)_2$. The flame
photometer now in use is the Meloy SA-285 although earlier work
involved the Meloy SA-185 and SH-202. The zero level of the
flame photometer is established by inserting a low pressure drop
filter between the diffusion stripper and the flame photometric
detector. A device for automatically switching the filter into and
out of the system has been constructed but not yet tested. The
final feature of the experimental set-up is the filtration system
which is used for collecting samples from the aerosol stream for
chemical analysis.

Aerosols of H_2SO_4 and $(NH_4)_2SO_4$ were generated and sampled
simultaneously by the flame photometer and by filtration. The
filter-collected sulfate was extracted in water and analyzed
by the flash volatilization-flame photometric method (14,21). In
our variation of this method 5 μl of the solution to be analyzed
are placed on a 0.025 cm diameter Pt wire connected to the terminals of a 0.3 F capacitor. The capacitor is charged to about
5V and discharged through the wire. The wire is rapidly heated to
greater than 1000°C, and any sulfur on the wire is converted to
SO_2. The SO_2 is drawn into the flame photometer, and the response
of the flame photometer recorded. Although the response should
depend only on the amount of sulfur deposited on the wire, the
observed response to $(NH_4)_2SO_4$ was approximately 15% greater than
for an equivalent amount of H_2SO_4 . Because of the desirability of
removing any systematic differences between H_2SO_4 and $(NH_4)_2SO_4$

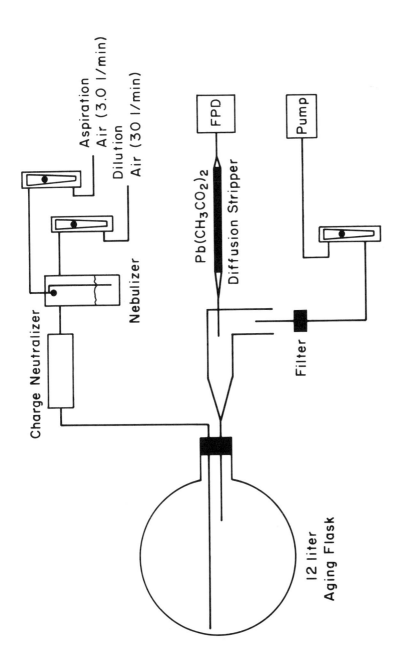

Figure 1.   Experimental set-up.   The flame photometric detector (FPD) samples polydisperse aerosol produced by the nebulizer.   The filter collects aerosol samples for chemical analysis.

analyses, all of the H_2SO_4 samples were converted to $(NH_4)_2SO_4$ by
bubbling NH_3 gas through them. This problem has since been satisfactorily solved however.

In Figure 2 the output of the flame photometer (operating in
the continuous mode) is plotted as a function of sulfur concentration in $\mu g(S)/m^3$ for H_2SO_4 and $(NH_4)_2SO_4$ aerosols. (Note that
the concentration refers to sulfur and not sulfate. Multiplication by a factor of 3 gives the sulfate concentration.) The error
bars are generally within the symbols. The most significant
aspects of Figure 2 are the differences in sensitivities of the
flame photometer to the two aerosols and in the slopes of the
lines (i.e., in the power law response). In this experiment the
flame photometer responded as $[S]^{1.9}$ for $(NH_4)_2SO_4$ and $[S]^{1.6}$
for $[H_2SO_4]$.

If these results are confirmed by future experiments, then
the implications for the monitoring of total particulate sulfur
by flame photometry are twofold. The differences in sensitivities
and power law responses yield an uncertainty in the actual sulfur
concentration which is zero at 2 $\mu g(S)/m^3$ (i.e., 6 $\mu g/m^3$ as $SO_4^=$)
and increases to $\pm 19\%$ at 10 $\mu g(S)/m^3$ (i.e., 30 $\mu g/m^3$ as $SO_4^=$).
Secondly, because the response to sulfur is usually linearized by
a log-antilog circuit, the difference in power law responses
precludes the use of a unique power in the linearization process.

A possible explanation for the difference in the response to
the two aerosols is the high temperature ($149°C$) at which the
flame photometer burner block is held. As discussed below, this
temperature is sufficiently high to vaporize the H_2SO_4 aerosol
given an adequate exposure time as the aerosol passes through
the block on its way to the burner. Because of its reactivity
the H_2SO_4 vapor is easily scavenged by surfaces. On the other
hand, $(NH_4)_2SO_4$ is stable with respect to sulfur loss up to about
$150°C$. If significant losses of H_2SO_4 are occurring, then this
could be the source of the difference between the two compounds.

If the lines of Figure 2 are extrapolated downward, the
response to 2 $\mu g(S)/m^3$ is about 10^{-10} A which is the peak to peak
noise level. This is illustrated in Figure 3A in which the
continuous response of the flame photometer to 2 $\mu g(S)/m^3$ as
H_2SO_4 is shown. For ease of presentation only the noise envelope
is given. In this figure the first six minutes correspond to the
instrument zero as established by inserting a glass fiber filter
between the diffusion stripper and the flame photometer, the next
seven minutes to the H_2SO_4 aerosol, and finally to the instrument
zero again. During this measurement the output of the flame
photometer was digitally averaged over 70 second periods by a
Hewlett-Packard 3370A integrator operating in the manual mode.

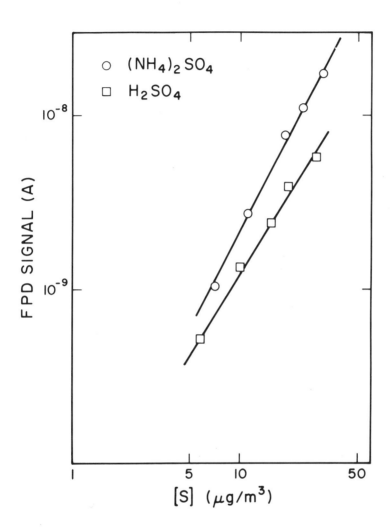

Figure 2.   Flame photometer output as a function of
sulfur concentration for H_2SO_4 and $(NH_4)_2SO_4$
aerosols.

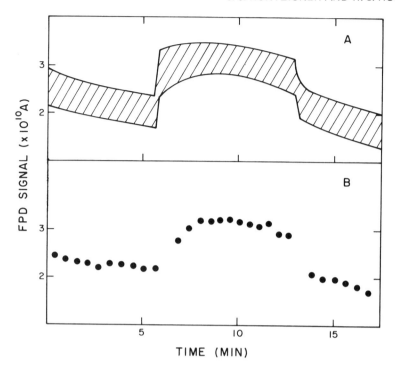

Figure 3. Flame photometer response to 2 μg/m³ of sulfur as
H₂SO₄ aerosol. A. Output of flame photometer with RC filter
time constant of one second. B. Output digitally averaged for
70 second periods. In both A and B the first six minutes correspond to the instrument zero, the next seven minutes to the H₂SO₄
aerosol, and finally to the instrument zero again.

The signal-averaged results are shown in Figure 3B, and the
dramatic improvement in signal to noise ratio is apparent.

Figure 3 also shows a slight downward drift in the baseline.
This can be caused by small changes in either the relative humidity
or in the flow rates of gases into the burner. For low level
measurements it is especially important that the baseline be
referenced often. In this regard the prototype instrument would
automatically switch between filtered and unfiltered air with each
cycle digitally averaged. Concentrations of sulfur-containing
aerosol below 0.5 μg(S)/m³ should be measurable with such an
instrument although the problem of compound-dependent response
must still be solved.

One possible solution to this problem is to measure the individual sulfur compounds in the aerosol separately. For the aqueous H_2SO_4-$(NH_4)_2SO_4$ system this was accomplished by heating the aerosol upstream of the diffusion stripper. Aerosols were produced by nebulizing aqueous solutions with the relative concentrations: H_2SO_4, $(NH_4)_2SO_4$, 1/2 $H_2SO_4 \cdot 1/2$ $(NH_4)_2SO_4$ (i.e., (NH_4HSO_4)), and 2/3 $H_2SO_4 \cdot 1/3$ H_2SO_4. The response of the flame photometer as a function of temperature for these systems is shown in Figure 4.

The response to the H_2SO_4 aerosol begins to decrease at about 70°C and by 135°C has completely disappeared. This may be understood as follows. The unheated H_2SO_4 droplet, which is initially in equilibrium with the relative humidity of the air, contains about 50% by weight water. As the droplet is heated, the water is evaporated, and the equilibrium vapor pressure of H_2SO_4 over the droplet increases considerably. This increase results in the transfer of H_2SO_4 from the liquid to the vapor phase with the subsequent loss of the H_2SO_4 vapor to the walls of the heater and diffusion stripper. This produces a decrease in the flame photometer output. The sharpness of the decrease is primarily dependent on heater design.

A similar effect is observed for $(NH_4)_2SO_4$ at about 150°C, but the actual process is more complex. At 100°C $(NH_4)_2SO_4$ loses NH_3 to form NH_4HSO_4, but with increasing temperature, $(NH_4)_3H(SO_4)_2$, $(NH_4)_2S_2O_7$, and NH_2SO_3H can also be formed. Above 200°C complete decomposition to SO_2, SO_3, and H_2SO_4 occurs (17). These gases are scavenged in the diffusion stripper and produce the observed decrease in the $(NH_4)_2SO_4$ signal.

The last two thermograms in Figure 4 correspond to the case of H_2SO_4 aerosol partially neutralized by NH_3. In these cases when the aqueous H_2SO_4-$(NH_4)_2SO_4$ droplet is heated to above 100°C, an amount of sulfuric acid sulfate equivalent on a molar basis to one-half the initial amount of NH_4^+ is precipitated as NH_4HSO_4. The remaining or excess H_2SO_4 is vaporized and scavenged in the stripper. If the $HSO_4^-/SO_4^=$ equilibrium is neglected, this process can be represented phenomenologically as:

$$H_2O + a(2H^+ + SO_4^=) + b(2NH_4^+ + SO_4^=)$$

liquid

$$\downarrow$$

T > 100°C

$$\downarrow$$

$$H_2O + 2b(NH_4HSO_4) + (a-b)H_2SO_4$$

vapor   solid   vapor

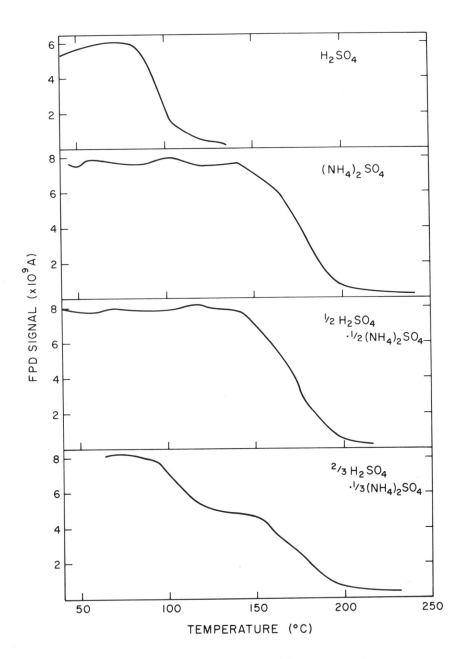

Figure 4.　Thermograms of sulfate aerosols

for a$\geq$b where a and b are stoichiometric coefficients. When
b>a, the resultant species are H_2O, NH_3, and NH_4HSO_4, which
decomposes as above upon heating above 150°C. Thus H_2SO_4 will be
vaporized only when there is an excess of H (either as H^+ or
HSO_4^-) with respect to NH_4^+ (i.e., a>b). As a result the decrease
in the flame photometer response at 70°C will only be observed
when the aerosol is more acidic than NH_4HSO_4. For less acidic
aerosols only the $(NH_4)_2SO_4$ thermogram will be observed. Similar
thermograms have also been observed by Husar (15) using temperature programmed nephelometry, and the same explanation applies.

Heterogeneous aerosol systems such as a mixture of aqueous
H_2SO_4 droplets and solid $(NH_4)_2SO_4$ particles have not yet been
investigated. It is likely, however, that the temperature
dependent response will be different from that of the homogeneous
system. A preliminary study of the thermal behavior of Na_2SO_4
aerosol indicated no decomposition or vaporization between room
temperature and 1100°C. This is consistent with the refractory
nature of Na_2SO_4.

Although these results are preliminary, the design of a
prototype instrument for separately measuring H_2SO_4, $(NH_4)_2SO_4$,
and metal sulfates is suggested. Aerosol would enter the diffusion stripper-flame photometer system through one of three tubes.
These tubes would be at room temperature, 140°C, and 250°C,
respectively. The difference in the flame photometer response
between the first two tubes would be a measure of the excess
H_2SO_4 concentration, and between the second two tubes a measure
of the $(NH_4)_2SO_4/NH_4HSO_4$ concentration. The response to the
aerosol transmitted through the 250°C tube would be a measure of
the concentration of metal sulfates. Of course, the feasibility
of this system depends on the temperature dependent response of
the flame photometer to the heterogeneous $H_2SO_4-(NH_4)_2SO_4$ system.

Summary

Preliminary results on the use of the sulfur specific flame
photometer for the continuous measurement of aerosol sulfur have
been presented. The most significant results are as follows:

1. Aerosol sulfur can be measured separately from gaseous
    sulfur by inserting a diffusion stripper upstream of the
    flame photometer.

2. Sensitivities to aerosol sulfur at the 0.5 $\mu g(S)/m^3$ level
    should be attainable if signal averaging techniques are
    used.

3. The sensitivity of the flame photometer to $(NH_4)_2SO_4$ aerosol is greater than the sensitivity to H_2SO_4 aerosol. This may result from the high temperature of the burner block.

4. The temperature dependent response to H_2SO_4 and $(NH_4)_2SO_4$ as indicated by the thermograms of Figure 4 suggest a possible method for the specific measurement of H_2SO_4, $(NH_4)_2SO_4/NH_4HSO_4$, and metal sulfates.

All of the results presented here are preliminary, and research is ongoing.

Acknowledgments

The financial support of the Environmental Protection Agency and the Northwest Air Pollution Center, a consortium of Northwest industries supporting the environmental work of the Oregon Graduate Center, is gratefully acknowledged. The participation of Lorne Isabelle and John Watson in the early stages of this work is appreciated. Mention of commercial products implies no endorsement on behalf of the Environmental Protection Agency, the Northwest Air Pollution Center, or the Oregon Graduate Center.

References

1 AMDUR, M. O.: Toxicological appraisal of particulate matter, oxides of sulfur and sulfuric acid. J. Air Pollut. Control Assoc. 19 (1969) 638.

2 CHARLSON, R. J., VANDERPOL, A. H., COVERT, D. S., WAGGONER, A. P., and AHLQUIST, N. C.: $H_2SO_4/(NH_4)_2SO_4$ background aerosol: optical detection in St. Louis region. Atmos. Environ. 8 (1974) 1257.

3 CHARLSON, R. J., VANDERPOL, A. H., COVERT, D. S., WAGGONER, A. P., and AHLQUIST, N. C.: Sulfuric acid–ammonium sulfate aerosol: optical detection in the St. Louis region. Science 184 (1974) 156.

4 CRAIG, N. L., HARKER, A. B., and NOVAKOV, T.: Determination of the chemical states of sulfur in ambient pollution aerosols by x-ray photoelectron spectroscopy. Atmos. Environ. 8 (1974) 15.

5 CRIDER, W. L.: Hydrogen flame emission spectrophotometry in monitoring air for sulfur dioxide and sulfuric acid aerosol. Anal. Chem. 37 (1965) 1770.

6 DAGNALL, R. M., THOMPSON, K. C., and WEST, T. S.: Molecular-
   emission spectroscopy in cool flames I. The behavior of
   sulphur species in a hydrogen-nitrogen diffusion flame and in
   a shielded air-hydrogen flame. Analyst 92 (1967) 506.

7 DURHAM, J. L., WILSON, W. E., and BAILEY, E. B.: Continuous
   measurement of sulfur in submicrometric aerosols. EPA-
   600/3-76-088 (1976).

8 ECKHARDT, J. G., DENTON, M. B., and MOYERS, J. L.: Sulfur
   FPD flow optimization and response normalization with a
   variable exponential function device. J. Chromatogr. Sci. 13
   (1975) 133.

9 ENVIRONMENTAL PROTECTION AGENCY: Health consequences of sulfur
   oxides: a report from CHESS, 1970-71. EPA-650/1-74-004.

10 ENVIRONMENTAL PROTECTION AGENCY: Position paper on regulation
   of atmospheric sulfates. EPA-450/2-75-007, September 1975.

11 HUNTZICKER, J. J., ISABELLE, L. M., and WATSON, J. G.: The
   continuous monitoring of particulate sulfate by flame
   photometry. Presented at the International Conference on
   Environmental Sensing and Assessment, Las Vegas, September 1975.

12 HUNTZICKER, J. J., ISABELLE, L. M., and WATSON, J. G.: The
   continuous measurement of particulate sulfur compounds by
   flame photometry. Paper 76-31.3 presented at the 69th Annual
   meeting of the Air Pollution Control Association, Portland,
   Oregon, June 27-July 1, 1976.

13 HUNTZICKER, J. J., ISABELLE, L. M., and WATSON, J. G.: The
   continuous monitoring of particulate sulfur compounds by flame
   photometry. Presented before the Division of Environmental
   Chemistry, American Chemical Society, New Orleans, Louisiana,
   March 20-25, 1977.

14 HUSAR, J. D., HUSAR, R. B., and STUBITS, P. K.: Determination
   of submicrogram amounts of atmospheric particulate sulfur.
   Anal. Chem. 47 (1975) 2062.

15 HUSAR, R. B.: In situ thermal analysis: detection of different
   sulfates. Unpublished report.

16 KITTELSON, D. B., MCKENZIE, R., LINNE, M., DORMAN, F., PUI, D.,
   LIU, B., and WHITBY, K.: Total sulfur aerosol detection with
   an electrostatically-pulse flame photometric detector system.
   Presented before the Division of Environmental Chemistry,
   American Chemical Soc., New Orleans, LA, March 20-25, 1977.

17 KIYOURA, R., and URANO, K.: Mechanism, kinetics and equili-
    brium of thermal decomposition of ammonium sulfate. Ind.
    Eng. Chem. Process Des. Develop. $\underline{9}$ (1970) 489.

18 LEAHY, D., SIEGEL, R., KLOTZ, P., and NEWMAN, L.: The separa-
    tion and characterization of sulfate aerosol. Atmos. Environ.
    $\underline{9}$ (1975) 219.

19 MUDGETT, P. S., RICHARDS, L. W., and ROEHRIG, J. R.: A new
    technique to measure sulfuric acid in the atmosphere. In
    "Analytical Methods Applied to Air Pollution Measurements"
    (R. K. Stevens and W. F. Herget, eds.), Ann Arbor Science
    Publishers Inc., Ann Arbor, Michigan (1974).

20 PIERSON, W. R., HAMMERLE, R. H., and BRACHACZEK, W. W.:
    Sulfate formed by interaction of sulfur dioxide with filters
    and aerosol deposits. Anal. Chem. $\underline{48}$ (1976) 1808.

21 ROBERTS, P. T., and FRIEDLANDER, S. K.: Analysis of sulfur
    in deposited aerosol particles by vaporization and flame
    photometric detection. Atmos. Environ. $\underline{10}$ (1976) 403.

22 STEVENS, R. K., MULIK, J. D., O'KEEFFE, A. E., and KROST, K.J.:
    Gas chromatography of reactive sulfur gases in air at the
    parts-per-billion level. Anal. Chem. $\underline{43}$ (1971) 827.

23 TANNER, R. L., and NEWMAN, L.: The analysis of airborne
    sulfate: a critical review. J. Air Pollut. Control Assoc.
    $\underline{26}$ (1976) 737.

24 WAGGONER, A. P., VANDERPOL, A. J., CHARLSON, R. J., LARSEN, S.,
    GRANAT, L., and TRÅGÅRDH, C.: Sulphate-light scattering ratio
    as an index of the role of sulphur in tropospheric optics.
    Nature $\underline{261}$ (1976) 120.

25 WEISS, R. E., WAGGONER, A. P., CHARLSON, R. J., and
    AHLQUIST, N. C.: Sulfate aerosol: its geographical extent in
    the midwestern and southern United States. Science $\underline{195}$ (1977)
    979.

26 WHITE, W. H.: Reduction of visibility by sulphates in
    photochemical smog. Nature $\underline{264}$ (1976) 735.

Discussion

SMITH – Did you do any measurements in the valley of the molecular emission? Sometimes the baseline moves around and if you don't measure the valley you don't know what the baseline is.

HUNTZICKER – In our instrument the baseline is determined by inserting a particulate filter between the diffusion stripper and the flame photometer. With this addition neither gaseous nor particulate sulfur enters the flame, and only the intrinsic emission of the flame itself is measured.

SMITH – It won't give you a spectral zero though.

HUNTZICKER – We don't need a spectral zero. All we need is the output of the flame photometer in the absence of sulfur. This corresponds to the flame emission in the 394 nm region of the spectrum.

SMITH – Have you considered using photodiodes rather than photomultiplier tubes?

HUNTZICKER – No. What we wanted to do was develop something which could be used as an add-on device to a commercially available sulfur flame photometer.

SMITH – They are much more stable than photomultiplier tubes.

HUNTZICKER – We haven't had too much trouble with the flame photometers as purchased. There has been considerable development which has gone into them, and the newest instruments on the market, the Meloy SA-285 and the Monitor Labs 8400, are reasonably stable.

FILBY – This is perhaps a very elementary question but when one thinks normally of sulfuric acid as an acid and not as an aerosol one would not expect it to dehydrate at around 100°C.

HUNTZICKER – What happens is that sulfuric acid in the atmosphere is an aqueous solution, the composition depending upon the relative humidity. For example, at 35% relative humidity, the composition of the sulfuric acid aerosol is about 50% water, 50% sulfuric acid by weight. Now as the temperature of the sulfuric acid is raised, water is evaporated until a constant boiling mixture containing about 98% sulfuric acid is obtained. At that point the vapor pressure of H_2SO_4 has risen to a point where it's non-negligible. The resultant H_2SO_4 vapor is continuously adsorbed on the heater walls, and because the H_2SO_4 vapor can never achieve its equilibrium pressure, the driving force is always in the direction of converting the H_2SO_4 liquid to vapor.

BROWN - What about competition reactions from, for example, the
reaction you have up there. What's the formation of sulfur oxides
under various conditions?

HUNTZICKER - The molecule SO may be formed in the flame in addition to S_2^*. If there are very high concentrations of sulfur in
the flame, quenching of the S_2^* excited state with itself or with
other species in the flame occurs. Typically the flame photometric
technique is useful up to about 1 ppm of SO_2 or H_2SO_4 in the flame.
Above this concentration the flame photometer is relatively insensitive to changes in the sulfur concentration.

BROWN - Would there be a difference between H_2SO_4 and $(NH_4)_2SO_4$
with respect to the relative amounts of S_2^* and SO formed in the
flame?

HUNTZICKER - That's a possible explanation for the difference.
Our approach to this problem, however, will be to convert any
H_2SO_4 to $(NH_4)_2SO_4$ by addition of NH_3 to sample air just upstream
of the flame photometer burner. Note added: Recent results have
shown that the diminished response of the flame photometer to
H_2SO_4 aerosol is a result of the high temperature ($150^{\circ}C$) of the
burner block. H_2SO_4 aerosol, when exposed to this temperature,
partially vaporizes, and the resultant vapor is removed at the
burner block wall before entering the flame. The addition of NH_3
to the H_2SO_4 aerosol proved to be an effective solution to this
problem.

LODGE - I wonder if the point made about taking a look at the
adjacent spectral valley by just rocking the interference filter
wouldn't be a good idea. I worked years ago on a sodium particle
counting flame photometer. We found that one of the basic limitations on sensitivity was the low level continuum emitted by particles not containing sodium. Rocking the interference filter
would take out that continuum background and would improve the
signal to noise ratio.

HUNTZICKER - That is a good point. We have looked at other ways
of trying to establish the baseline but haven't considered this
one in detail.

SMITH - You could use two photomultiplier tubes with one slightly
off-angle.

HUNTZICKER - That could be done. There was a prototype instrument
developed a couple of years ago in which two interference filters
were used. One was centered on the 394 nm peak and the other on
the spectral valley. The filters were alternately rotated into
and out of the optical system. It was a very nice concept.

SMITH — May I draw what I had in mind? I think it's simpler than that.

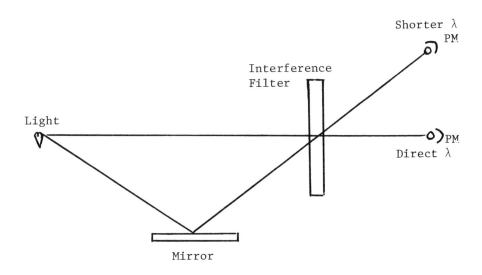

There are no moving parts. This will get bluer light.

HUNTZICKER — Yes, something like that is certainly feasible.

FILBY — Let us consider a completely different detection system, having instead of a flame photometer, a solution gathering device that would use a sulfate sensitive electrode and ammonium ion sensitive electrode. What is the sensitivity of this?

LODGE — Worse.

FILBY — Is it?

HUNTZICKER — There are other configurations that one can think of? For example, if the aerosol were heated to $1000^{\circ}C$, any H_2SO_4 or $(NH_4)_2SO_4$ would be converted to SO_2, and any SO_2 monitoring instrument could be used as the detector. We tried this in our early work and the results were promising.

CAHILL – Much data are being taken now, and yet the final data are usually averaged over an hour. I'm wondering just what kind of response time one really ought to have in a case like this. Ideally a few seconds response is lovely, but we think that particulate properties are slowly varying functions of time. Thus you might be able to give up some of your fast response time and use longer averaging.

HUNTZICKER – If the full speciation capabilities of the instrument are utilized, only four measurement cycles per hour would be possible.

CAHILL – It probably would be adequate.

FILBY – It seems to me if you could have that same time response on sensitive electrodes, maybe just three or four measurements an hour. That isn't very fast.

HUNTZICKER – Yes, however, there are other problems associated with such a system–such as the collection efficiency of the bubbler for aerosols.

TORIBARA – A kind of naive question but maybe there are others who also wonder why you have to differentiate between the species?

HUNTZICKER – The current thinking on the toxicity of these compounds is that they depend upon the chemical nature of the sulfur. There is quite a bit of investigation going on now into the role of sulfuric acid, the role of ammonium sulfate, and the role of other sulfur containing aerosols in respiratory irritation. It's certainly not a closed case, but some of the earlier works suggest that sulfuric acid is a more important irritant than ammonium sulfate and ammonium sulfate is more important than sodium sulfate. As was mentioned yesterday, it may be a moot point because most of the sulfuric acid may be neutralized by ammonia in the respiratory tract.

WEST – I would like to comment on this. To a large extent it is possible that the health effects are not due to sulfate at all; they are due to protons. Sulfuric acid, of course, is an insult to any physiologic system. Ammonium sulfate hydrolyzes to produce protons wherever there is moisture. The work of Mary Amdur substantiates the general conclusion that protons and not the sulfate are the principal problem. Thus speciation refers really to how many protons can be obtained from a given sulfate species.

LODGE – Except Phil, that that wouldn'nt account for the anomalous, by that theory, irritancy of zinc ammonium sulfate which seems is worse than sulfuric acid according to Mary.

WEST - Yes, I talked to Mary about that. I think this needs to be investigated further because zinc has some peculiarities of its own. I don't think however, that it is sulfate that is significant in the zinc ammonium sulfate. I think this may be a synergistic effect, the proton plus the zinc.

TORIBARA - Since you were down in Pasadena in the Los Angeles area, could you tell us what are the relative amounts of sulfuric acid and ammonium sulfate as compared to the other sulfates in the Los Angeles smog for example.

HUNTZICKER - The ACHEX study of a few years ago showed an almost stoichiometric balance between ammonium ion and the sum of sulfate (as $(NH_4)_2SO_4$) and nitrate which would imply the absence of any acid sulfates. However, these samples were not protected from ammonia in the laboratory and ammonia in the air after they were taken. It is my understanding that the Caltech and University of California, Riverside groups feel that $(NH_4)_2SO_4$-not H_2SO_4- is the most important sulfate in the Los Angeles area.

TORIBARA - I think this would point out very much the value of taking a field instrument and getting instantaneous readings.

LODGE - I think the one thing you can say about that is that there is very little evidence that much of the sulfate or nitrate is present as the metal salts.

HUNTZICKER - That's right. Friedlander has shown that sea salt is a minor contributor to the Los Angeles aerosol, and $MgSO_4$, which is a small fraction of sea salt, would therefore be present only in small amounts. There are no kraft mills in Los Angeles, and Na_2SO_4 would not be expected. It's almost certainly sulfuric acid or its neutralized forms by ammonia.

Note added: The complete version of this paper has been published in Atmospheric Environment, January 1978.

TUNABLE DIODE LASER DETECTION OF AIR POLLUTANTS

E. D. Hinkley*

Laser Analytics, Inc.

Lexington, Massachusetts 02173

ABSTRACT

Tunable diode lasers can be tailored to emit coherent radiation over a wide portion of the infrared "fingerprint" region of
the electromagnetic spectrum. This, coupled with relatively large
absorption cross sections for many molecular species and high
specificity have resulted in several important applications of
these devices to gaseous pollutant detection. In addition to providing the fundamental spectroscopic data for a number of pollutants, diode lasers have been used for low-pressure sampling, in
situ source monitoring, long-path ambient-air monitoring, and as
tunable local oscillators for infrared heterodyne detection. There
have been several recent developments in the capabilities of diode
lasers for the detection of air pollutants, and these will be
discussed.

INTRODUCTION

Laser techniques based on resonance or differential absorption
of infrared, visible or ultraviolet radiation are currently under
development in several laboratories throughout the world. Resonance absorption occurs when the wavelength of electromagnetic
radiation coincides with the center frequency of a spectral line
of an atomic or molecular species. According to Beer's law, the

---

*Present address: Jet Propulsion Laboratory, California Institute
of Technology, Pasadena, California 91103.

average transmittance over a path of length L(cm) is given by
exp[-NσL], where N is the species concentration (cm^{-3}), and σ the
absorption cross section (cm^2). Representative absorption cross
sections are shown in Table I for a variety of gases. Tunable
lasers have made it possible to scan the characteristic absorption lines of such species with unsurpassed resolution, allowing
the full cross sections to be used, while permitting the selection
of wavelengths which minimize potential interferences. Of these
lasers, the tunable diode has proved to be the most versatile since
it can reach any absorption line in the infrared "fingerprint"
region between 2.7 and 30 μm. Thus, diode lasers can be used to
perform fundamental laboratory spectroscopy required for any monitoring scheme based upon differential absorption; and, moreover,
they can be employed in fieldable instrumentation.

TABLE I.   Absorption Cross Sections for Various Molecules at
Atmospheric Pressure.

Molecule	λ (μm)	ν (cm^{-1})	σ (10^{-18} cm^2)
NO	0.2265	44150	1.3
O_3	0.2536	39425	12.0
OH	0.2826	35386	12.0
SO_2	0.3001	33330	1.0
NO_2	0.4358	22946	0.3
I_2	0.5895	16963	4.6
CH_4	3.270	3058	2.0
CO	4.709	2124	2.8
NO	5.263	1900	0.6
NH_3	9.220	1085	3.6
O_3	9.506	1052	0.9
C_2H_4	10.526	950	1.7
$C_2H_3C\ell$	10.638	940	0.4
$CC\ell_2F_2$	10.860	921	11.0
$CC\ell_3F$	11.806	847	4.4
$CC\ell_4$	12.610	793	4.8
C_2H_2	13.891	720	9.2

TUNABLE DIODE LASERS

Tunable diode lasers are fabricated from Pb-salt semiconductor material whose composition is chosen to produce laser emission in a desired region of the infrared (1). An individual laser can be tuned by varying its temperature or electrical excitation current, which indirectly affects the temperature through Joule heating. Until recently, it was necessary to maintain the operating temperature of the diode lasers below approximately 20K, a requirement that restricted tunability to around 30 cm^{-1}. A diode laser with significantly wider tunability was made possible in 1974 by developments of Groves et al (2) who produced lasers which operated continuously at temperatures as high as 80K and tuned over nearly 280 cm^{-1}. Using a simpler technique of compositional interdiffusion (3), lasers have now been developed that have operated at temperatures as high as 130K, with tunability of over 400 cm^{-1} in some cases. Even with the use of liquid nitrogen to achieve the operating temperature, it is still possible to obtain 100 cm^{-1} tuning by varying the diode current.

These widely-tunable lasers are now available commercially (4). They permit a large number of gases to be analyzed spectroscopically with a single laser, as well as making it possible to develop laser monitoring systems which are capable of nearly simultaneous monitoring of several pollutants. Using one of these new devices, infrared spectroscopy of ethylene, vinyl chloride, ozone, ammonia, Freon-11, Freon-12, and ethyl alcohol was performed in the 9-12 μm region (3).

DIODE LASER SPECTROSCOPY OF POLLUTANT MOLECULES

Tunable laser spectroscopy has been carried out in the laboratory to establish the absorption coefficients, linewidths, and lineshapes of important transitions of a variety of molecules. These laboratory studies provide a basis for selecting the most appropriate absorption lines for monitoring, based primarily on line strength and susceptibility to interferences from other molecules which may be present.

Figure 1 illustrates a diode laser scan of the P(5) line of carbon monoxide at 2123.7 cm^{-1}, broadened with one atmosphere of air. The dashed line is a theoretical Lorentzian fit to the experimental scan using a peak absorption constant of 7.1 x 10^{-5} $cm^{-1}ppm^{-1}$ (which corresponds to a cross section of ~3 x 10^{-18} cm^2) and FWHM (full width at half-maximum absorption coefficient) of 0.10 cm^{-1}. There is excellent agreement between the experimental lineshape and the theoretical Lorentzian curve, except in the wings where the line appears to absorb more strongly than predicted.

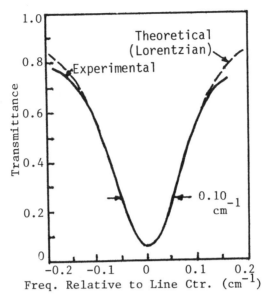

Fig. 1. Diode laser scan of the CO P(5) line centered at 2123.7 cm^{-1} for a 1316 ppm mixture of CO:air at atmospheric pressure in a 30-cm cell.

In order to uncover any potential interferences, all of the carbon monoxide lines between 2111.5 and 2139.4 cm^{-1} [P(8) to P(1)] were scanned in samples of raw automobile exhaust.

Spectroscopic measurements of this type are being used to establish the optimum lines for monitoring pollutant gases as well as the related fundamental parameters. For certain reactive species, such as chlorine monoxide, laser spectroscopy appears to be the only answer (5) because of the low pressure needed for a stable mixture and relatively small absorption cross sections.

AMBIENT AIR MONITORING

The essential components of a diode laser system for monitoring are shown in Fig. 2. The diode laser is mounted in a closed-cycle refrigerator, and its emission is collimated by a parabolic mirror. The beam is transmitted to a remote retrore-flector (hollow corner-cube) which reflects it back towards the paraboloid, which refocuses it onto the infrared detector situated behind a calibration cell. In order to minimize the effects of atmospheric turbulence on system sensitivity, a derivative spec-troscopic technique is employed in which the laser is frequency-modulated at 10 kHz at the pollutant gas absorption line of interest (6).

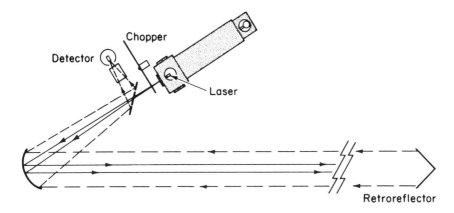

Fig. 2.   Diode laser system for long-path monitoring.

Calibration is performed by placing a known mixture of the gas to be monitored in the 10-cm cell. If the monitored path outside is 1,000 meters long, a calibration gas of 1,000 ppm mixture in a 10-cm calibration cell produces the same signal as 100 ppb over the long path. Linearity may be confirmed by using several different volumetric mixtures in the calibration cell. By proper controls, a measurement accuracy of around 5% can be achieved.

During the summers of 1974, 1975 and 1976, long-path measurements were made of atmospheric carbon monoxide in the St. Louis, Missouri area in conjunction with the Regional Air Pollution Study (RAPS) of the U.S. Environmental Protection Agency (7). One purpose of RAPS was to produce mathematical models for pollutant transport and modification. The diode laser measurements were compared with those of point monitors located near the path. Although there were, at times, marked differences between the long-path and point measurements, these generally occurred during conditions for which the pollutant concentration would be expected to be nonuniform.

Figure 3 shows one example of excellent agreement between the laser long-path (380 meters to retroreflector) and point monitor (Beckman 6800) at a farm site in Granite City, Illinois. The source of the carbon monoxide plumes could not be established, but both instruments responded similarly both in time and concentration.

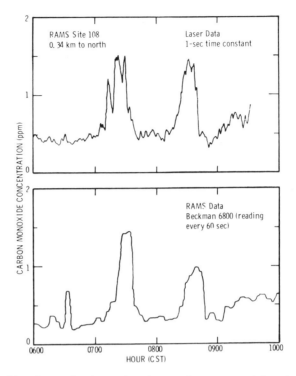

Fig. 3. Monitoring of atmospheric carbon monoxide at site 108 of
the Regional Air Pollution Study in Granite City, Illinois on
2 August 1975. Comparison between long-path diode laser data and
point-sampling instrumentation.

IMPROVEMENT IN SENSITIVITY

At atmospheric pressure, for pathlengths of up to a few hundred meters, it has been possible to measure changes in transmission of 0.3% using a derivative/ratio technique involving frequency-
modulation of a tunable diode laser (6). Recently, Reid and
Shewchun were able to detect changes in transmission as small as
0.001% in a multipass cell at a reduced gas pressure of around
10 torr, using a second-harmonic processing of the signal from a
frequency-modulated diode laser (8). This advance greatly improves
the sensitivity for tunable laser monitoring, not only for reduced-
pressure sampling, but for monitoring of the stratosphere where the
gas pressure is also low.

Figure 4 shows a diode laser scan (second harmonic) around
8.75 μm of a sample of air from Hamilton, Ontario, reduced in
pressure to 10 torr in a 200-m cell. The sulfur dioxide lines are

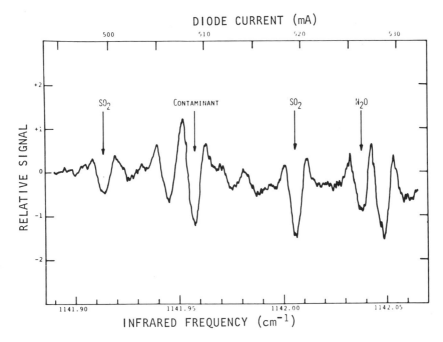

Fig. 4. Diode laser scan of air sample from Hamilton, Ontario.
[after Reid and Shewchun (8)]

clearly evident, and correspond to a concentration of 40 ppb.
Also within the tuning range is an identifiable absorption line of
nitrous oxide (N_2O) and a few unknown lines due to other species.
Although the signal-to-noise ratio for SO_2 detection is adequate,
it is worth noting that if the laser were tailored to operate in
the 7.5 µm region, the sensitivity would be an order of magnitude
better. Although water lines generally preclude use of this region, the low ambient pressure minimizes potential interferences.

REMOTE HETERODYNE DETECTION

Because of the effects of pressure broadening on a spectral
absorption line, it is possible to use lineshape information to
determine the concentration of a pollutant gas as a function of
altitude. In Fig. 5(a) the technique of solar heterodyne radiome-
try is illustrated, in which the sun serves as the source of radiation and a laser local oscillator is used to measure the amount of
the transmitted signal. By tuning the laser through a known absorption line of a pollutant present in the region between the
receiver and the sun, a dip in the transmission will be observable.
The same analysis applies over the two-way path of Fig. 5(b) involving a satellite reflector.

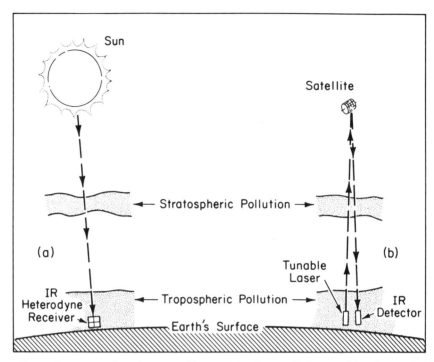

Fig. 5. Diagrams for long-path vertical profile monitoring of the atmosphere using (a) passive system with sun as radiation source and tunable-laser heterodyne detection, and (b) active, ground-based system employing cooperative satellite reflector.

Two laser measurements based on the heterodyne technique of Fig. 5(a) have been performed during the past year. In early 1976, Menzies and Seals used several lines of a discretely-tunable CO_2 laser to determine the height profile of ozone (9). More recently, Frerking and Muehlner of M.I.T. have performed the same measurement using a continuously-tunable semiconductor diode laser as local oscillator (10). Their apparatus consisted of a diode laser, beam splitter, infrared detector, and optics all attached to the cold head of a <u>liquid nitrogen</u> dewar, operating at 77K. Figure 6 shows their results for ozone in the frequency range 1010.9 to 1011.8 cm^{-1}. Trace (a) is a fully-resolved (0.0001 cm^{-1} resolution) direct absorption spectrum of ozone measured by transmitting the diode laser radiation through a laboratory gas cell. The heterodyne absorption signal from the sun is displayed in (b), which shows the same ozone absorption lines due to the presence of the gas in the stratosphere. The instrumental resolution in this case is 0.006 cm^{-1} (180 MHz), limited by the bandwidth of the rf amplifier system. This result represents the first use of a tunable

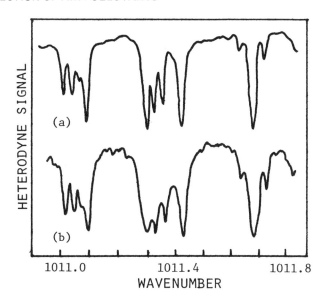

Fig. 6.   Ozone spectra measured with a tunable diode laser.
Trace (a) obtained in laboratory by direct absorption spectroscopy.
Trace (b) was a measurement over the same spectral region using
heterodyne detection with the sun as source of radiation, in the
manner illustrated in Fig. 5(a).   [Adapted from Frerking and
Muehlner].

laser heterodyne receiver to trace out the continuous absorption
profile of a molecule over a broad (0.8 cm^{-1} or 24 GHz) spectral
range, as well as the first wide-scan spectroscopy of an atmospheric constituent with a tunable heterodyne receiver.

CONCLUSION

With the commercial availability of tunable semiconductor
diode lasers, there has been a remarkable series of developments
and advances in the applications of these devices to environmental
monitoring and related spectroscopy. Other device improvements
need to be made, such as mode quality, power, stability, etc. and
are being vigorously pursued; and it appears that these devices
will prove to be even more useful in the years to come. The purpose of this paper is to point out the recent developments in diode
laser technology as applied to air pollution monitoring, and
indicate potential applications for the future.

REFERENCES

1 CALAWA, A.R.: J. Lumines. 7 (1973) 477.

2 GROVES, S.H., NILL, K.W., and STRAUSS, A.J.: Appl. Phys.
   Lett. 25 (1974) 331.

3 HINKLEY, E.D., KU, R.T., NILL, K.W., and BUTLER, J.F.: Appl.
   Optics 15 (1976) 1653.

4 Model SDL-3, available from Laser Analytics, Inc., Lexington,
   Massachusetts.

5 MENZIES, R.T., MARGOLIS, J.S., HINKLEY, E.D., and TOTH, R.A.:
   Appl. Optics, March 1977.

6 KU, R.T., HINKLEY, E.D., and SAMPLE, J.O.: Appl. Optics 14
   (1975) 854. For a comprehensive review of laser monitoring
   techniques, see Laser Monitoring of the Atmosphere, edited
   by E.D. Hinkley (Springer-Verlag, Heidelberg) 1976.

7 KU, R.T. and HINKLEY, E.D.: "Long-Path Monitoring of Atmos-
   pheric Carbon Monoxide," Technical Report to the National
   Science Foundation (RANN) and U.S. Environmental Protection
   Agency, M.I.T. Lincoln Laboratory, April 1976.

8 REID, J. and SHEWCHUN J.: Applied Optics (to be published).

9 MENZIES, R.T. and SEALS, JR., R.K.: Science (to be published).

10 FRERKING, M.A. and MUEHLNER, D.J.: Appl. Optics 16 (1977) 526.

AN EVALUATION OF LASERS FOR

AMBIENT AIR POLLUTION MEASUREMENT

Lucian W. Chaney

The University of Michigan

Ann Arbor, Michigan

ABSTRACT

Laser based air pollution measurement techniques which might possibly be applied to future air pollution monitors are examined. Four specific systems for which prototypes are currently being developed are reviewed from the standpoint of a user. The known problems are discussed. The four systems depend on either resonance absorption or resonance fluorescence. The first system is a long measurement path step tuned CO_2 laser designed to measure ozone, methane, and ethylene. The second system is a solid state tuned diode laser which is designed to measure carbon monoxide and nitric oxide. By continuously modulating the laser frequency a derivative signal is obtained which is ratioed with the transmission signal. The resulting normalized derivative signal is essentially independent of fluctuations in laser power or any broad band attenuation. This technique represents a fundamental improvement over the first system. The third system has been termed the opto-acoustic technique which measures resonance absorption directly by focusing the beam through a small cell. The laser beam is chopped by a variable speed chopper. The laser beam in passing through the gas in the cell heats the gas which produces a pressure variation. The pressure variation is detected by a microphone. Measurements have been made by this technique of concentration below one part per billion. The fourth technique utilizes resonant fluorescence and has been demonstrated to measure NO_2 with a noise equivalent concentration of 4-5 ppb.

Introduction

Lasers have been under continuous development for the past 17 years and every year new lasers are announced and new and different applications are introduced. The application of lasers to atmospheric measurements in general and to air pollution measurements in particular is firmly established. Commercial laser radar (lidar) (3) systems are now available to measure the height of industrial plumes and the height of the urban mixing layer by the back scattered radiation from atmospheric aerosols.

The four systems to be discussed in this paper have not yet advanced to the point of being developed commercially. However, they have all been previously reported as research techniques. The common denominator for these systems in addition to the fact that each uses a laser is that they were selected by the Environmental Protection Agency (EPA) to be funded for the development of a demonstrable prototype. The usual criteria for EPA funding is that (1) the end result should be a usable monitoring system, and (2) eventual commercial development be a possibility. The second criterion has, because of cost, eliminated from consideration some previously reported systems, such as those which employ Raman scattering (11) or use a spin-flip (12) laser.

We are all aware that most new techniques are subject to unanticipated problems. Sometimes very promising techniques remain in the development stage for a long time in a struggle to overcome an unanticipated problem. At other times the development must be laid aside until newer solutions become available. The purpose of this paper is to look at the problems which have been uncovered as a result of the prototype developments.

The development of any measurement technique which incorporates a laser as a radiation source is an attempt to take advantage of the laser's unique characteristic of generating essentially monochromatic radiation. The laser's monochromatic characteristic can lead to a number of advantages compared to broad band sources used in conventional spectroscopy. The beam can be almost perfectly collimated. Mirrors only a few inches in diameter are usable at distances up to several kilometers. The beam can be passed through small cavities with volumes as small as one square centimeter. Still another advantage is the inherent high energy density both spatially and spectrally. The laser line width can be less than $.001$ cm^{-1} and the beam diameter can be a few millimeters.

As a result of these advantages many investigators have applied lasers to previously existing techniques. The techniques to be discussed in this paper depend on resonance fluorescense or resonance absorption. In both instances the selected laser line is nearly coincident in frequency with a molecular absorption line of

the target gas. The difference between the two phenomena are in
the mode of the reradiated energy. In the case of resonance absorption the energy reappears in translation of the molecule or
thermal energy. In resonance fluorescence the energy is reradiated as photons at a different wavelength.

Conventional spectroscopy is based on resonance absorption.
A broadband energy source is resolved spectrally by a grating and
a pair of slits. The transmission is measured. The difference
between the on-line and the off-line transmission is due to resonance absorption of the transmitted energy by the target gas. If
the path length and the absorption coefficients are known then
the concentration can be calculated. A laser can be used in precisely the same way if a pair of lines can be found such that one
is coincident with a target gas absorption line and the other is
unattenuated. The laser thus becomes a spectroscopic tool and the
first system to be discussed uses this procedure exactly. A second
system which is tunable measures the derivative of the transmission
and again knowing the path length and the absorption coefficient
the concentration is calculated.

One of the laser's principal advantages over broadband sources
is that the beam can be accurately collimated making long path
measurements with small optics possible. Even with a total transmitted power of less than a milliwat, it has been possible to make
measurements over an optical path of 2 km.

The third system to be discussed also depends on resonance
absorption, but in this case the absorption is measured directly
instead of measuring it indirectly by taking the difference in two
transmission signals. The laser beam is passed through a small
sample chamber, sometimes less than one cubic centimeter. The beam
is chopped by a variable speed chopper and the change in pressure
inside the sample chamber is measured. The target gas absorbs some
of the laser energy and when the gas returns to the quiescent state
the energy is released as translational energy which increases the
gas pressure. The technique is now termed opto-acoustic (OA) (10,
12, 19). The principle has been used in the past and is the basis
for the Golay cell which was commonly used as an infrared detector
a few years ago.

The fourth system utilizes resonance fluorescence (1,4,5). In
this case the laser energy absorbed by the target gas is reradiated
as photons rather than appearing as translational or thermal energy.
As in the OA system the beam is passed through a sample chamber.
The wavelengths used in the first three systems are in the infrared.
Here a molecular transition of the target gas (NO_2) at a visible
wavelength, 488 nm, is excited by an argon ion laser. The NO_2,
when deexcited, emits photons in a band between 700 nm to 810 nm.
The intensity of the emitted radiation is a measure of the pollutant

in the sample chamber.

This paper will not include a detailed description of any of
the four systems. The tunable diode laser (6,7,13) has been described at this conference in a previous paper by Hinkley. A review
by Hodgeson (8) et al describes the step tuned laser and the laser
induced fluorescence technique. A recent paper by McClenny (16)
describes the opto-acoustic technique as well as reviewing the long
path and the laser induced fluorescence technique. This paper will
evaluate the techniques from the standpoint of a potential user.
The two long path laser systems(2,14) and the laser induced fluorescence system (1) were tested as a part of the recently completed
St. Louis Regional Air Pollution Study (RAPS).

The cost and sophistication of any of these systems are such
that in the forseeable future none is apt to be employed for routine
monitoring of air pollution. The most likely use is as research
tools to answer basic questions regarding the nature of air pollution.

Long Path Monitoring

Long path monitors refer to those monitors which have optical
paths which cannot be contained in the basic instrument enclosure.
Such paths are generally greater than 20 meters and may be as large
as 10 km. There are at least three reasons for wanting to use long
optical paths to measure gaseous atmospheric pollutants. The first
reason is to increase the total pollutant concentration in the path.
It should be remembered that although some pollutants are present
in the parts per million (ppm) range most are in the parts per billion (ppb) range and some, such as the fluorocarbons, are in the
parts per trillion (ppt) range.

A common method of testing a long path system is to place the
source on one building and the receiver an appropriate distance away
on another building. This of course leads to communication problems,
requires at least two operators, and lacks flexibility. An alternate
technique is to use a multipass mirror system. A practical limit to
the number of passes is 50-100 with base lengths from 10-20 meters
which restricts the path to 0.5 km - 2.0 km. A second reason for
long path or open path monitoring is to maintain the sample integrity.
The handling of the gas to get it into and out of a sample chamber
can be extremely vexing in the ppb range especially for reactive
materials such as ammonia and hydrogen chloride. The third reason
and the motivation behind the two systems to be discussed is to make
integrated average measurements along specified paths. The EPA funding for these developments came from the Regional Air Pollution Study
(RAPS) program which had as its objective the development of urban
air pollution models. One of the problems in the testing of air
quality models is to relate point measurements to area wide averages.

The long path monitor was seen as an aid to making this correlation.

Discrete Laser Wavelength System

The first system to be discussed was developed by the General
Electric Company (9,20) and is a CO_2 laser system which measures
the infrared absorption of the laser radiation. General Electric
(G.E.) has called their monitor ILAMS (Infrared Laser Atmospheric
Monitoring System). The development of the system began in 1966 but
in the past three years EPA has funded G.E. to modify the system
to measure ozone and to install the system in a mobile trailer with
beam steering 360° in azimuth and $\pm$ 10° in elevation. A sketch of
the basic system is shown in Fig. 1. The laser beam is directed
toward a retroreflector approximately 25 cm. in diameter which can
be placed at any distance up to one kilometer away. The beam is
returned to the same optics. The returned beam is directed to the
detector by the beam splitter.

The CO_2 laser is capable of lasing on about 60 different lines.
Hence, the basic idea is to find a pair of laser lines which are
matched and unmatched with the target gas absorption line. Ideally
one line would be near the maximum absorption and the other would
be completely off an absorption line. The ideal condition cannot
be met and in addition there are other absorbing gases present. In
fact, many of the common gaseous pollutants as well as water vapor
have vibration bands which might interfere with the measurement in
this spectral region, 9 to 10 µm. Hence, the solution developed was
to use a set of six selected lines to measure ozone, ammonia, and
methane. Different weighting functions were assigned to each line
depending on the target gas to be measured and a matrix calculation
is performed to determine the concentration. A built-in small computer (CDC, pdp-11) is used to make the required real time computation.

The energy received at the signal detector for each wavelength
can be written in the form

$$I = I_0' \, kd \, \exp\,[-\alpha(\nu)CL] \, \exp\,[-\beta(t)L].$$

Where $\alpha(\nu)$ is the target gas absorption coefficient, a function of
the wave length (ν). $\beta(t)$ is the atmospheric extinction coefficient
due to atmospheric turbulence and aerosol scattering and generally
not related to the target gas, but time dependent. C is the target
gas concentration and L is the total path length. I_0 is the beam
intensity monitored at the reference detector. The output beam
intensity (I_0') is a fraction d of I_0. The factor k accounts for the
attenuation due to optical surface losses and the overfilling of
optical elements.

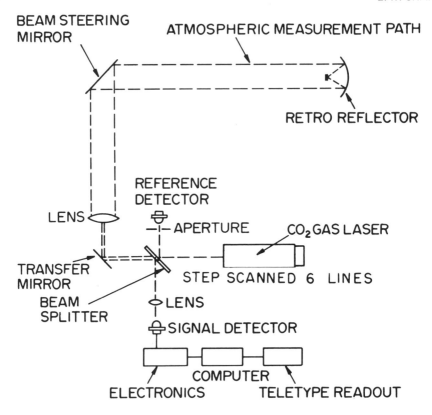

Figure 1

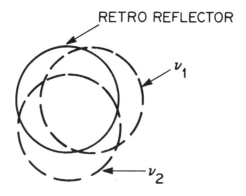

Figure 2

If we consider the two wavelength systems only, the concentration can be determined from the following equation.

$$C = [(\alpha(\nu_1)-\alpha(\nu_2))L]^{-1}[\ln(T_1 d_2 k_2/T_2 d_1 k_1)-(\beta_2(t)-\beta_1(t))]$$

where $T = I/I_o$, which was obtained by ratioing the two received signals. If the two wavelengths are close enough to each other then we assume that $\beta_1(t)=\beta_2(t)$, $d_1=d_2$, and $k_1=k_2$. The expression for the concentration thus reduces to

$$C = [(\alpha(\nu_1)-\alpha(\nu_2))L]^{-1}\ln[T_1/T_2]$$

This requires in addition to a measurement of the transmission an accurate knowledge of the value of the absorption coefficients and the optical path length. Since both of these parameters can be determined to a high accuracy it would seem that the concentration could be detected with the anticipated precision. A set of tests conducted in the fall of 1973 at General Electric produced a minimum detectable signal using a 20 second time constant of 2.5 ppb. This was considered quite good and a field test of the equipment was conducted in the summer of 1974 in St. Louis. A review of the collected data indicated that the assumption that $k_1=k_2$, $\beta_1=\beta_2$, and $d_1=d_2$ was not valid for $\nu_1 \neq \nu_2$. The closer that $\nu_1 \rightarrow \nu_2$ the better the assumption. There are many possible wavelength effects which occur as a result of reflection, diffraction, and refraction at optical surfaces and aperture stops. The beams for each wavelength may not be co-axial (Fig. 2) and a change in the index of refraction of the air due to a temperature gradient could reposition the beam and change the ratio of the signal returned for each wavelength. This represents a change in the k_1/k_2 ratio which appears as a change in concentration. It is also possible that the beam shifts in other portions of the system such as the receiving optics or at the reference detector aperture stop might lead to changes in the d_1/d_2 ratio.

General Electric has now contracted with EPA to deliver a prototype instrument which has optimized the wavelength selection in order to minimize the possible changes in k or d. The prototype is scheduled to be tested in the late summer of 1977.

Semi-Conductor Diode Laser System

An alternate approach to a long path laser system is the semiconductor diode laser pioneered by Hinkley at Lincoln Laboratory and described in the previous paper. The inherent advantage of the semiconductor diode is its tunability. The laser line can be tuned to the best possible match with the selected molecular absorption line. The problems which arose in the step tuned system due to $\nu_1 \neq \nu_2$ are virtually eliminated since the maximum wavelength difference is $\Delta\nu$,

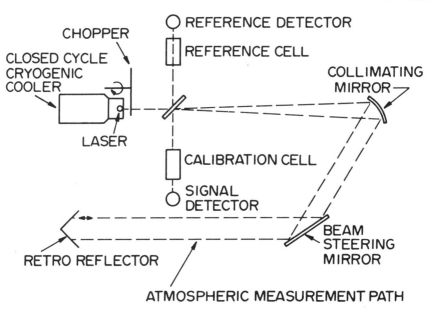

Fig. 3. Schematic Diagram of Tunable
Semiconductor Diode Laser System Optics

the modulation.

The complete system (Fig. 3) adjusted to measure carbon monoxide was housed in a General Motors motor home and extensively
field tested (2,14) in connection with the St. Louis RAPS program.
The measurement path is like the ILAMS bistatic with a 13 cm diameter retroreflector placed at the far end of the measurement path.
Data were collected during the St. Louis test with this reflector
located at a distance of one kilometer.

The same basic equation used by ILAMS applies to the transmission measurement by this system.

$$I(\nu) = kI_o \exp[-\alpha(\nu)CL] \exp[-\beta(t)L]$$

There are, however, significant differences. The parameter,
d, does not enter the calculation. But the most important difference is that by frequency modulating ν it is possible to obtain
the derivative signal. This is done by applying a small amplitude
10 kHz signal to the diode. The expression for the derivative is

$$\frac{dI}{d\nu} = -kI_o CL \exp[-\alpha(\nu)CL] \exp[-\beta(t)L]\frac{d\alpha(\nu)}{d\nu} .$$

By using two lock-in amplifiers and a ratio meter it is possible to obtain a voltage equal to

$$dI/d\nu/I = CLd\alpha(\nu)/d\nu.$$

The parameters k, I_o, and β drop out of the expression and if L and $d\alpha(\nu)/d\nu$ are constant then the voltage is directly proportional to C. The distance to the retroreflector is obviously a constant and under many conditions, but not all, $d\alpha(\nu)/d\nu$ is a constant.

The system was extensively field tested in St. Louis during the RAPS program and when the CO concentrations were low, less than 1 ppm, the agreement with point monitoring systems was excellent. However, ν, the laser operating frequency varies with temperature which cannot be controlled absolutely. Secondly, as the concentration increases the absorption becomes non-linear and $d\alpha(\nu)/d\nu$ decreases so that further increases in C do not produce proportional increase in the output signal. Among the major pollutants the problem of non-linearity only occurs with CO which can vary in concentration from 50 ppb to over 50 ppm in the urban atmosphere. An additional third problem is that under most conditions the diode laser has more than one output frequency or oscillates simultaneously in more than one mode. These are the essential system problems for which we believe there are adequate solutions.

The closed cycle cryogenic cooler cools the diode laser down to an operating temperature of about 10°K. However, the precise temperature of the cold finger on which the laser is mounted is maintained by a feedback circuit incorporating a temperature sensor near to but physically removed from the laser. The set point temperature can be controlled to an accuracy of 0.001°K and it is by this means that the frequency of the diode is made coincident with a selected absorption line. However, if the ambient temperature changes the diode will also change temperature slightly even though the set point remains constant. The solution to this problem has been to place a beam splitter in the laser path and divert a portion of the transmitted energy through a reference cell containing pure CO. The amplitude of the transmitted signal can be monitored and variations in amplitude from a reference voltage are incorporated into another feedback loop. The requirement is an extremely accurate reference voltage which has been difficult to establish. An alternate technique is to use as a reference an AC signal derived from the chopper and fed to the phase-lock amplifier before detection. A different suggestion has been to use a second derivative signal which will be zero at the first derivative maximum. The feedback loop would be designed to minimize the second derivative. The advantage would be that the system would lock on a very specific frequency determined by the gas in the reference cell. The specific frequency could be moved towards the wing of the line by increasing the concentration in the reference cell.

The solution to the problem of non-linearity due to the increase in the concentration of the target gas has to be to either move the monitoring frequency to the wing of the absorption line or to use a nearby isotopic line. However, since $C^{13}O^{16}$ is two orders of magnitude less abundant than $C^{12}O^{16}$ it would be desirable to use the wing of the line if adequate frequency control can be established.

The two basic approaches to the problem of multiple modes is either to not generate or to remove them. At the present state of the art the diode can be made to oscillate in a single mode but not necessarily at the frequency selected. The present approach is to incorporate into the system a tunable Fabry-Perot etalon which will only pass a single mode. The modified system will be tested later this year.

The conclusion regarding long path monitoring is that it remains to be demonstrated that a discrete frequency system can be made sufficiently immune to atmospheric turbulence to be useful as a monitoring tool. The diode laser system by ratioing the derivation's signal by the transmission has effectively minimized the atmospheric turbulence problem, but has multimode and frequency stability problems for which there seem to be solutions.

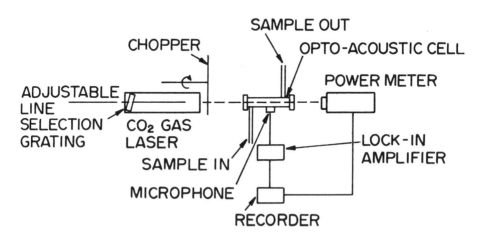

Fig. 4.   Schematic Diagram of EPA Opto-acoustic
Ammonia Monitor

Opto-Acoustic System

An opto-acoustic system specifically designed to measure ambient concentrations of ammonia is being developed as an in-house project by the EPA Environmental Sciences Laboratory. Preliminary results from the project have recently been reported by McClenny (16).

The experimental system (Fig. 4) consists of a commercial grating tuned CO_2 laser; a variable speed chopper; a 15 cm long by 0.95 cm diameter OA cell with Irtran 2 windows; a laser power meter; and a lock-in amplifier. The sample containing the ammonia is continuously flowed through the cell at 60 ml min^{-1}. The basic configuration is the same as that reported by other investigators. A noise equivalent sensitivity of 0.5 ppb in air is now being reported (17). This compares favorably with the theoretical minimum detectable signal of 0.1 ppb determined by Max and Rosengren (15). The reported sensitivity has been achieved by tuning the chopper frequency to take advantage of acoustic resonances set up inside the cell. Signal enhancements or Q's greater than 100 have been reported (15) using this feature. Further design improvement in this area should be expected.

The problems associated with the opto-acoustic system are primarily associated with background noise. The background noise can come from the gas flowing into the cell, or ambient noise coupled back through the sample lines. It is also possible for ambient noise to be coupled directly to the microphone. The chopper itself can also couple mechanical vibrations to the microphone. The vibrations from this source are bothersome since they occur at some sub-multiple of the chopping frequency.

New references citing the use of the OA technique are appearing regularly in the literature. In addition to being able to measure gaseous pollutants, suspended aerosols, liquids, and solids (18) can be detected. In fact a major problem in reducing the minimum detectable signal is the signal generated by the entrance and exit windows which must be balanced out. The major advantage of the technique is that it is possible to integrate the re-radiated energy both in space and time. The further development of the OA technique as a research tool and probably as commercial instrumentation should be expected.

Laser Induced Fluorescence

A laser induced fluorescence monitor for the detection of NO_2 has been developed by Aerospace Corporation and described by Birnbaum (1) et al. The basic optical configuration is shown in Fig. 5. The argon laser generates a 488 nm beam which traverses the sample chamber. The photomultiplier tube used to detect the fluorescence

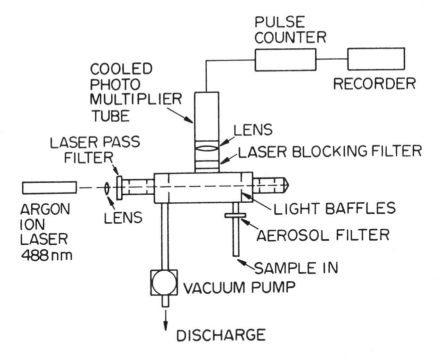

Fig. 5.   Schematic Diagram of NO_2
Fluorescence Monitor

is mounted perpendicular to the beam to minimize the possibility
of collecting any of the laser radiation. The exit filter is designed to completely block the 488 nm laser radiation. The fluorescence occurs in the band from 710-800 nm which is passed by the
filter. The careful design and selection of the exit filter is
very necessary to the success of this technique. The careful match-
ing of the laser line and the filter to the target gas decreases
the flexibility of the system. The vacuum pump is required to maintain a lower than atmospheric pressure in the sample chamber. The
lower pressure permits sharper, more specific lines, but it also
decreases the concentration of the pollutant under study.

The photons counted by the photomultiplier tube are a direct
measure of the NO_2 concentration. Using an argon ion laser with
a power of 100 milliwatts the standard deviation of the detectable
signal is 4-5 ppb. This is roughly equivalent to commercially
available chemiluminescent (CL) monitors. However, the laser fluorescence monitor has the advantage of measuring the NO_2 directly
in contrast to the CL monitor which must first measure the NO in
the sample, convert the NO_2 to NO and determine the NO_2 by the difference in the NO measurements. The basic difficulty with the CL

monitor procedure is that other nitrogen compounds such as PAN, HNO_3, and formic acid can also be converted to NO. In fact, a principal reason for the EPA sponsorship of the LF monitor development was to ascertain the magnitude of the interference from other nitrogen compounds. The St. Louis RAPS comparison of a standard CL monitor and the LF monitor showed that the CL monitor invariably read higher, but the difference was often less than the standard deviation of the differences. The standard deviation of both monitors was $\pm$ 5 ppb and the standard deviation of the difference was $\pm$ 7 ppb. Recent reports (21) on research CL monitors are that standard deviations less than 1 ppb are achievable. The conclusion from the St. Louis test was that more data and testing would be required to separate the difference attributable to the atmosphere from those due to the instrumentation. However, the basic question was answered that the differences were less than the standard deviation in presently available commercial equipment.

Following the St. Louis test the equipment was and still is being used to measure the NO_2 in a smog chamber. For this purpose the equipment is satisfactory and the conclusion is that it measures NO_2 unambiguously. However, there is little prospect that the system at this time will be developed commercially because of the cost, size, and lack of flexibility to measure other pollutants.

In passing it might be of interest to note that a commercial fluorescence monitor which measures SO_2 in concentrations from 0.1 to 1600 ppm is now available. The fluorescence is induced by a 15 watt resonance lamp; zinc (213.8 nm) and cadmium (228.8 nm).

Conclusion

The laser has already proved useful as a building block in the development of air pollution monitors. However, without tunability its usefulness as a spectroscopic device is limited. As compensation a large number of lines must be available such as the 60 lines in a CO_2 laser. On the other hand tunability without frequency stability is a poor alternative. The best prospect for a long path laser system is the diode laser plus a mode filter and a satisfactory technique for frequency stabilization.

The opto-acoustic technique holds a great deal of promise for the future. It is still in the stage of vigorous development and how far it may go remains to be seen. The laser induced fluorescence development has resulted in one useful instrument, but because of cost and specificity, the possibilities for future developments seem remote at the present time.

REFERENCES

1   BIRNBAUM, M. and TUCKER, A.W.: NO$_2$ Measuring System, EPA Report
    650/2-74-059, Final Contract Report for EPA Contract No. 68-02-
    1225 Aerospace Corporation, May 1974.

2   CHANEY, L.W., McCLENNY, W.A., and KU, R.T.: Long Path Laser
    Monitoring of CO in the St. Louis Area, paper 75-56-6, 68th
    Annual Meeting of APCA, Boston, Mass, June 15-20, 1975.

3   COLLIS, R.T.H.,: Appl. Opt. 9 (1970) 1782.

4   GELBWACHS et al, Opto-Electronics 4 (1972) 155.

5   GELBWACHS, J. and BIRNBAUM, M.: Appl. Opt. 12 (1973) 2442.

6   HINKLEY, E.D. and KELLY, P.L.: Science 171 (1971) 635.

7   HINKLEY, E.D., J. Opto. Electronics 4, (1972) 69.

8   HODGESON, J.A., McCLENNY, W.A. and HANST, P.L.: Science 182
    (1973) 248.

9   JACOBS, G.B. and SNOWMAN, L.R.: IEEE J. of Q.E., QE-3 (1967)
    603.

10 KERR, E.L. and ATWOOD, J.G.: Appl. Opt. 7 (1968) 915.

11 KOBAYASI, T. and INABA, H.: Appl. Phys. Letter 17 (1970) 139.

12 KREUZER, L.B., and PATEL, C.K.N.: Science 173 (1971) 45.

13 KU, R.T., HINKLEY, E.D., and SAMPLE, J.O.: Appl. Opt. 14,
    (1975) 854.

14 KU, R.T. and HINKLEY, E.D.: Long Path Monitoring of Atmospheric
    Carbon Monoxide, Report submitted to National Science Foundation
    by Lincoln Laboratory, Contract No. NSF/RANN/IT/GI-37603 (April
    1976).

15 MAY, E. and ROSENGREN, L.-G., Optics Communications 11, (1974)
    422.

16 McCLENNY, W.A. and RUSSWURN, G.J., paper presented to the 8th
    Materials Research Symposium, National Bureau of Standards (1976).

17 McCLENNY, W.A.: U.S. Environmental Protection Agency, Research
    Triangle Park, N.C., private communication March 1977.

18 ROSENCWAIG, A.: Science 181 (1973) 657.

19 SCHNELL, W. and FISHER, G.: Rapport de la Société Suisse de
    Physique 26 (1975) 133.

20 SNOWMAN, L.R., MORGAN, D.R., ROBERTS, D.L. and CRAIG, S.E.:
    Development of a Gas Laser System to Measure Trace Gases by
    Long Path Absorption Techniques. EPA Report 650/2-74-046-A
    General Electric Final Report Contract No. 68-02-0757, June 1974.

21 STEDMAN, D.: Chemistry Department, University of Michigan,
    Ann Arbor, private communication, March 1977.

Acknowledgement

The information concerning the opto-acoustic technique and
many reference papers on the long path systems were furnished by
Dr. W. A. McClenny of the Environmental Protection Agency. This
was very helpful and appreciated in the preparation of the paper.

The evaluation studies were supported by the Environmental
Protection Agency Grant R-803399. The contents of this paper do
not necessarily reflect the views and policies of the Environmental
Protection Agency.

Discussion

ETZ - If you use the laser technique developed by Lincoln Laboratory to monitor atmospheric gases, how well can you match up the
reference and signal detector? Is that a problem?

CHANEY - In the Lincoln Laboratory system, this is not a problem.
The signal detector is used to measure the amplitude of the returned signal, whereas the reference detector is used in a feedback network to control the laser operating frequency.

SMITH - The last slide that you showed reminded me of work that
Dr. Gary Hieftje is doing at Indiana University. He uses a multi-
frequency mode argon laser to excite fluorescence. Then he does
a wave analysis of the fluorescence signal from which he obtains
a time decay. I wonder if you could specify the particular nitrogen oxide by the time decay.

CHANEY - The fluorescence monitor which I described is specific
for NO_2. The problem of separating the nitrogen oxides is in
the chemi-luminescence monitor which is specific for NO.

SMITH - Could you do a multi-analysis by measuring the time decay
signal?

CHANEY - I don't know, but that is a suggestion which would be
worth investigating.

CAHILL - About two years ago, a group at the University of California at Berkeley including Owen Chamberlin was working with a
known scattering atmosphere and high-energy lasers. Is anyone
now using such lasers and measuring the emitted fluorescence radiation over long paths?

CHANEY - Dr. Hinkley can probably answer that question better
than I can. The projects with which I am familiar, sponsored by
Environmental Protection Agency, do not include such devices. It
sounds as if it would be beyond their normal funding capabilities.

CAHILL - I think there was a problem in specifying the laser power
in units of Audubons - How many birds do you knock down with one
of those things?

GEISS - Referring to the tunable diode laser, you stated that the
selected operating frequency depended on the expected maximum concentration of CO. Does that mean that the maximum signal is
changing as a function of concentration?

CHANEY - No it doesn't. The total range of ambient CO concentrations is such that if the system is adjusted for minimum

concentrations, it will become non-linear and saturate for larger concentrations. This condition can be compensated for by changing the laser operating frequency to the wing of the same line or switching to a weaker line.

GEISS - In area of low concentration, the absorption line is sharp, but in areas of high concentration in the same path, the absorption line may be flatter. Do you correct for this?

CHANEY - Yes we do. As I previously noted, the operating frequency is determined by the reference cell and reference detector in a feedback arrangement. In the last version of the system, the operating frequency is located at that position on the frequency scale where the second derivative of the transmission signal is equal to zero and the first derivative is positive. This point can be moved towards the wing of the line by increasing the reference cell pressure. In moving towards the wing of the line, the measurement is less sensitive, but remains linear to a higher concentration.

HUNTZICKER - Dr. Hinkley, regarding one of the last slides you showed of the Canadian work about nitrous oxide, did they make an independent measurement of the concentration of N_2O?

HINKLEY - No. They measured it themselves. That was pure N_2O in a reference cell.

HUNTZICKER - I would have expected to see the N_2O lines in the ambient concentration. Does anyone know the nominal N_2O concentration?

LODGE - At least a quarter of a part per million. Rasmussen is reporting 0.3 ppm.

GEISS - Has anyone used the data from the long path monitors as inputs for air pollution models?

CHANEY - I am not aware that any of the data has been used. Some of the data that we collected was placed in the Regional Air Pollution Study data bank. However, there was not enough data available to believe that it could be used to actually verify models.

GEISS - Regarding the measurement of low concentrations of very strange pollutants, do you see any advantage to the long path?

CHANEY - Yes, it has been used very successfully. Dr. Phillip Hanst from EPA has conducted several studies designed to identify the unknown constituents in Los Angeles smog. He was the first to identify peroxy acetyl nitrate (PAN) and later formic acid,

by using a multipass cell totaling 0.5 km and a Fourier Transform
Spectrometer (FTS). He has aided the University of California at
Riverside in setting up a new system which has a folded 2-km path
and an FTS monitor.

GEISS - Of course, you have the complication of the FTS instrument
which Hanst uses. But there is the advantage of the larger spectral
range and you don't have to change lasers.

CHANEY - If you are examining a broad spectral region, the FTS
system is certainly preferred. There is also a frequency identification problem if the laser is scanned very far.

GEISS - Is there an advantage to the laser's high resolution at
atmospheric pressure?

HINKLEY - It is definitely an advantage compared to a two wave
number instrument. It gives you higher specificity and higher
sensitivity. Of course, if it is used in the stratosphere where
the pressure is low, there is a big advantage to the high resolution.

GEISS - I recently participated in a seminar where the relative
advantages of laser techniques were discussed and we came to the
following conclusion. The diode laser has two advantages; one,
it can be used over a small frequency band and two, it is not
dangerous when used over a long path. However, it cannot compete
with the differential absorption technique for measuring the concentration at a distance.

HINKLEY - This is using lasers?

GEISS - Yes. They use two pressure-tuned CO lasers. One is on
the absorption line and one just off the line. It is just a matter
of electronic treatment to obtain the concentration at a given
distance.

HINKLEY - I agree. The next best type of laser compared to the
tuned diode is either the capillary CO or CO_2 laser. These high
pressure lasers are being developed for nuclear fusion. Nevertheless, you are still talking about a restricted wavelength region.
You do not have the flexibility to select the wavelength you wish
to use. But I agree that differential absorption lidar as it is
called, is a very nice technique.

SESSION IV:

HIGH SPATIAL RESOLUTION

MICROPROBE METHODS

METHODS OF MICROPROBE ANALYSIS

Kurt F. J. Heinrich

Institute for Materials Research
Analytical Chemistry Division
National Bureau of Standards
Washington, DC 20234

Introduction

Microanalytical techniques were first developed by Pregl [1] for the determination of elementary composition of organic materials. Such procedures provided a means of characterization for materials such as products of biological processes which were available in small amounts only. The observation of precipitation reactions under the microscope [2] permitted the scaling down of qualitative analytical procedures for inorganic materials, and the formation of colors and precipitates was used for qualitative microanalysis in Feigl's "Tüpfelanalyse" (spot analysis)[3]. However, the separation methods of chemical quantitative ("wet") analysis were difficult to adapt, mainly because of the strong capillary effects in the manipulation of liquids in the microscopic domain.

The sampling of components of complex solid specimens such as individual mineral grains or inclusions in metals requires painstaking mechanical or etching procedures, with danger of alteration or contamination. The problem is avoided in the probe techniques, where a physical reagent of microscopic cross-section (microbeam or microprobe) elicits a signal from a small selected region at or near the surface of a solid object [4] (figure 1). This beam can be used either in a static mode or can be made to scan in a raster over a small area at the specimen surface [5]. Such techniques are also very successful in the analysis of individual small particles [6].

242 K. F. J. HEINRICH

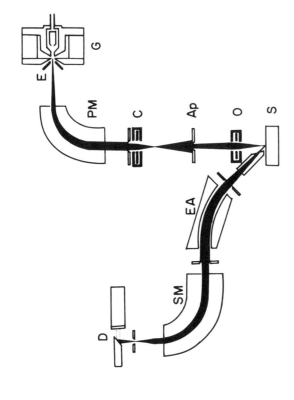

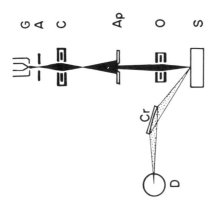

1 Electron and Ion Probe Microanalyzers. Left: Electron Probe Microanalyzer. The primary electron beam, generated in the electron gun (G) and focused by electron lenses, impinges on the specimen (S). The emitted x-rays are analyzed by a curved-crystal spectrograph (Cr). In energy-dispersion analysis, this detector is replaced by a solid state, Si(Li), detector. For cathodoluminescence, an optical spectrometer is substituted, and an electron spectrometer for Auger and backscattered electrons. Right: Ion Probe Microanalyzer: A duoplasmatron gun (G), ion (electrostatic) lenses, and a mass spectrograph are substituted. A magnet in the primary beam selects the ion to be used.

Microanalysis and Microscopy

A single microscopic region of matter or single particle is seldom of sufficient interest to warrant the effort of the analyst and price of the equipment (exceptions occur, mainly in failure analysis of microcircuits). Therefore, a question of representativeness of sampling arises. In some experiments such as in the tracing of concentration changes through a single- phase region of diffusion, credibility of the results is established by internal consistency - points will fall on a smooth line - but in most cases statistical boundaries are established from a number of measurements of different regions or particles that "look alike" in the microscope (optical or electron). In either case it is important that the sampling is not random; sampling should be largely determined by an initial working hypothesis, and/or a microscopic criterion such as shape, color, cathodoluminescence, etc. Therefore, microprobe analysis is closely related to microscopy as well as to analytical chemistry: not only is a microscope invariably used for the selection of areas to be analyzed, but characteristic shapes of analyzed phases or particles, and the local distribution of the components, are frequently essential parts of the information provided by microprobe analysis. Microscopes are, therefore, built into most microprobe instruments. Probe techniques are also often extended into microscopy by scanning a probe in the form of a raster over the specimen surface.

Standardization

Microprobe analysis of a large homogeneous specimen will be of no practical interest; homogeneity is, however, a most important requirement for standards used in microprobe analysis, particularly if quantitative information is expected. The microscopic region should have the same composition as the standard aliquots which were characterized by macroscopic analytical procedures. Moreover, the presence of interelement effects in many microprobe procedures justifies the requirement that the standard also be homogeneous within the microscopic region used in the standardization of the microprobe analysis. This requirement is often disregarded by necessity, negligence or ignorance. It is not licit, however, to assume that the sum of signals obtained from parts of a heterogeneous specimen is equal to the signal that would be obtained from a microscopically homogeneous specimen of identical average composition. The resultant error will be the larger the more severe the "interelement" or "matrix" effect is. This fact has caused the extensive effort in predicting theoretically the matrix effect observed when the standards differ considerably from the specimen in composition, in techniques such as electron and ion probe microanalysis. It should be noted that

Standard Reference Materials (SRM's) intended and certified for use in macroanalysis are not necessarily suited to microprobe analysis. The National Bureau of Standards has issued SRM's of tested micro-homogeneity [7], and efforts are under way to extend such tests to other SRM's.

Classification and Characteristics of Probe Techniques

The techniques of microprobe analysis can be classified according to the nature of the exciting agent and of the signal, as shown in table 1.

The spatial resolution of photon probes is limited by the available focusing optics, particularly when x-ray photons are employed. A further limitation in spatial resolution arises from the deep penetration of x-rays, and, for many specimens, of photons in the visible range. Particle optics, especially electron optics, can be designed to focus very fine beams, but they require high vacuum and the equipment is usually more expensive.

Before discussing in more detail specific methods, it is useful to cite the parameters and characteristics which must be considered in their choice for a given problem (table 2). The basic parameters are the level and range of application. The level indicates the type of answers the method can provide. It can be isotopic, atomic, valence, molecular, structural, or shape. Molecular probe analysis is, of all levels, the most recent and most promising, being, at present, represented only by laser-probe Raman microanalysis. This method also gives structural information which is otherwise mainly obtained by x-ray and electron diffraction; shape (habit) is determined by the microscopic techniques.

The range of application of a method can be fairly narrrow, as for instance, in cathodoluminescence which is not observed in most types of specimens, or universal, e.g., in ion probe microanalysis, which can be applied to the detection of all isotopes and elements in all solid specimens. Specificity, sensitivity and accuracy are parameters of interest in any chemical method of analysis. A specific signal can be related uniquely to the presence of a given component in the specimen. Interferences by similar signals limit the specificity, and hence the applicability of the method. The specificity of a signal such as radiation (line of a spectrum) depends on the resolution (i.e., capacity of separation from adjacent lines) of the spectroscope. This, and other parameters of a method, are not fully determined by the method, but also depend on the characteristics of the instrument and the specimen, and conditions of the measurement (beam intensity, optical aperture, length in time of operation, speed of spectral scan, etc.).

Table 1. Types of Probes

Reagent	Signal	Methods
Photons	Photons	X-Ray Diffraction Laser Probe Laser Raman Microprobe X-Ray Fluorescence Probe
Photons	Particles	(ESCA)
Particles	Photons	Kossel-line Electron Probe Scanning Electron Microprobe:X-Rays (SEM-EDX) Proton-Excited X-Rays (PIXE) Cathodoluminescence Infrared Cathodoluminescence
Particles	Particles	Electron Diffraction Auger Spectroscopy Scanning Electron Microscopy Ion Probe Electron Probe Gas Analysis

Table 2

Significant Parameters

Levels: Isotopic
           Atomic
           Valence
           Molecular   Basic
           Structural
           Shape

Range of Application

Specificity
Relative Sensitivity   Analytical
Accuracy

Absolute Sensitivity
                                  Microscopic
Localization

The (relative) sensitivity is expressed by the lowest detectable
concentration of a sample component. It is important to distinguish in this context microanalysis (i.e., detection of small
amounts) from trace analysis (i.e., detection of low concentrations). For instance, electron probe analysis can lead to the
detection of 10^{-15}g, but, since this is only over a very small
volume, this corresponds to about 0.01 percent of an element in a
sample. This technique is thus microanalysis, but not trace
analysis. The definition of limits of detection is somewhat
controversial; for a good discussion of the subject see reference
[8]. Sensitivity depends on the efficiencies of signal generation
and detection and on the level of background noise. The accuracy
is the capability of a method to measure quantitatively a property
such as the concentration of a component. Its definition includes
consideration of possible systematic errors in the determination.
Hence, it must not be confused with the statistical precision
(i.e., repeatability) of the measurement. The absolute sensitivity (expressed in units of mass, or moles) can also be defined
for a macrotechnique; but it is of special interest in microanalysis. It is equal to the product of the relative sensitivity by
the total mass of specimen excited. Concerning localization,
the dimensions of the region excited in a large specimen from
which the observed signal is emitted are of utmost importance. It
is useful to distinguish in the excited regions the "width" of
excitation in the plane of the specimen surface, and the "depth of
penetration" in the normal to the specimen plane. Either dimension
can be determined by the physical size of the specimen, e.g., in

the analysis of thin films or small particles. If the specimen is
larger than the excited region both in width and depth, the width
of the excited region is determined by the diameter of the probe
and the lateral diffusion of the exciting agent. The effective
depth of excitation is determined by the penetration of the
exciting beam and/or by the maximum depth from which the signal
can emerge, without excessive attenuation or, in Raman probe
microanalysis, is efficiently collected. These limitations vary
greatly among methods.

Comments on Specific Techniques

We will here comment on specific probe methods. Some parameters of the more important ones are contained in table 3 which
should be used in conjunction with these notes.

X-Ray Diffraction: This is not used as a microprobe method since
the volume required to scatter x-rays efficiently is too large for
a microanalytical technique.

Laser Microprobe [9]: In this instrument a portion of the specimen is evaporated leaving behind a crater. The vaporized material
is excited to emit atomic spectra in the visible range which are
analyzed in a conventional light spectrograph. The advantages of
the technique are the low cost of the equipment, the possibility
of analysis with the specimen in air, and the fact that the
specimen can be an electrical insulator. Disadvantages are the
following: the size of the evaporated portion depends on the
heat conductivity of the specimen and is difficult to regulate.
Accuracy and localization are inferior to those of electron probe
analysis. The technique is destructive of the analyzed site.

Laser Raman Microprobe [10,11] (figure 2): A laser beam, typically
a moderate-power gas laser, after filtering out non-lasing light,
is focused into a beam a few μm wide, with the aid of a light
microscope objective. The specimen, should be a thin film, or a
small particle on a lithium fluoride or alumina substrate, is
placed into the beam. The Raman scattered radiation emanating from
a volume of the order of few picoliters is directed for spectral
analysis to a monochromator. The specimen must not strongly
absorb the primary radiation; otherwise, it may be destroyed by
heating. Molecular spectra, arising from the characteristic
vibrational modes of the basic chemical units can be obtained from
many organic or inorganic compounds; thus, unknown particles of
micrometer size often can be identified.

X-Ray Fluorescence Probes [12]: These probes provide atomic
spectra of specimens without need for preparation or evacuation.
The spatial resolution, however, is poor ($\sim$0.1 mm).

Table 3. Signal Characteristics of Microanalytical Techniques

Method	Count rates for pure elements or compounds [a]	Ratio of line to background [a]	Limits of detection [a]	Beam widths [a]	Depth of sampling	Depth of sampling limited by
X-R-D	10^4-10^5 cts/sec (single crystal)	$\sim10^3$	.1%	available	10-100 μm	specimen thickness x-ray absorption
Laser Microprobe	photons not usually counted typically plates used	10^3-10^4	.01-.1%	10-100 μm	10-100 μm	heat transfer
Laser-Raman probe	~5000	100-5000	.1-10%	2-10 μm	sample thickness	sample thickness
X-Ray Fluor probes	100-10^4	100-1000	.05%	.1-1 min	~100 μm	x-ray absorption
Electron probe	5000-50,000	100-1000	50-500 ppm	3000 Å	1-5 μm (for thick samples)	electron stopping or sample thickness
SEM and Si(Li)	1000-10,000	20-200	0.02-0.2%	300 Å	1-5 μm (for thick samples)	electron stopping or sample thickness
Auger probe	5000-50,000 [b]	.01-.1	.1 at %	500-2000 Å	5-20 Å	emergence of Auger electrons
Ion probe	1000-10^6	10^3-10^5	1-100 ppm	1-10 μm	20-200 Å	ion emergence sputtering

[a] Typical available instruments and typical conditions.
[b] Counting techniques not usually employed.

The data for detection limits are valid for static beam conditions. For scanning images, the detection limits will be typically a factor of 10 or more higher.

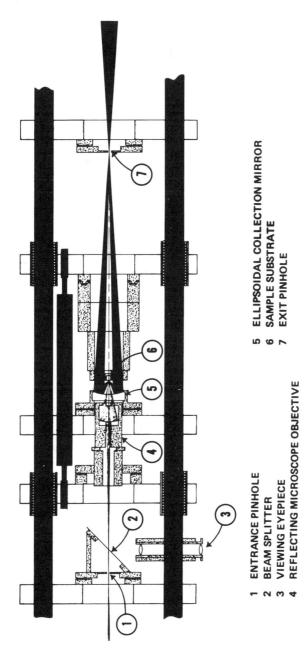

1 ENTRANCE PINHOLE
2 BEAM SPLITTER
3 VIEWING EYEPIECE
4 REFLECTING MICROSCOPE OBJECTIVE

5 ELLIPSOIDAL COLLECTION MIRROR
6 SAMPLE SUBSTRATE
7 EXIT PINHOLE

2 Central part of the Laser-Raman Microprobe, in which the focused laser beam (from left) impinges on the specimen (position 6), and the Raman scattered signal is collected by the mirror (position 5) and sent to the monochromator (at right).

Electron Probe Microanalysis: This technique [13], as well as the
analysis of x-rays produced in scanning [5] or transmission [14]
electron microscopes by means of solid-state detectors [15] are
well-documented in the literature. Atomic spectra of elements of
atomic number above 4 are obtained. Some information on valency
can be obtained from the shifts of low energy x-ray lines [16].
The main advantage of this technique is the high accuracy achievable ($\sim$2% absolute). The sampling volume is a hemisphere of the
order of 1 μm radius, depending on the beam energy and the matrix.

Kossel Diagrams [17]: They are generated by diffraction in the
specimen of x-rays produced at the point of impact of an electron
beam. They provide x-ray diffraction patterns which characterize
the lattice of a region of about 20 μm diameter. The technique
is useful in studies of the properties of metals, but it is not
at present widely employed.

Probes of Protons and Other Heavy Particles [18]: They provide an
x-ray spectrum of lower background than electron probes. The high
cost and large dimensions of the particle sources have limited the
use of such devices.

Cathodoluminescence [19]: This phenomenon is the generation of
light from specimens irradiated by electrons. In appropriate
media, the color and intensity of cathodoluminescence can be
influenced by very low concentrations of some elements. However,
relatively few substances emit this radiation, and the interpretation of cathodoluminescence is difficult. Infrared cathodoluminescence has been used in the study of semiconductors [20].

Electron Diffraction [21]: This technique is an excellent adjunct
of other electron beam techniques, providing information about
the crystallographic structure of a microscopic specimen. Its
application is limited to specimens of small depth, and an electron
acceleration of 50-100 kV is required in most practical applications (e.g., for the identification of asbestos fibers).

Auger-Spectroscopy [22]: Auger emission is closely related to
electron-excited x-ray emission (electron probe microanalysis)
but, due to the small depth of emergence of the Auger electrons
generated within the specimen, the Auger spectrum reflects the
state of a few atomic layers close to the specimen surface. Due
to the high resolution of available electron spectrometers, Auger
spectroscopy not only provides information at the atomic level,
but also on valence states and binding energies. On the other
hand, the large signal absorption in the specimen precludes an
analytical accuracy similar to that of x-ray microanalysis.

Electron Probe Gas Analysis: The presence of Argon in minerals
was shown by a technique in which the gas is liberated from the
solid specimen under an electron beam impact and analyzed in a
small mass spectrometer (residual gas analyzer) [23]. Other gas-

liberating processes of this type are also of potential interest
in a diverse branches of science and technology, and this gas
analysis technique should be further explored.

Ion Probe Analysis [24]:   This technique is based upon secondary
mass spectrometry. A stream of accelerated atomic ions, generated
in a duo-plasmatron and, in some instruments, reduced to a single
ion species by a primary mass spectrograph, is focused into a
small spot on the specimen, and the secondary ions sputtered from
the specimen are mass-analyzed. The advantages of the method are
the feasibility of measuring isotopic ratios, the high sensitivity
in the detection of many elements, the shallow sampling of the
signal, and the possibility of investigating the distribution in
depth of elements in a specimen by observing the signal variations
as the sputtering proceeds. A serious disadvantage lies in the
large matrix effects, which presently render impossible any
attempts at accurate quantitation in complex matrices.

Scanning Techniques

    Probe techniques can be adapted to the investigation of small
areas, by scanning the exiting beam in a raster over the specimen
surface or scanning the specimen mechanically. Such procedures
are practiced in electron and ion probe analysis, scanning electron
microscopy [25], cathodoluminescence spectroscopy, certain
techniques of electron diffraction, and laser Raman microanalysis.
The gain in information concerning spatial distribution is offset
to some extent by a loss of quantitativeness and scanning techniques thus are intermediate between analytical spectrometry and
microscopy. Multiple exposures can be combined in color addition
procedures [5]. In scanning electron microscopy, the observation
of the intensity variation of secondary electron emission provides
information on specimen surface topography and on microscopic
magnetic domains [26]. The absorbed current [27] or backscattered
electrons [28] are also used for image formation.

Related Techniques

    Images similar to scanning micrographs can also be obtained
by related techniques which do not involve, however, the formation
of a focused beam. These include contact techniques such as
contact auto-radiography [29], and projection techniques such as
transmission electron microscopy [30], ion microscopy [31], and
the formation of images by photoemission electron microscopy [32].

References

[1] Pregl, F., Die quantitative organische Mikroanalyse, Berlin, J. Springer, 3rd ed., (1930).

[2] Chamot, E. M. and Mason, C. W., Handbook of Chemical Microscopy, Vol 2, J. Wiley and Sons, New York, N.Y., (1940).

[3] Feigl, F. and Anger, V., Spot Tests in Inorganic Analysis, 6th ed., Elsevier Publ. Co., New York (1972).

[4] Hillier, U.S. Patent 2, 418029, filed 1943.

[5] Heinrich, K. F. J., in Adv. Opt. El. Microscopy, 6, R. Barer and U. E. Cosslett, Eds., Academic Press, London p 275 (1975).

[6] Armstrong, J. T. and Buseck, P. R., Anal Chem., 47, 2178 (1975).

[7] NBS, Available from Author.

[8] Currie, L. A., Anal Chem., 40, 586 (1968).

[9] Moenke, H., Moenke-Blankenburg, L., Mohr, J. and Quillfeldt, W., Mikrochim. Acta, 1154 (1970).

[10] Rosasco, G. J. Etz, E. S. and Cassatt, W. A., Appl. Spectroscopy, 29, 396 (1975).

[11] Rosasco, G. J., Etz, E. S., R. and D. Mag. to appear in June 1977 issue.

[12] Heinrich, K. F. J., Adv. in X-Ray Analysis, 5, 516 (1961).

[13] Reed, S. J. B., Electron Microprobe Analysis, Cambridge University Press, Cambridge (1975).

[14] Duncumb, P., in The Electron Microprobe, T. D. McKinley, K. F. J. Heinrich and D. B. Wittry, Eds., John Wiley and Sons, New York, N.Y., p 490 (1966).

[15] Energy Dispersion X-Ray Analysis, ASTM Special Technical Publ. 485, J. C. Russ, Ed., ASTM, Philadelphia, PA. (1971).

[16] White, E. W., Chapter 4 of Microprobe Analysis, C. A. Anderson, Ed., J. Wiley and Sons, New York, N.Y. (1973).

[17] Yakowitz, H., in The Electron Microprobe, T. D. McKinley, K. F. J. Heinrich and D. B. Wittry, Eds., John Wiley and Sons, New York, N.Y. p 417 (1966).

[18] Pierce, T. B., J. Microscopie Biol. Cell., 22, 349 (1975).

[19] Kinseley, R. N., and Laabs, F. C., Chapter 9 of Microprobe Analysis, Anderson, C. A., Ed., J. Wiley and Sons, New York, N.Y. (1973).

[20] Wittry, D. B., Chapter 4 of Microprobe Analysis, Anderson, C. A., Ed., J. Wiley and Sons, New York, N.Y. (1973).

[21] Alderson, R. H., and Halliday, J. S., Chapter 15 of Techniques for Electron Microscopy, D. Day, Ed., Blackwell Scient. Publ., Oxford, UK (1965).

[22] Palmberg, P. W., in Electron Spectroscopy, Shirley, D. A., Ed., North-Holland Publ. Co., Amsterdam, p 835 (1972).

[23] Zähringer, J., Earth a. Planetary Science Letters, 1, 20 (1966).
[24] Secondary Ion Mass Spectrometry, NBS Special Publication 427 (1975), K. F. J. Heinrich and D. E. Newbury, Eds.
[25] Wells, O. C., Scanning Electron Microscopy, McGraw-Hill, New York, N.Y. (1974).
[26] Newbury, D. E. and Yakowitz, H., Chapter 5 of Practical Scanning Electron Microscopy, Goldstein, J. I. and Yakowitz, H., Eds., Plenum Press, New York, N.Y. p 149 (1975).
[27] Newbury, D. E., in Scanning Electron Microscopy, (Part 1) IITRI, Chicago, Ill., p 111 (1976).
[28] Wells, O. C. in Scanning Electron Microscopy, IITRI, Chicago, Ill, p 1 (1974).
[29] Kennan, J. A. and Larrabee, G. B., Chapter 11 of Characterization of Solid Surfaces, Kane, P. F. and Larrabee, G. R., Eds., Plenum Press, New York, N.Y. (1974).
[30] Techniques for Electron Microscopy, Kay, D., Ed., Blackwell Sci. Publ., Oxford, UK (1965).
[31] Morrison, G. H. and Slodzian, G., Anal. Chem., 47, 932A (1975).
[32] Pfefferkorn, G., Weber, K. Schur, K. and Oswald, H. R., in Scanning Electron Microscopy, 1976 (Part 1) IITRI, Chicago, Ill., p 129 (1976).

ANALYTICAL TRANSMISSION ELECTRON MICROSCOPY AND

ITS APPLICATION IN ENVIRONMENTAL SCIENCE

D. R. Beaman

The Dow Chemical Company

Midland, Michigan

ABSTRACT

In a transmission electron microscope equipped with an energy dispersive spectrometer it is possible to obtain the high resolution morphology, crystal structure, and elemental composition of sub-micron particulates and unsupported thin films. A limitation on quantitative x-ray spatial resolution which is in the range of 200-1000Å is imposed by elastic scattering of the incident electron beam. Smaller regions may be examined qualitatively. Quantitative accuracy and spatial resolution are degraded by the excitation of regions lying outside the perimeter of the primary electron beam and can only be optimized by recognition of the source of the excitation and appropriate electron column modifications. The quantitative accuracy is not well established, but limited results indicate that ±10% of the amount present can be expected when using relative sensitivity factors which may be experimentally measured and, in some cases, calculated. When using standards of known thickness and composition, better accuracy is attainable. It is sometimes necessary to correct the measured concentrations for absorption and secondary fluorescence effects, even in the case of apparently thin materials. The mass sensitivity is in the range of 10^{-17}-10^{-18} grams, and the minimum mass fraction detectable can be as low as 1-3 wt%. It has been possible to measure concentrations of asbestos fibers in environmental samples at the ppt-ppb level with a precision of ±30%.

INTRODUCTION

The need to characterize small particles and thin films has continuously intensified. This encompasses such problems as: the study of grain boundary regions, fine phases, and precipitates in metallurgy; elemental distributions and concentrations on the cellular level in biology; particulate identification in environmental science; and thin film composition and homogeneity in solid state electronics. These needs have served as the motivation for the development of the analytical transmission electron microscope (ATEM) which consists of a conventional transmission electron microscope (CTEM) equipped with energy dispersive spectroscopy (EDS) and scanning transmission electron microscopy (STEM) capabilities. In such an instrument high resolution microscopy is possible in the transmission mode ($<8\text{Å}$ resolution). All elements with atomic numbers above ten can be easily and rapidly detected with the EDS. The electron beam diameter can be reduced to 1000Å in the TEM mode and less than 50Å in the STEM mode making sub-micron particle and phase analysis possible. Identification on the basis of crystal structure is possible using selected area electron diffraction (SAED) in the CTEM mode or micro-micro diffraction in the STEM mode. When identification is based on the nearly simultaneous determination of three quantities--morphology, elemental composition, and crystal structure--the reliability of the analysis is significantly improved because the individual modes sometimes yield ambiguous information. The major limitation in ATEM is that the sample must be thin and transparent to the electron beam, which can lead to significant difficulties in sample preparation.

This paper describes present capabilities in regard to ATEM resolution and quantitation and indicates the difficulties that presently limit the utility of the technique. Experimental determinations of the x-ray spatial resolution are compared with theoretical values predicted by Monte Carlo calculations and Rutherford elastic scattering. The ability to perform standardless quantitation using relative elemental sensitivity factors is demonstrated. The quality of the results are dependent on instrumental parameters, and the need to correct for matrix effects and contamination is clearly established. The sensitivity and accuracy of the technique are shown to be reasonably good for the extremely small volumes of material being analyzed. The application of ATEM to the quantitative determination of asbestos fiber concentrations in liquids is described. This is the only technique which provides reliable data in measuring asbestos contaminations at the ppt-ppb level.

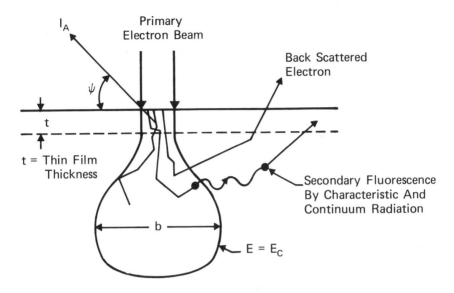

Figure 1. Schematic representation of electron scattering in a low atomic number matrix. The broadening, b, establishes the spatial limit for quantitative x-ray resolution and decreases as the sample thickness decreases.

ELECTRON SCATTERING AND X-RAY SPATIAL RESOLUTION

The primary scattering event occurring when moderate energy electrons impinge on a solid is elastic scattering of the incident electrons by the atomic nucleus. There are various stages of elastic scattering (single high angle, plural, multiple and diffusion) but the x-ray resolution will be determined by those electrons scattered through the largest angles. The probability of elastic scattering through a relatively large angle, ϕ, is well described by the Rutherford expression:

$$P = 9.76 \ 10^9 \ \frac{z^2 \ \rho \ t}{A \ E^2} \ \cot \frac{\phi}{2} \qquad (1)$$

where z = atomic number, A = atomic weight, ρ = density in g/cm^3, t = film thickness or depth in cm, and E = electron energy in ev. The effect of this type of scattering is to cause a change in direction of the incident beam sufficient enough, in some cases, to cause backscattering (see Figure 1). As the electron is changing its direction through elastic scattering, it will also lose energy in inelastic scattering events, the most frequent of which involves

interaction with the atomic electron cloud and a small energy loss
of a few ev per event. Infrequently an electron will lose energy
by inner shell ionization of an atom with subsequent x-ray photon
or Auger electron emission. Another infrequent, but important
inelastic event occurs when the electron loses its energy through
continuous slowing down in the coulombic field of the atomic
nucleus yielding a photon in the continuous spectrum (Bremstrahlung
or white radiation). The overall effect of these interactions is
the creation of a volume of x-ray excitation as shown in Figure 1.
The perimeter of the volume is defined by the point in the range
of the electron where its energy is equal to the excitation
potential, E_C , of the atomic level (K,L,M) of interest. The
continuum and characteristic radiation generated by the electron
beam can expand the volume of primary x-ray generation through
secondary fluorescence of atoms outside the primary volume. From
the point of view of quantitative x-ray microanalysis in bulk
materials these scattering and x-ray generation phenomena have two
primary consequences:

    1.   The elastic scattering limits the spatial resolution
defined by the broadening, b, to about 1 μm at best, no matter how
small the incident electron beam.

    2.   The emitted x-ray intensity is related to composition in
a complex manner(2) such that: $C_A = (I_A^{'}/I_A^{o})(ZAF)$ where C_A is
the weight fraction of A in the sample; $I_A^{'}$ and I_A^{o} are the emitted
x-ray intensities from the unknown and pure element standard
respectively; ZAF is a correction factor that accounts for the
differences existing between the standard and unknown with respect
to "Z"-x-ray generation and electron backscattering, "A"-x-ray
absorption, and "F"-secondary fluorescence.

    Raymond Castaing(12), recognized in 1951 that improvements in
spatial resolution could be realized by analyzing thin films (see
Figure 1). To achieve such an improvement, the film has to be
sufficiently thin so that most of the electrons are transmitted
through the film with negligible energy loss (less than 5%) and,
therefore, little elastic scattering or beam broadening. In
addition, the ZAF corrections should be markedly reduced, particularly in the case of the absorption correction which generally pre-
dominates in bulk materials(3). An added benefit of thin film
analysis is less specimen damage because so little of the incident
beam energy is absorbed in the sample.

    In the early years of microprobe analysis, x-radiation was
detected using wavelength dispersive spectrometers based on Bragg
diffraction and described in the preceding article. These high resolution spectrometers suffer from poor overall collection efficiency.
The solid angle, Ω, (crystal area/(crystal to target distance)^2)
intercepted is generally less than 0.01 steradian because of the

large spectrometer diameter. The spectrometer efficiency, 100 $\Omega/4\pi$, is less than 0.1% and, because crystal diffraction efficiency is less than 20%, the overall collection efficiency is 0.02% at best. Thus, early attempts at thin film analysis were characterized by very low x-ray intensities making the analysis difficult.

Energy dispersive spectrometers consist of relatively large, (30 mm^2), Si(Li) crystals which can be positioned close to the x-ray source, therein intercepting a relatively large solid angle (0.3 steradians) and providing a high spectrometer efficiency (2.4%). Because the detector efficiency is 100% over a wide range of energy (2-15 kv), the collection efficiency of an EDS is generally several hundred times greater than that of a WDS. This sensitivity has rekindled interest in areas of analysis involving low x-ray count rates such as: the analysis of thin films and sub-micron particulates where the signal is low because of the small excited x-ray volume; in electron beam instruments where high spatial resolution is achieved through the use of small and consequently low (<1 na) current electron beams; and in situations where a low current density is required to minimize electron beam induced damage or artifacts, as in biological tissue or other organic materials, and in the study of diffusible ions (K, Na). All of these analyses can be performed in the ATEM which was first described by Peter Duncumb(16) in 1962 who designed a combination instrument called EMMA (electron microscopy and microanalysis) using a WDS. Bender and Duff(6) installed an EDS on a CTEM in 1970. Because small volume analysis is the aim of analytical transmission electron microscopy, it is essential to determine just how small of a region can be examined.

Goldstein et al.(22) have used the Rutherford expression (Eq. 1) to calculate the electron beam broadening in thin films due to elastic scattering alone. Assuming that the region of quantitative excitation contains 90% of the incident electron trajectories (P = 0.1) and the scattering configuration shown in Figure 2, Eq. 1 yields:

$$b = 625\left(\frac{\rho}{A}\right)^{1/2}\frac{z}{E_0}t^{3/2} \qquad (2)$$

b is the broadening in cm, ρ is the film density in g/cm^3, A is the atomic weight, z is the atomic number, E_0 is the acceleration potential in keV, and t is the film thickness in cm. The broadening predicted by Eq. 2 is plotted for Al, Cu, and Au in Figure 2. The most significant observation is that elastic scattering limits the ultimate spatial resolution achievable in films that are easy to

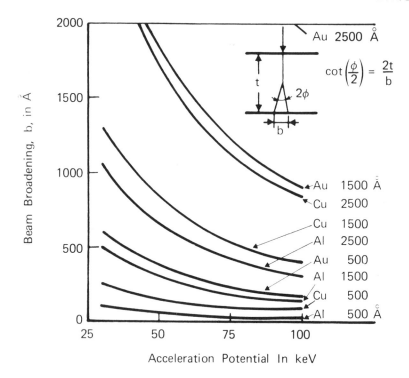

Figure 2. Electron beam broadening, b, due to elastic scattering
predicted by the Rutherford equation. The broadening is plotted
as a function of primary electron beam energy for various thicknesses (500, 1500, and 2500Å) of Al, Cu, and Au. The schematic
inset illustrates the relationship between broadening and thickness
suggested by Goldstein, et al.(22).

---

produce (t>1000Å) and analyze in CTEMs operated in the normal range
of 50-100 keV. Resolution degrades rapidly as the thickness and
atomic number increase and acceleration potential decreases.
Resolution of better than 200Å will be difficult to achieve.
Table I compares some experimental measurements and Monte Carlo
calculations with the values predicted by the simple model of Eq. 2.
Monte-Carlo calculations would be expected to provide the best
results. The relatively good agreement between the simple
scattering model results and the Kyser and Geiss(27) Monte Carlo
results tends to validate Eq. 2. In general, the experimental
broadening is greater than that predicted by Rutherford scattering.
This is at least partially due to beam and sample drift and other
instrumental difficulties.

Table I. Calculated and Experimental Values for Quantitative X-ray Resolution in Thin Films

investigator and reference	material	acceleration potential, keV	thickness in Å	Resolution in Å		
				experimental	Monte Carlo Calculation	Rutherford scattering
Bolon, et al. (10)	Au	20	500		~ 1300	750
Rao & Lifshin (34)	stainless steel	200	near edge	400		
Goldstein et al. (22)	Fe Co Ni	100	1000	250		210
Kyser & Geiss (27)	Cu	100	1000		160	200
Kyser & Geiss (27)	Cu	50	1000		320	430
Faulkner et al. (18)	TiC in steel	100	1500±200	970±100		350
Faulkner et al. (18)	Fe	100	1500		1000*	350
Faulkner et al. (18)	TiC in steel	200	1500±200	525±200		170
Faulkner et al. (18)	Fe	200	1500		500*	170

*using a simplified scattering model

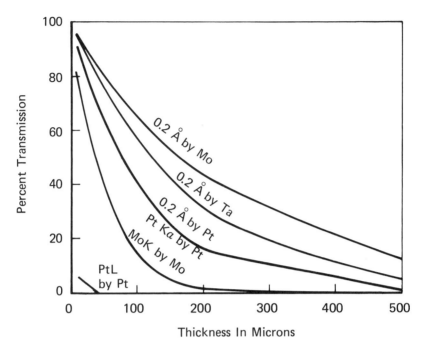

Figure 3. The percentage of radiation for various wavelengths transmitted as a function of thickness of Pt, Ta and Mo.

INSTRUMENTAL LIMITATIONS

Instrumental problems arise because CTEMs were never intended to be used in such a quantitative manner and ATEMs have been constructed by retrofitting EDS and STEM capabilities to existing systems. There are two prime sources of the problem: 1) The EDS is not a focusing spectrometer and is insensitive to the location of the x-ray source and thus, will detect all generated x-rays within a line-of-sight path to the detector(3); 2) In a typical CTEM column there is, within a confined space, a high density of hardware such as pole pieces, apertures, anti-contamination surfaces, support grids, sample holders and associated clips. These two features combine to cause remote x-ray generation, i.e., x-radiation originating from regions outside of the volume excited by the primary electron beam. This causes 1) spectral peaks unrelated to the sample to appear in the EDS spectra leading to quantitative inaccuracy and errors in identification, 2) increased background radiation which raises the detectability limits, and 3) a loss in spatial resolution. The sources of the problem are

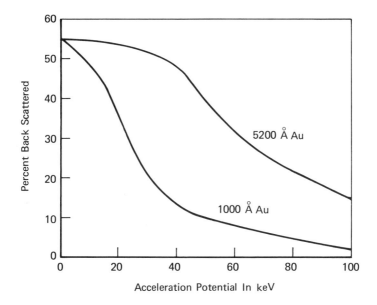

Figure 4. The percentage of backscattered electrons as a function of incident electron energy for two different thicknesses of Au. The data are from Philibert and Tixier(33).

secondary fluorescence induced by characteristic and continuous radiation generated in the column apertures, backscattered electrons from the sample and its support, and scattered primary electrons.

The use of high voltages to penetrate thin samples and retain good spatial resolution leads to the generation of characteristic and continuous radiation in the CTEM column apertures. The second condenser (C_2) variable aperture, which is the last aperture above the sample, poses the most serious problem. The maximum in the generated continuum at 100 keV and the PtKα characteristic radiation both have wavelengths of about 0.2Å and are readily transmitted by thin Pt apertures as shown in Figure 3. Over 40% of the 0.2Å Pt radiation is transmitted by an 100 µm thick Pt aperture. Mo and Ta apertures transmit even greater portions of the radiation. Most of this radiation will be dissipated by absorption in the column but any that does reach the sample area can generate secondary fluorescence at and near the sample which is unrelated to primary electron beam excitation.

Because most primary electrons are transmitted by thin films and small particles, the backscattered electron fraction is small,

as indicated for Au films in Figure 4(33). If the beam voltage
is high and the sample thin, less than 5% of the incident electrons
will be backscattered. Any electrons that are backscattered
toward the detector can penetrate the 7.5 μm Be window of the EDS
because they will, for the most part, have energies close to the
incident beam energy. About 50% and 80% of the electrons with
energies of 60 and 100 keV respectively can penetrate 7.5 μm of Be,
and, in so doing, lose less than 5% of their energy. Many backscattered electrons do not reach the detector because they are
confined by the strong objective lens field. They can, however,
excite remote positions of the sample, support grid, and sample
holder.

Scattered electrons in the column cause electron beam tailing
(11) which leads to excitation of areas in the sample immediately
adjacent to the region of primary beam excitation. This effect is
due to improper alignment and scattering by column components and
increases in severity as the beam voltage is lowered.

The following list indicates some steps that may be taken to
alleviate these instrumental problems. The magnitude of the problem
and therefore the effectiveness of these alterations will vary
appreciably from one instrument to another because of differences
in electron optical configurations, alignment procedures, column
cleanliness, aperturing (sizes, materials, thicknesses and location), and operating mode (TEM) vs. STEM).

I.   Reduce the Generation in and Transmission of Radiation by
Column Apertures

a)   Use thick apertures(39).

b)   Use Pt apertures rather than Mo or Ta(11, 25).

c)   Use column inserts somewhere between C_2 and the sample(45).

d)   The use of low acceleration potential reduces these
problems but promotes tailing, back scattering and absorption
effects.

e)   Determine if performance depends upon emission current
for the instrument being used and the type of sample being studied.

II.   Reduce the Excitation of Material Remote to the Sample.

a)   Specimen holders, specimen clamps, and support grids
should be made of low atomic number materials (Be, graphite, or
polymer) or coated with such materials(2, 32, 39).

b)   Use support grids with maximum open area(39).

c) Coat components near the specimen such as anti-contamination devices and sample support rods with low atomic number materials (Aquadag®).

d) The objective aperture must be removed during EDS data acquisition.

e) The sample support film should be as thin and have as low an atomic number as possible.

f) Operate at as low of a tilt angle as will provide adequate EDS intensities (less area of grid exposed to excitation).

III. Optimize the EDS Detector Configuration.

a) Use the greatest Si(Li) crystal to sample distance that will provide adequate count rates(21).

b) Collimate the detector with a low atomic number material.

c) The collimator should be thick enough or shielded with sufficient material (high z) to absorb any stray radiation(27).

IV. Minimize Electron Scattering

a) Use a small (100 μm) condenser aperture(25).

b) Operate at high acceleration potential.

c) Have the column clean and properly aligned.

These effects of extraneous radiation can best be examined by comparing spectra obtained on and off the edge of a thin film or by comparing the spectra obtained with the beam positioned in a hole (hole-count)(11) with spectra obtained on the sample. In performing on- and off-film measurements on a Sn-Cu-Cr film, 3% of the Cr intensity was attributable to a Cr plating on the sample hold-down clip while the Cu TEM grid was responsible for 15% of the Cu signal. Insertion of an aperture just beneath the variable C_2 aperture in a Philips EM300 operated in the TEM mode increased the Cu peak-to-background ratio and reduced the off-film Cu count by 35%. Kyser and Geiss(27) have found that operation in the STEM mode reduces the extraneous background by about a factor of two.

In quantitative work, even when the precautions and modifications listed above have been taken, it is advisable to substract in-hole, off-film, or off-particle spectra from the sample spectra. If one fails to make these modifications and corrections, the quantitative x-ray resolution may be several hundred microns rather than several hundred angstroms. In an unmodified CTEM, Joy and Maher(25) detected Si when a 0.1 μm beam was placed at the center of a 300 μm hole in a Si disc.

QUANTITATIVE ANALYSIS

The quantitative analysis of thin films and small particles is not a new endeavor. In ATEM analysis thin generally means that the films are transparent to the electron beam and are unsupported or positioned on thin (∿100Å) polymer (collodian, Formvar) substrates that do not effect the analysis. The analysis of supported thin films on bulk substrates is best performed in an electron probe using the Monte Carlo methods described by Kyser and Murata(26). Small particle analysis in the ATEM should be restricted to sub-micron particulate. Larger particles which absorb most of the incident beam energy can be analyzed in the electron probe or scanning electron microscope in the manner described by Armstrong and Buseck(1).

Philibert and Tixier(33, 43) established the foundations for accurate analysis of unsupported thin films, and found that the ratio of concentrations in a thin film could be written as:

$$\frac{C_A}{C_B} = \frac{(P - B)_A'}{(P - B)_A^0} (\rho t)_A^0 \bigg/ \frac{(P - B)_B'}{(P - B)_B^0} (\rho t)_B^0 \tag{3}$$

where C is the composition in weight percent, P is the peak intensity, B is the background intensity, ρ is the density, t is the thickness, ρt is the mass thickness, and the superscripts prime and zero represent the unknown and pure material respectively. Hall(24) developed a method for use in biological thin sections in which

$$C_A \propto \frac{(P - B)_A'}{\beta'} \bigg/ \frac{(P - B)_A^{std}}{\beta^{std}} \tag{4}$$

The proportionality is determined from the standard used for A. β is the total integrated intensity from a portion of the contin-uous spectrum free of characteristic peaks and is assumed to be proportional to the total mass of the sample. While both of these methods have provided respectable results, there has been consid-erable effort directed toward developing a standardless method of quantitation because of the difficulties involved in obtaining homogeneous thin standards of known density, composition, and thickness. Suzuki et al.(42) have proposed a Hall-like method without standards in which:

$$C_A = (P - B)'_A \Big/ \text{Total EDS Spectral Count} \qquad (5)$$

This would not be expected to yield good results if the sensitivities for the elements in the sample varied appreciably. Faulkner et al. (18) simply normalize the net EDS intensities and propose:

$$C_A = (P - B)_A \Big/ \sum_{i=A}^{n} (P - B)_i \qquad (6)$$

This applies only if the atomic number variation in the sample is small, and the line energy exceeds 3 keV.

The most successful approach has involved the following expression originally proposed by Duncumb(17) and pursued by Cliff and Lorimer(13) and Russ(40).

$$\frac{C_A}{C_B} = S_{AB} \frac{I_A}{I_B} = S_{AB} \frac{(P-B)_A}{(P-B)_B} \qquad (7)$$

where I is the net peak intensity corrected for background and peak overlap and S_{AB} is a relative sensitivity factor, i.e., the ratio of the detected intensities (I_B°/I_A°) for two pure, thin standards of the same mass thickness. Absorption, secondary fluorescence and backscattering effects must be negligible for Eg. 7 to be applicable. S_{AB} is most easily measured on multi-element thin standards of known composition. There is not much experimental data available and the bulk of what is available has been published by Cliff and Lorimer(13) and Sprys and Short(41). S_{AB} can be calculated in the same manner as has been used in the past to calculate the intensities expected from bulk standard(3). For thin films the situation is markedly simpler than for bulk materials because absorption, fluorescence, and backscattering are often negligible(33).

Let dn = ionizations per unit path length dx

$\rho N_o/A$ = no. of atoms per unit volume where ρ = film density
N_o = Avogadro's no. and A = atomic weight

Q(E) = ionization cross-section of atom for electrons with energy E

dn = Q(E) ρ N_0 dx/A

For thin film let dx = t and assume a negligible energy loss as
the primary electrons penetrate the film so that Q(E) is constant.

C = weight fraction

n = number of ionizations per incident electron

$n = Q\, C\, \rho\, N_0\, t/A$

Ω = solid angle defined by the detector

ω = fluorescence yield = no. photons/no. ionizations

G = fractional emission in line of interest, e.g.,

$G(K\alpha_{12})$ = $K\alpha_{12}$ intensity/($K\alpha_{12}$ intensity + Kβ intensity)

$$\text{I generated intensity} = Q\,\omega\,G\,C\,\rho\,N_0\,t\,\Omega/A \tag{8}$$

Neglecting matrix effects, the only reduction of the generated
intensity takes place as the radiation passes through the Be
window (7.5 μm thick), Au contact surface (0.01 μm thick), and Si
dead layer (0.1–0.2 μm thick) of the solid state detector.
Absorption in the Be window, which is the predominant effect, is
corrected for by using Beer's law so:

$$\text{I detected intensity} = \frac{Q\,\omega\,G\,C\,\rho\,N_0\,t\,\Omega}{A}\; \exp\left(-\left.\frac{\mu}{\rho}\right|_{Be}^{line} \rho_{Be}\, t_{Be}\right) \tag{9}$$

where μ/ρ is the mass absorption coefficient for the line of
interest by Be, ρ_{Be} is the density of Be, and t_{Be} is the thickness
of the window. Because $S_{AB} = (C_A/C_B)\,(I_B/I_A)$, using Eq. 9 for I_B
and I_A gives:

$$S_{AB} = \frac{Q_B\,\omega_B\,G_B\,A_A \exp\left(-\left.\frac{\mu}{\rho}\right|_{Be}^{B} \rho_{Be}\, t_{Be}\right)}{Q_A\,\omega_A\,G_A\,A_B \exp\left(-\left.\frac{\mu}{\rho}\right|_{Be}^{A} \rho_{Be}\, t_{Be}\right)} \tag{10}$$

Russ(40) and Goldstein et al.(22) have derived similar expressions
for S_{AB} and Russ(40) has also considered absorption by the Si dead
layer. Goldstein et al.(22) recommend the Q value of Green and

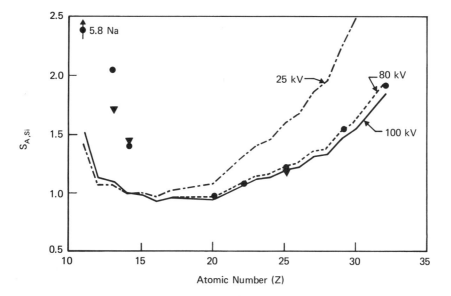

Figure 5. Relative sensitivity factors, $S_{A \; Si}$, for $K\alpha$ radiation as a function of the atomic number of element A. The curves are calculated from Eq. 11 and the points are experimental values from Cliff and Lorimer(13); ▼ from this report.

Cosslett(23) (Q = const. $\ln U/E_0 E_c$) for K series lines because it provides the best agreement with the experimental values of Cliff and Lorimer(13). Using $U = E_0/E_c$, $t_{Be} = 7.5 \; \mu m$, $\rho_{Be} = 1.85 \; g/cm^2$ and $\omega = z^4/(10^6 + z^4)$ (Ref. 44), Eq. 10 becomes:

$$S_{AB} = \frac{A_A \left(\dfrac{z_B^4}{10^6 + z_B^4}\right) \; G_B \; \ln\left(\dfrac{E_0}{E_{C,B}}\right) E_{C,A} \; \exp\left(-\left.\dfrac{\mu}{\rho}\right|_{Be}^B \; 13.9 \times 10^{-4}\right)}{A_B \left(\dfrac{z_A^4}{10^6 + z_A^4}\right) \; G_A \; \ln\left(\dfrac{E_0}{E_{C,A}}\right) E_{C,B} \; \exp\left(-\left.\dfrac{\mu}{\rho}\right|_{Be}^A \; 13.9 \times 10^{-4}\right)} \tag{11}$$

Table II. Calculated and Experimental Values of the Relative
Sensitivity Factor, $S_{A\,Si}$ for $K\alpha$ Radiation

investigator and conditions	Experimental $S_{A\,Si}$ values					
	S_{NaSi}	S_{MgSi}	S_{AlSi}	S_{TiSi}	S_{FeSi}	S_{CuSi}
Cliff & Lorimer (13) EMMA-4 100kv $\Theta=0°$ $\psi=45°$ amphibole particles	5.77	2.07	1.42	1.08	1.27	1.58
Beaman & File (2) EM300 80kv $\Theta=39°$ $\psi=26°$ asbestos fiber=0.1 μm		1.7 ±0.2	1.4 ±0.2		1.25	
Sprys & Short (41) EM300 100kv silicide particles			1.22	1.08	1.30	
Morgan et al. (3) EM300 80kv $\psi=42°$ 3 μm iso-atomic drops	3.92	1.55	1.16	1.13	1.38	
Suzuki et al. (42) JEOL 100C 40kv $\Theta=0°$ mineral fibers		1.7	1.3		2.5	
Calculated $S_{A\,Si}$ Values						
Goldstein et al. (22) 100 kv	1.66	1.25	1.12	1.16	1.33	1.59
this report Eq. 11 100 kv	1.52	1.13	1.09	1.07	1.22	1.46
Russ (4) 100kv	2.01	1.39	1.12	0.95	1.22	1.34

Θ = tilt angle ψ = x-ray take-off angle

Note that this expression shows no dependence on the instrumental configuration. However, S_{AB} values determined in different
instruments may differ from each other and from the theoretical
values because: 1) the contribution of secondary fluorescence,
backscattering, and beam tailing may be vastly different in different instruments; 2) the Be window thickness and detector
efficiencies may be different and, in some instances, the Si dead
layer and Si crystal thickness may be significant; 3) the samples
used to measure S_{AB} may not be thin.

Figure 5 compares the calculated values (Eq. 11) obtained using
the Reed and Ware(36) values for G with the experimental values of
Cliff and Lorimer(13); the ratios are relative to Si i.e., B = Si.
As originally shown by Goldstein et al.(22) the agreement is poor
below 2keV and good above 2keV. Table II also compares calculated
and experimental S_{AB} values. For $S_{Mg\ Si}$, $S_{Al\ Si}$, $S_{Ti\ Si}$, and
$S_{Fe\ Si}$, the agreement in the experimental values is better than
13% (fractional standard deviation or coefficient of variation)
notwithstanding the variation in the experimental configurations.
With the exception of the $S_{Na\ Si}$ and $S_{Mg\ Si}$, the agreement between
theory and experiment is better than 15%. The $S_{Mg\ Si}$ value
determined from eight different mineral fiber standards using the
data of Beaman and File(2) was 1.7 ± 0.2 (±14%). This variation
is primarily due to inaccuracies in the bulk chemical values for
the mineral fibers. The situation is not nearly as favorable for
L radiation as shown in Table III.

If $\Sigma C = 1$ and the S values are all relative to Si,

$$C_A = S_{A,Si} I_A / \sum_{i=A}^{n} S_{i,Si} I_i \qquad (12)$$

Other relative sensitivity factors can be calculated from the Si
values because

$$S_{AB} = S_{CB}/S_{AC} \qquad (13)$$

If the S values are not relative to Si

$$C_A = I_A / (I_A + \sum_{i=B}^{n} S_{i,A} I_i) \qquad (14)$$

We measured the composition of a 3000Å-thick Cu-Sn-Cr thin
film on a Cu TEM grid using a Philips EM300 CTEM at 80keV and a
Cameca electron probe operated at 25keV. The results in Table IV
are compared with results obtained using neutron activation analysis. The ATEM results are seriously degraded by secondary
fluorescence and electron scattering as evidenced by the high Cu

272   D. R. BEAMAN

Table III. Relative Sensitivity Factors, $S_{A\,Si}$, for L and M
Radiation at 100 keV

| element A and atomic number | Experimental $S_{A\,Si}$ | | | Calculated $S_{A\,Si}$ | |
| | Sprys and Short(41) EM300 | Sprys and Short(41) EM300 | Cliff and Lorimer(13) in EMMA-4 | Goldstein et al.(22) | Russ(40) |
	M Line	L Line	L Line	L Line	L Line
Sr 38		2.35			
Mo 42		2.10			
Ag 47			2.3	3.0	1.4
Ba 56		1.70	3.0	3.0	
Ta 73	3.25				
W 74		2.0	3.2	3.1	2.0
Au 79			4.2	3.1	2.2
Pb 82	2.7	2.9	5.2	3.2	2.4
Th 90	1.5				

value resulting from the use of a Cu TEM grid. Off-film spectra
were subtracted from the film measurements. The Cr/Sn ratio is in
good agreement with the chemical data (relative error = 11%). The
Cu grid was used to demonstrate the difficulties associated with
quantitation in the ATEM. As indicated previously, the results
will be improved by using low atomic number grids and grids that do
not contain any of the elements present in the sample. The results
obtained in the electron probe, where scattering problems are minimized by the instrumental configuration and the use of low acceleration potential, are excellent (relative error <10%), and there
is good agreement between the results obtained using sensitivity
factors and those obtained using pure element standards and a ZAF
correction. The results obtained by normalizing net peak intensities as suggested by Eq. 6(18) are poor for the ATEM and electron
probe. From these limited data we conclude that the thin film
model of Eq. 7 is valid and capable of providing relative errors

Table IV. Experimental Composition of a 3000Å
Thick Cu-Sn-Cr Film

method	Composition in Wt. %			Cr/Sn
	Cu	Sn	Cr	
neutron activation	14.6	77.6	7.8	0.101
ATEM at 80keV with S_{AB} values	27	67	6	0.090
electron probe at 25keV with S_{AB} values and absorption correction	15.6	76.7	7.6	0.099
electron probe at 25keV with S_{AB} values but no absorption correction	16.4	76.3	7.3	0.096
electron probe at 25keV using pure standards and ZAF correction	15	78	7	0.090
normalize the net peak values as in Eq. 6.	electron probe at 25keV	23	61	16	0.26
	ATEM at 80keV	40	48	12	0.25

of less than 10% when using experimentally determined S_{AB} values.
This represents reasonably good performance when compared with the
5% relative error attainable with EDS systems and bulk samples(4).
However, this will only be possible in CTEMs after taking the precautions described previously. The accuracy will be best when
measuring concentration ratios. The presence of oxide films or
organic contamination on the surface and the tendency for real
surface segregation to occur complicates and degrades quantitative
thin-film or small-particle analysis. Geiss and Huang(20) have
reported relative errors of less than 2% in Co-Ho and Ni-Fe alloys
using Eq. 3 and standards of known thickness.

CORRECTIONS TO THE DATA

It has often been assumed that when structure is visible in the TEM image the sample will be sufficiently thin so that the only consideration necessary in quantitative analysis is the variation in x-ray generation by the primary electron beam. This may not be the case when the acceleration potential is increased to high levels (>80keV) or the instrument is operated in the STEM mode to obtain contrast in thicker samples.

The loss of ionization through backscattering will be negligible if the film is transparent to electrons. From Figure 4, it is seen that for 1000Å of Au, a strong scatterer, the voltage could be as low as 50 keV and the backscattered fraction still below 10%, whereas over 50% would be backscattered by a bulk material. At 100keV over 3000Å of Au would be required to backscatter 10% of the primary electrons.

Philibert and Tixier (33) have found that continuous fluorescence is negligible and that characteristic fluorescence will be negligible if $(\mu/\rho)_{alloy}^{B\ line} t \ll 1$. μ/ρ is the mass absorption coefficient for the exciting radiation by the material. Philibert and Tixier (33) derived the following expression to correct for characteristic fluorescence in thin films, where I_A^f and I_A^P are the A line intensities resulting from characteristic fluorescence by element B and the primary electron beam respectively.

$$\frac{I_A^f}{I_A^P} = \gamma = 2\ \omega_B\ C_B\ \frac{r_A - 1}{r_A}\ \frac{A_A}{A_B}\ \frac{\mu}{\rho}\bigg|_A^{B\ line}\ \frac{\mu}{\rho}\bigg|_{AB}^{B\ line}\ \frac{E_{C,A}}{E_{C,B}}\ t^2 \quad (15)$$

where r(A) is the absorption jump ratio for element A and the other terms have been defined previously. The measured net intensity in Eq. 7 is corrected for characteristic fluorescence using $I_A = I_A$ (measured)$/(1 + \Sigma\ \gamma)$ with γ evaluated for each element that could cause significant fluorescence of element A(37). It is not presently clear how necessary the characteristic fluorescence correction is for thin films because the limited accuracy of the analyses in most CTEMs obscures the effect of characteristic fluorescence.

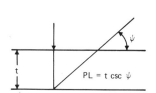

Figure 6. Schematic drawing showing the absorption path length, PL, in a thin film.

$PL = t \csc \psi$

In order to make any corrections to the data, it is necessary to know the thickness which certainly complicates the analysis and detracts from the simplicity of standardless correction.   For particles and fibers, the thickness can often be estimated from the TEM image.   Thin sections of film cross-sections are generally easy to prepare and useful when the film thickness is uniform.   The separation of carbon contamination deposits on the top and bottom of the film can be measured by tilting the sample (22).   Hall (24) uses a portion of the continuum spectrum to measure the mass thickness.

If the absorption path length (PL in Figure 6) is sufficiently short so that 90% of generated intensity will be transmitted by the film, x-ray absorption effects should be negligible in view of the existing capabilities with respect to accuracy.   In this case

$$I/I_o = 0.9 = e^{-(\mu/\rho)\ \rho t \csc \psi} = e^{-\chi \rho t}, \text{ where } \chi = \mu/\rho \Big|^{\text{line}}_{\text{film}} \csc \psi.$$

μ/ρ is the mass absorption coefficient of the x-ray line by the film composition and ψ is the x-ray take-off angle.   For $I/I_o > 0.9$, $t < 0.1/\chi\rho$.   This is a guideline to predicting if absorption will be a problem, but it does not offer a means of correcting for absorption in Eq. 7.   We first noticed absorption affects in the analysis of thin materials when studying asbestos fibers (2). Figure 7 shows the dependence of the ratio I_x/I_{Si} on fiber size for various minerals.   The ratio of intensities at one fiber radius (r_1) to that at another fiber radius (r_2) can be determined from Beer's law.

$$\frac{R_1}{R_2} = \frac{\left(\dfrac{I_x}{I_{Si}}\right)_{r_1}}{\left(\dfrac{I_x}{I_{Si}}\right)_{r_2}} = \frac{\left(\dfrac{I_x}{I_{Si}}\right)^0_{r_1}}{\left(\dfrac{I_x}{I_{Si}}\right)^0_{r_2}} \frac{\dfrac{\exp\left(-\dfrac{\mu}{\rho}\Big|^x_m \rho_m\ r_1\ \csc \psi\right)}{\exp\left(-\dfrac{\mu}{\rho}\Big|^{Si}_m \rho_m\ r_1\ \csc \psi\right)}}{\dfrac{\exp\left(-\dfrac{\mu}{\rho}\Big|^x_m \rho_m\ r_2\ \csc \psi\right)}{\exp\left(-\dfrac{\mu}{\rho}\Big|^{Si}_m \rho_m\ r_2\ \csc \psi\right)}} \qquad (16)$$

where μ/ρ_m is the mass absorption coefficient for x or Si radiation by the mineral, ρ_m is the mineral density, and $(I_x^\circ/I_{Si}^{\circ\prime})_{r_1}/(I_x^\circ/I_{Si}^\circ)_{r_2}$ is the ratio of the generated intensities which is independent of r. The intensity is assumed to be generated at the center of the fiber. Rearranging this expression gives

$$\ln \frac{R_1}{R_2} = \rho_m \ \mathrm{csc}\ \psi\ (r_2 - r_1)\ \left(\left.\frac{\mu}{\rho}\right|_m^x - \left.\frac{\mu}{\rho}\right|_m^{Si}\right) \tag{17}$$

This expression(2) provides a satisfactory fit (±10%) to the experimental data in Figure 7 except in the case of severe contamination at very small fiber diameters. Eq. 17 illustrates that it is the difference between the mass absorption coefficients that determines the magnitude of the absorption effect(2). When $\mu/\rho\left|_{mineral}^{Si}\right. \gg \mu/\rho\left|_{mineral}^{x}\right.$, a decrease in I_x/I_{Si} occurs with decreasing size because the relative increase in emission will be greater for the element with the larger absorption coefficient. Thus, in grunerite there is a greater relative increase in Si emission ($\mu/\rho\left|_{grunerite}^{Si}\right. = 1455$) than in Fe emission ($\mu/\rho\left|_{grunerite}^{Fe}\right. = 65$) and a subsequent 25% decrease in I(Fe)/I(Si) as the diameter decreases from 1.5 to 0.15 µm. When $\mu/\rho\left|_{mineral}^{Si}\right. \ll \mu/\rho\left|_{mineral}^{x}\right.$,

I_x/I_{Si} increases with decreasing size because the relative increase in emission is greater for x than for Si. Thus, in grunerite where $\mu/\rho\left|_{grunerite}^{Mg}\right. = 3460$ and $\mu/\rho\left|_{grunerite}^{Si}\right. = 1455$, there is a greater relative increase in Mg emission and a subsequent 50% increase in I(Mg)/I(Si) as the size decreases from 1.5 to 0.15 µm. The easiest way of correcting for such effects is to use calibration curves of the type shown in Figure 7.

Combining Eqs. 7 and 16 shows that $(S_{AB})_{t_1}/(S_{AB})_{t_2} = R_2/R_1$

where t is the film thickness (r = t/2). In the case of a very thin film or fiber, taking the limit in Eq. 17 as t approaches zero gives:

$$\ln \frac{S_{AB}\,(\text{not-so-thin})}{S_{AB}\,(\text{thin})} = - \ \rho_{film}\ \mathrm{csc}\ \psi\ \frac{t}{2}\left(\left.\frac{\mu}{\rho}\right|_{film}^B - \left.\frac{\mu}{\rho}\right|_{film}^A\right) \tag{18}$$

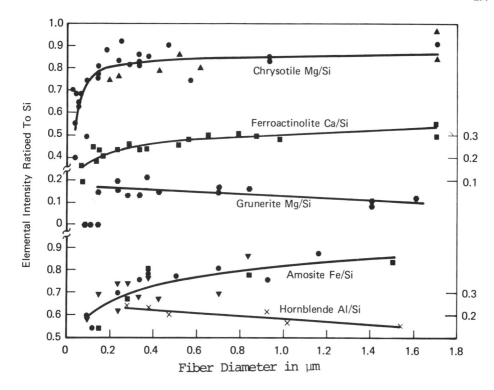

Figure 7. Elemental intensities ratioed to the Si intensity as a function of mineral fiber diameter. The scales for chrysotile, grunerite, and amosite are on the left and on the right for ferroactinolite and hornblende.

which is in accord with the expression published recently by Goldstein et al.(22). The $S_{Cu\ Si}$, $S_{Sn\ Si}$ and $S_{Cr\ Si}$ values used to calculate the Cu-Sn-Cr values were corrected for absorption using S_{AB} (not-so-thin) values from Eq. 18, and in all cases the relative error in concentration decreased as shown in Table IV. Figure 8 can be used as a guide to determine when an absorption correction is advisable. When the absorption coefficient difference for a given particle radius or film thickness is above the line, the absorption effect will be greater than 10% and should be taken into account. Many of the amphibole fibers with diameters of 0.2 µm and over require absorption corrections(2).

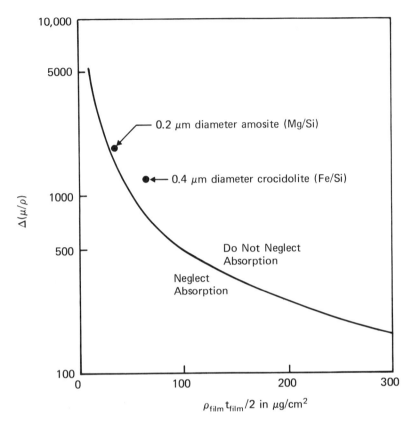

Figure 8. $\Delta(\mu/\rho) = \mu/\rho\Big|_{film}^{B\ line} - \mu/\rho\Big|_{film}^{A\ line}$ (ρt) film = film mass thickness. When the value of $\Delta(\mu/\rho)$ for a particular film thickness is below the line, the absorption correction will be less than 10%. The values shown for amosite and crocidolite indicate that the absorption correction is significant for relatively thin fibers.

INSTRUMENTALLY-INDUCED CONTAMINATION

Superimposed on the effects just described is the sample contamination which occurs when the hydrocarbons from the vacuum pump fluids are decomposed by the electron beam and deposited on the sample surface(3). The deposited thickness can, in time, represent an appreciable portion of the total sample thickness. The magnitude of the problem depends upon: 1) the cleanliness of the vacuum system, 2) the electron beam current density, 3) the duration of

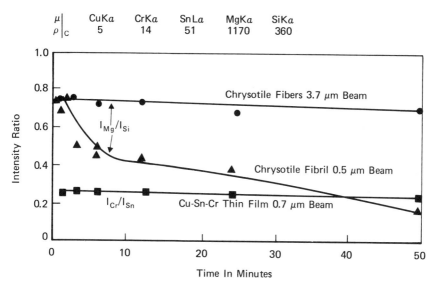

$\dfrac{\mu}{\rho}\Big|_C$	CuKα	CrKα	SnLα	MgKα	SiKα
	5	14	51	1170	360

Figure 9.   Elemental intensity ratios as a function of the duration of electron bombardment in an ATEM operated at 80keV. I_{Mg}/I_{Si} and I_{Cr}/I_{Sn} are plotted for chrysotile asbestos fibers and a Cu-Sn-Cr thin film respectively. The beam diameter for each analysis is indicated on the curves. The values of the mass absorption coefficients for the indicated radiation by carbon are also shown.

analysis, and 4) the difference in absorption by carbon for the x-ray lines of interest. The magnitude of the latter effect can be estimated from the following expression:

$$\ln \frac{(I_x/I_{Si})_{\substack{\text{with} \\ \text{contamination}}}}{(I_x/I_{Si})_{\substack{\text{without} \\ \text{contamination}}}} = \rho_c\, t_c\ \csc\psi\left(\frac{\mu}{\rho}\Big|_C^{SiK} - \frac{\mu}{\rho}\Big|_C^x\right) \qquad (19)$$

where ρ_c is the density of carbon and t_c is the thickness of the carbon deposit in cm. Figure 9 shows the observed variation of I_{Mg}/I_{Si} in chrysotile with time for different current densities. The analysis of small (300-400Å) chrysotile fibers often requires a small electron beam (higher current density) and a longer analysis time (>5 minutes) to generate credible counting statistics. Even though $\mu/\rho\Big|_C^{Si} - \mu/\rho\Big|_C^{Mg}$ is 800, the rapid decrease in I_{Mg}/I_{Si} at small fiber diameter can only be partially accounted for by contamination implying other electron beam induced effects. When the

Table V. Experimental Determinations of the Acceleration Potential
         Providing the Maximum Peak-to-Background Ratios in the ATEM

investigator	instrument and mode	x-ray line	E_O in keV for maximum peak to background
this report	EM300-TEM	CuK	60
this report	EM300-TEM	SnL	40
Russ (39)	EM300-TEM	FeK	50
Joy & Maher (25)	JEOL 100B-STEM	MgK	100
Mizuhira (29)	JEOL 100C-TEM	Na-ClK	20-40
Galle et al. (19)	Cameca-TEM	AlK, Au	20
Kyser & Geiss (27)	EM301-STEM	Fe and CuK	100

difference in absorption coefficients is small, contamination is not
a serious problem as indicated in Figure 9 for the Cu-Cr-Sn film.

OPTIMUM CONDITIONS OF ANALYSIS

Eq. 9 shows that the detected characteristic intensity should
vary with E_O just as the ionization cross section, Q, which maximizes
and then drops slowly ($Q \alpha \ln U/E_O E_c$). The background intensity
(I continuum) from a thin film decreases with increasing F_O
according to Reed (35) as $I_{cou} = 420z\rho tdE/E_O E$. The peak to background ratio, I_{char}/I_{cou}, then would be expected to vary approximately as $\ln U$ with E_O, increasing rapidly at low E_O and then more
slowly. This is not always observed experimentally as shown in
Table V. The failure to increase continuously with voltage is, in
part, due to the background contribution from extraneous radiation
which varies from instrument to instrument. In our EM300 the
highest peak to background ratios have been achieved using a lead
column insert (1 mm ID x 2.57 mm OD x 3 mm thick) in the lower end
of the vacuum tube through which the variable C_2 aperture passes.
The superiority of the STEM (vs. TEM) configuration is indicated in

Table VI. Values of the Mass Sensitivity and Minimum Mass Fraction Detectable Using ATEM

investigator and configuration	beam diameter Å	acceleration potential keV	sample thickness Å	acquisition time sec.	mass sensitivity grams	mass fraction wt.%
Joy & Maher (25) STEM	100	100	1000	100	3×10^{-19}	0.2
Goldstein (22) et al. STEM	300		1000		1×10^{-18}	1
Joy & Maher (25) STEM	1000	100	1000	100	1×10^{-18}	3
Beaman & File (2) CTEM*	1000**	80	340	300	3×10^{-18}	3 Si

*Experimental measurement on chrysotile fibrils

**Fiber length

Table VII. Methods of Preparing Liquids for ATEM Analysis

method →	Jaffe-Fusion	Jaffe-wick	Condensation washing
reference	31, 46	14	2, 5
filter medium	0.22 μm Millipore	0.1 μm Nuclepore	0.22 μm Millipore
pre-treatment	fused in acetone vapor for 5-10 min.	none	none
fiber fixation by vacuum evaporation of carbon	yes	yes	no
pre-conditioning	none	10 μl droplet of solvent onto sample positioned on grid	acetone wetting of grid without filter
extraction	filter section on grid on polyurethane in enclosed petri dish	filter section on grid on wire mesh on several layers of filter paper in enclosed petri dish	filter section on grid on cold finger in reflux column
solvent	acetone	chloroform	acetone
duration of extraction	12 hrs.	10-24 hrs.	10-50 min.

Table V where the two SEM instruments give their best peak-to-background ratios at the highest voltage; unfortunately, fiber or particle counting in the STEM mode is not practical(5). When column modifications are completed in each instrument, the optimum operating conditions should be experimentally determined. Note that low voltage operation will enhance absorption and backscatter effects.

SENSITIVITY AND DETECTION LIMITS

The ATEM operated in either the CTEM or STEM mode offers remarkable mass sensitivity and modest mass fraction detection limits as shown in Table VI. The mass sensitivity is approximately 10^{-18}g with minimum mass fractions of less than 3 wt.% for less than a 1000Å of material and beam diameters between 100-1000Å. These limits will be further improved as high brightness sources, improved detector configurations, and ultra-high-vacuum systems before more widely available.

THE ATEM AND ENVIRONMENTAL ANALYSIS

The ATEM has vast utility in environmental analysis where the accurate identification of sub-micron particulate is required in water and air pollution studies. The ATEM is widely used in the measurement of asbestos fiber concentrations in liquids and air. The fundamental difficulty with such a determination is that the asbestos concentration constitutes an almost infinitesimal fraction of the total solids present in these samples and the particle diameter is generally less than 0.5 µm. Bulk methods lack the needed sensitivity and selectivity.

In the ATEM the electron beam size can be reduced and localized on the fiber of interest to avoid contributions from the surrounding and interfering debris. EDS spectra which are sufficiently unique to permit chemical characterization(2) can generally be acquired in 30-300 seconds even on the smallest fibrils such as chrysotile (300-400Å). The identification of fibers on the basis of their EDS spectra is not without its difficulties. A reliable means of identification is to compare the net elemental intensities ratioed to Si with the ratios determined in the same instrument for asbestos standards(2). The fiber size must be measured in order to account for the absorption effects discussed previously and illustrated in Figure 7. Possession of the same elemental ratios as a standard is a necessary, but not always sufficient, condition for positive chemical identification, e.g., hedenbergite (clinopyroxene) could be mistaken for ferroactinolite (amphibole). Fayalite and amosite have similar EDS spectra as do grunerite and iron-rich Minnesotaite(2,5).

Table VIII.   Experimentally Measured Asbestos
              Concentrations(2)

sample	concentration millions of fibers per liter	mass of asbestos in parts per billion by weight
Midland, Mi, Tap Water[*]	0.6	0.01
waste water effluent[*]	10-400[***]	0.2-10
50% NaOH[*]	50-5000[***]	0.5-40
Duluth Tap water[**]	25	25

*Chrysotile
**Amphibole
***Range for a variety of samples from different sources

     The reliability of the chemical identification can be improved
by examining the selected area electron diffraction (SAED) pattern
of the fiber and comparing it with standard patterns. SAED is
also not without problems primarily because the quality of the
pattern is affected by fiber size;very thin fibers provide
inadequate diffracted intensity and thicker fibers are opaque to
the electron beam. Morphological appearance is distinct in the
ATEM but can be compromised by the presence of other solids,
electron beam induced alterations, and the environmental history of
the sample. In the ATEM, it is possible to resolve most of the
ambiguities associated with a single mode of identification and
improve the overall reliability of the analysis by basing the
identification on morphology, structure, and elemental composition.
When attempting to characterize unknown specimens, the ATEM, which
is most useful for obtaining quantitative concentrations, should be
used in conjunction with definitive electron (28, 38), and x-ray
diffraction(8, 14).

     Water samples are prepared by vacuum filtration through
0.22 μm Millipore® filters which are dried and then cut into 3mm-
diameter filter sections(2, 5). TEM grids with carbon-coated
Formvar films are positioned on the Ni support screen of the cold
finger in a condensation washer. A piece of Whatman filter paper
placed between the TEM grid and the Ni support screen has been
shown to reduce fiber loss during solvent extraction(7). The grids
are preconditioned by application of a few drops of acetone

beneath the Ni support screen to prevent warping of the filter section. The filter sections are placed, sample side down, on the TEM grid immediately following preconditioning. The Millipore is removed in 10-50 minutes of acetone vapor extraction. The sample is then ready for fiber counting and identification in the ATEM. The complete procedure and sources of errors are described elsewhere(2, 5).

One of the sources of error in the analysis is the fiber loss which occurs during sample preparation. Condensation washing is a popular method of preparation, but it introduces variability in the results and yields higher fiber losses (45% for amphiboles) than Jaffe-type methods (<10% for amphiboles) (2). While some investigators have obtained good results with condensation washing there are a sufficient number of technique problems so that serious differences occur in inter-laboratory comparisons. In Jaffe extraction the filter paper is removed by long-time (generally over 10 hours), solvent extraction at room temperature. The three most common methods of sample preparation are shown in Table VII. The short pretreatment in acetone in the Jaffe-fusion technique destroys the structure of the Millipore and therein avoids the formation of a replicated network structure during C coating which would seriously interfere with fiber counting. Other methods, such as micropipetting(9) and low temperature ashing(15), also have merit but have not been extensively evaluated. There are problems to overcome with each preparation method, but sufficient progress has been made so that inter-laboratory reproducibility in measuring fiber concentrations in relatively clean samples is presently in the range of ±20 to ±50%. Using the ATEM our variation in multiple determinations has been ±20% for relatively clean samples such as 50% NaOH. We consider the method good to a factor of two for typical environmental samples (river water). This is reasonably good performance for the small amount of material involved, as indicated in Table VIII (10 ppt to 40 ppm by weight).

ACKNOWLEDGEMENTS

The author wishes to thank H. J. Walker of The Dow Chemical Company for his assistance with the experimental measurements, and R. H. Geiss and D. F. Kyser of IBM, San Jose, California, and H. M. Baker and R. G. Asperger of the Dow Chemical Company for their helpful discussion and critical review of the manuscript.

REFERENCES

1. ARMSTRONG, J. T., BUSECK, P. R.: Anal. Chem. 47 (1975)
2178.

2. BEAMAN, D. R. and FILE, D. M.: Anal. Chem. 48 (1976)
101, also in Proceedings Microbeam Analysis Society, 10th Annual
Conference (1975) paper 31.

3. BEAMAN, D. R. and ISASI, J. A.: Electron Beam Microanalysis, STP506, American Society for Testing and Materials,
Philadelphia (1972).

4. BEAMAN, D. R. and SOLOSKY, L. F.: in Proceedings
Microbeam Analysis Society, 9th Annual Conference (1974) paper 26.

5. BEAMAN, D. R., and WALKER, H. J.: in FDA Symposium on
Electron Microscopy of Microfibers, (Aug. 1976) in press.

6. BENDER, S. L. and DUFF, R. H.: in Energy Dispersion X-ray
Analysis: X-Ray and Electron Probe Analysis, STP485, American
Society for Testing and Materials, Philadelphia (1971) 180.

7. BENEFIELD, D.: The Dow Chemical Co., Freeport, Texas,
private communication (1977).

8. BIRKS, L. S., FATEMI, M., GILFRICH, J. V. and JOHNSON,
E. T.: Naval Research Laboratory Report 7874 (Feb. 28, 1975).

9. BISHOP, K. and RING, S.: Minnesota Dept. of Health
Section of Analytical Services Report (July 26, 1976).

10. BOLON, R. B., LIFSHIN, E., and CICCARELLI, M. F.: in
Practical Scanning Electron Microscopy; Electron and Ion Microprobe
Analysis, J. I. Goldstein and H. Yakowitz Ed., Plenum Press, New
York (1975) 299.

11. BOLON, R. B. and McCONNELL, M. D.: in Scanning Electron
Microscopy/IITRI/SEM/76 Part 1 (1976).

12. CASTAING, R.: Thesis University of Paris (1951); also
publication No. 55 Office National d'Etudes et de Recherches
Aeronautiques ONERA, Chatillon-sous-Bagneux (Seine).

13. CLIFF, G. and LORIMER, G. W.: J. Microscopy 103 (1975) 203.

14. COOK, P. M., RUBIN, I. B., MAGGIORE, C. J., and NICHOLSON,
W. J.: in Proceedings of International Conference on Environmental
Sensing and Assessment, Section 34-1 I.E.E.E., Las Vegas (1976).

15. CUNNINGHAM, J.: in FDA Symposium on Electron Microscopy of Microfibers (Aug. 1976) in press.

16. DUNCUMB, P.: in Proceedings of the 5th International Congress on Electron Microscopy, Academic Press, New York (1962) kk4.

17. DUNCUMB, P.: J. de Microscopie 7 (1965) 581.

18. FAULKNER, R. G., HOPKINS, T. C., and NORRGARD, K.: X-ray Spectrometry 6 (1977) 73.

19. GALLE, P., CONTY, C., and BOISSEL, A.: in Proceedings Microbeam Analysis Society, 11th Annual Conference (1976) paper 71.

20. GEISS, R. H. and HUANG, T. C.: J. Vac. Sci. Technol. 12 (1975) 140.

21. GEISS, R. H. and HUANG, T. C.: X-ray Spectrometry 4 (1975) 196.

22. GOLDSTEIN, J. I., COSTLEY, J. L., LORIMER, G. W. and REED, S. J. B.: in Scanning Electron Microscopy, IITRI/SEM/77 1 (1977) 315.

23. GREEN, M. and COSSLETT, V. E.: Proc. Phys. Soc. 78 (1961) 1206.

24. HALL, T. A.: in Quantitative Electron Probe Microanalysis, K. F. J. Heinrich Ed. National Bureau of Standards Special Publication 298 (1968) 269; also in Physical Techniques in Biological Research, G. Oster Ed., 1A 2nd Edition, Academic Press, New York (1971) 157.

25. JOY, D. C. and MAHER, D. M.: in Scanning Electron Microscopy, IITRI/SEM/77 1 (1977) 325.

26. KYSER, D. E. and MURATA, K.: IBM Journal of Research and Development 18 4 (1974) 352.

27. KYSER, D. F. and GEISS, R. H.: in Proceedings Microbeam Analysis Society, 12th Annual Conference (1977) paper 110; also private communication with R. H. Geiss (1977).

28. LEE, R.: in FDA Symposium on Electron Microscopy of Microfibers (Aug. 1976) in press.

29. MIZUHIRA, V.: Acta Histochem. Cytochem. 9 1 (1976) 69.

30. MORGAN, A. J., DAVIES, T. W. and ERASMUS, D. A.: J. Microscopy 104 (1975) 271.

31. ORTIZ, L. W. and ISOM, B. L.: in 32nd Annual Proceedings of EMSA (1974) 554.

32. PACKWOOD, R. H., LAUFER, E. E. and ROBERTS, W. N.: in Proceedings Microbeam Analysis Society, 12th Annual Conference (1977) paper 115.

33. PHILIBERT, J. and TIXIER, R.: in Physical Aspects of Electron Microscopy and Microbeam Analysis, B. M. Siegel and D. R. Beaman, Eds. John Wiley and Sons, New York (1975) 333.

34. RAO, P. and LIFSHIN, E.: in Proceedings Microbeam Analysis Society, 12th Annual Conference (1977) paper 118.

35. REED, S. J. B.: Electron Microprobe Analysis, Cambridge University Press, Cambridge (1975) 326.

36. REED, S. J. B., and WARE, N. G.: X-ray Spectrometry 3 (1974) 149.

37. REED, S. J. B.: British J.A.P. 16 (1965) 913.

38. ROSS, M.: in FDA Symposium on Electron Microscopy of Microfibers (Aug. 1976) in press.

39. RUSS, J. C.: in Scanning Electron Microscopy, IITRI/SEM/77 1 (1977) 335.

40. RUSS, J. C.: in Proceedings Microbeam Analysis Society, 8th Annual Conference (1973) paper 30; also in Edax Editor 5 1 (1975) 11; also J. Submicr. Cytol. 6 (1974) 55.

41. SPRYS, J. W. and SHORT, M. A.: in Proceedings Microbeam Analysis Society, 11th Annual Conference (1976) paper 9; also private communication with J. W. Sprys (1977).

42. SUZUKI, M., AITA, S., and HAYASHI, H.: JEOL News 13e 1 (1975) 7.

43. TIXIER, R. and PHILIBERT, J.: in 5th International Congress on X-ray Optics and Microanalysis, G. Mollenstedt and K. H. Gaukler eds. Springer-Verlag, Verlin (1969) 180.

44. WENTZEL, G.: Zeit Phys. 43 (1927) 524.

45. ZALUZEC, N. J. and FRASER, H. L.: in Proceedings Microbeam Analysis Society, 11th Annual Conference (1976) paper 14.

    46. ZUMWALDE, R.: in FDA Symposium on Electron Microscopy of Microfibers (Aug. 1976) in press.

Discussion

NEWBURY - I would like to make a general comment. I believe that people with scanning electron microscopes and
electron probes who have never carried out any experiments
to look for stray radiation might be grossly shocked because the problem is similar to what we are finding in
the analytical transmission electron microscope. In fact,
some SEM people have found as much as 20% stray radiation
generation from a region as large as a centimeter in diameter. We always advise people when they have an EDS
system on their SEM to make stray radiation measurements.

BEAMAN - The data in Table IV illustrate the importance
of this problem. The remote X-ray generation in an unmodified ATEM column was responsible for the 85% error
in the measured Cu concentration. In the absence of
scattering problems, the error was reduced to 7%. Incorporating the suggestions listed on pages 10-11 should
eliminate most of the problem. An excellent description
of the scattering problems encountered in the scanning
electron microscope and the means of detecting them can
be found in the paper by Bolon and McConnell (reference
11).

NEWBURY - For the benefit of outsiders would you care
to explain your acronym, in hole?

BEAMAN - A Faraday cup may be produced by press fitting
a Mo electron microscope aperture with a 10-20 μm
opening into a blind hole drilled in a Ni block (see
reference 11). The ratio of the EDS intensities (Mo
and Ni) measured with the beam centered in the aperture
to the same intensities measured with the beam located
on the metals is a measure of the scattering problem
and is distressingly high in some instruments. This
procedure is called an in-hole measurement of the stray
X-ray generation. Similar measurements can be made on
and off the edges of thin films. The ratio for Mo is
primarily due to electron beam tailing while the Ni
ratio is primarily due to secondary fluorescence. Unfortunately, the measurements do not demonstrate the
effects of electron backscattering from the analyzed
volume. Fortunately, the problems can be alleviated
using the procedures outlined on pages 10-11. The
problem is less severe in SEM's than in the ATEM because of the lower acceleration potentials used and the
presence of less material in the immediate vicinity of
the sample.

CAHILL - I would like to comment on the use of active
filtration as a way to improve the sensitivity of
analysis with the types of samples described by the
first two speakers. These samples have a lot of low
atomic number material, and only small amounts of high
atomic number materials. Thus the low Z elements emit
a large number of X-ray photons although in a low resolution region of the spectrum, while the high Z elements
emit smaller numbers of X-ray photons but in a high
resolution region of the spectrum. In the detector,
resolution is lost in the regions of the detector around
the edges; thus by shielding these regions from the soft
X-rays from low Z elements, one can increase resolution.
Since with these samples, one has plenty of soft X-rays,
there is no sacrifice of sensitivity. This shielding
is accomplished by placing in front of the detector, a
thin Mylar sheet in the center of which has been drilled
a hole equal to about one-tenth the area of the detector.
Soft X-rays pass through the hole and are detected with
higher resolution though lower efficiency. Hard X-rays
from high Z elements pass through the Mylar sheet, and
thus the detector has effectively a ten-fold greater
solid angle for these X-rays. The result is higher
energy resolution for low Z elements and about a three
fold increase in sensitivity for high Z trace elements.
It is about as close as one can get to something for
nothing. It has an additional advantage in that it
prevents accidental damage to the detector from such
devices as screwdrivers and thumbs.

BEAMAN - Selective filtering of this nature is useful
in certain cases, e.g., where you have low intensity
at high energy in the presence of high intensity at
low energy. The filtering of the low energy radiation
allows the electron beam current to be increased, thereby increasing the signal at high energy, without increasing the spectral count rate to a level where the
detector resolution would be degraded. In this case,
the detection sensitively for the high energy radiation
would be improved. For the case of recent vintage
systems, I don't believe you would observe a significant
change in detector resolution at low energy because the
resolution dependance on crystal area is primarily a
function of crystal thickness and capacitance which are
unaffected by selective filtering. The resolution at
low energy might improve with filtering if 1) the detector was poorly collimated allowing radiation to
reach the very edges of the crystal where the dead layers are thicker or 2) the system was an older one in

which baseline restoration and pile-up rejection did
not work well at energies below 1.5 kev.

In using selective filtering, it is necessary to use
pure standards in quantitative analysis because some
low energy radiation will be transmitted and some high
energy radiation will be absorbed by the filter.   There
may be difficulty in defining the effective X-ray
take-off angle for the low energy radiation.   Selective
filtration is not advisable in ATEM analyses involving
low levels of X-ray generation, e.g., in analyzing thin
films or submicron particulate or when using low beam
currents to reduce specimen damage or improve spatial
resolution.   In these cases, the primary problem is
one of achieving sufficient X-ray intensity to perform
a reliable analysis.

COLEMAN - How thick is the plastic window?

CAHILL - We've been using about two thousand micro-
grams per cm. squared of plastic foil just a half mil
to 2 mil foil. You govern it by what you're looking
at and you get an efficiency curve that corresponds.

COLEMAN - About how thick a coating of carbon were you
getting on those fibers that cut down your detection?

BEAMAN - Some rather rudimentary measurements made by
tilting the sample after analysis and measuring the
thickness the carbon contaminations spots on a variety
of samples indicated that the carbon thickness was in
the range of 0.1-0.2 µm.   This thickness is not suffi-
cent to account for the decrease in the I_{Mg}/I_{Si} intensity for chrysotile shown in Figure 5, implying
that some other electron beam induced phenomena is ocur-
ring.

COLEMAN - And that's with an effective cold finger
around the detector.

BEAMAN - Yes.   The contamination in some TEM's is quite
serious.   When the beam diameter is reduced to less
than 0.3 µm, we have seen cases where the sample becomes opaque to electrons in less than five minutes due
to carbon deposition.

FILBY - On the last slide, you had 25 million asbestos
fibers per liter for Duluth drinking water.   What's
the health significance of that?

BEAMAN - That's a controversial subject, that I'm not really qualified to discuss and best left to the experts performing epidemiological studies. Asbestos workers with industrial exposure do show increased incidences of gastro-intestinal cancer, but I do not know of any work showing such affects in non-occupational exposures. Several government agencies are conducting extensive feeding studies from which better answers to your question will be forthcoming in the next year or two. Meanwhile, Duluth has installed a filtration system which, I believe, has removed over 90% of the fibers in the tap water.

NEWBURY - How do you concentrate your asbestos fibers when they are in suspension in water. What's your procedure in getting them into the microscope?

BEAMAN - The water samples are prepared by vacuum filtration through 0.22 μm Millipore filters. After drying, 3 mm diameter sections are cut and placed, sample side down, on TEM grids in an acetone condensation washer. These grids have previously been covered with carbon coated Formvar films. The Millipore is removed in 10-50 minutes of acetone vapor extraction, after which the sample is already for ATEM examination. Some fiber losses do occur during sample preparation but they are generally less than 20% for chrysotile.

DAHNEKE - What is the cause of the plus or minus 30% reproducibility? Is it that some of the fibers are lost in the matrix of the millipore?

BEAMAN - The $\pm$ 30% variation in our analyses is primarily due to the following: 1) variable fiber losses occur during sample preparation, 2) a very small area is examined and the distribution of asbestos fibers on the grid is not always uniform, 3) other solids present in the sample can interfere with the identification so that all the fibers observed are not accurately classified, 4) asbestos fibrils may be clumped, bundles or otherwise entwined to such an extent as to make an exact fiber count difficult. This variation is not objectionable for an analyses being carried out at the ppt-ppb level, particularly in view of the lack of data concerning the toxicity of ingested fibers.

DAHNEKE - Can you check to see if any of the fibers are actually being transmitted through the filter? Is there a possibility that fibers with such shapes might be

hydrodynamically oriented during flow, and thus pass
through the filter by virtue of the size of their
diameters being about the same size or smaller than the
pores? For aerosol fibers with similar shapes, we find
we must use filters with pore sizes smaller than the
diameter of the fiber.

BEAMAN - With 0.22 μm Millipore, we have found that over
99.9% of the fibers are retained. Those fibers trapped
in the pore network are also deposited on the TEM grid
during extraction in the condensation washer. Because
fibers must have an aspect ratio exceeding three to be
counted the fibril length of chrysotile is over 0.1 μ.
It is difficult for a fiber of this length to penetrate
the irregular network of pores in a 0.22 μm Millipore
filter.

HEINRICH - I have a question myself. We have now at
least three instruments for microanalysis in which we
use the solid state detector-the electron probe, scanning electron microscope, and transmission microscope.
Would you give a brief synopsis of the usefulnesses and
limitations of these instruments in particulate analyses.

BEAMAN - The ATEM is preferred in the analysis of unsupported thin films and sub-micron particulate, particularly when high resolution electron microscopy and/or
selected area diffraction are needed for visualization
and identification. The ATEM is also preferred in fiber
or particulate counting because of its superior imaging
capabilities and suitability for searching at high
magnifications (> 20000x). The image in the TEM is not
distorted by sample movement as is the acquired scanning
image in the SEM. Video scan rates may be used to alleviate this problem with an accompanying loss of image
quality. The tendency to lose focus with sample movement is minimized by the eucentric goniometer stage of
the TEM. The excellent brightness and contrast in the
TEM are easily obtained, require only simple adjustment
for photographic documentation, and make the detection
of the smallest fibers straightforward.

The electron microprobe is preferred for larger particulate and substrate supported thin films because in
these cases it provides excellent imaging (secondary)
electron, backscattered electron, and specimen current),
lower detection limits, superior quantitation, and the
analysis of elements with low atomic numbers (C, O, N).

ELECTRON ENERGY LOSS SPECTROSCOPY: A NEW MICROANALYTICAL TECHNIQUE?

Michael Isaacson

Department of Physics and
The Enrico Fermi Institute
Univ. of Chicago, 5642 S. Ellis Avenue
Chicago, Illinois 60637

I. Introduction

Out of all the techniques which will be discussed in this conference concerning microanalysis, the one which I will present here is the newest and cannot as yet be thought of as a viable technique. That is my main reason for putting the question mark in the title. However, performing microchemical analysis by analyzing the energy of electrons transmitted through thin specimens may have great potential at very high spatial resolution and high sensitivity and that is the reason that there is interest in the technique.

What I hope to cover in this short paper is to describe the fundamentals of the technique, explain in more detail why there is interest in it in the first place, cover some of the problems in the technique which have to be solved before it becomes a really viable technique for microchemical analysis and then give some examples of the use of the technique. Although this is a conference on environmental pollutants, none of my examples are taken from that arena. However, I think the obvious extension will be clear.

First let us see what happens when fast electrons (i.e., electrons of kinetic energy greater than 10 keV) pass through a material sample. If the sample is thin enough so that most of the electrons do not get absorbed in it then it is convenient to classify the electrons which have been transmitted into three groups. This is illustrated schematically in figure 1. In the first group (Figure 1A) are those electrons which pass

through the sample without interacting at all. These are the
unscattered electrons, and the angular distribution of these
electrons is (of course) that of the incident electron beam.

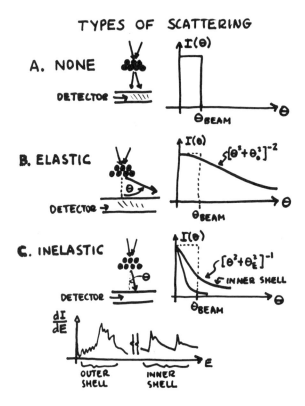

Figure 1.   Schematic illustrations of the three groups of electrons
            which are transmitted through thin samples. $I(\theta)$ is
            the intensity at an angle θ with respect to the forward
            direction.

        In the second group (figure 1B) are those electrons which
have interacted with the atomic nuclei of the atoms comprising
the specimen and have been elastically scattered. These interactions are billiard ball type collisions and the transmitted
electrons essentially lose no energy in the process, but are
merely deflected through angles of several degrees from their
incident direction. The angular distribution of this group is
different from that of the unscattered electrons and varies as
$1/[\theta^2 + \theta_o^2]^2$ where θ is the scattering angle and $\theta_o = \lambda/2\pi a$
where λ is the electron wavelength and a is a characteristic
atomic radius which is slightly dependent on atomic number

(e.g. 1). This distribution is approximately correct for free
atoms, or if the sample consists of amorphous-like material.
If it is crystalline, the angular distribution is peaked at
angles satisfying the Bragg condition. The main point to be
emphasized here is that that characteristic angle of scattering
(θ_0 or the Bragg angle) is generally larger than the angular
distribution of the unscattered electrons and so at a plane
far from the sample, these two groups are separated in space.
Since the probability of elastic scattering is approximately
proportional to the three-halves power of the atomic number (2)
this group can be used for crude elemental discrimination. It
is not the aim of this paper to discuss how to extract information from the sample using these two groups and the interested
reader is referred to references 3-5.

The third type of electrons which are transmitted through
a thin sample are those that have been inelastically scattered
by the electrons within the sample. These electrons lose
energy in traversing the sample and are deflected through
relatively small angles (compared to the angles for elastic
scattering) [see figure 1C]. The angular distribution of these
electrons varies as $1/[\theta^2 + \theta_E^2]$ where θ_E = E/Pv, E is the energy
lost by the incident electron of momentum P and velocity v
(e.g. 6). Thus, electrons which lose more energy tend to be
scattered through larger angles than those which have lost less
energy. It is, however, the fact that electrons lose different
amounts of energy in different materials (i.e., the distribution
of energy losses is material dependent) that makes this group
of electrons useful for chemical analysis. And it is this third
group which we will concern ourselves with in this paper. I
should point out that the other techniques utilizing electron
beams for chemical determination utilize secondary products of
inelastic interactions (such as emitted X-rays or Auger electrons)
rather than the basic interaction itself.

II. What is ELS?

The general method of electron energy loss spectroscopy is
quite simple. A beam of monoenergetic electrons is incident on
a thin sample. Many of the electrons lose energy in traversing
the specimen by exciting or ionizing atoms within the sample.
With the aid of a device (a spectrometer) that analyzes the
energy (or momentum) of the transmitted electrons, one can
obtain a distribution of energy losses (the energy loss spectrum)
within the sample. This spectrum is characteristic of the area
being irradiated and so, in principle, one can identify the
sample. One typical spectrum of an organic material is shown
in figure 2 (7, 8). We will discuss this in more detail later.

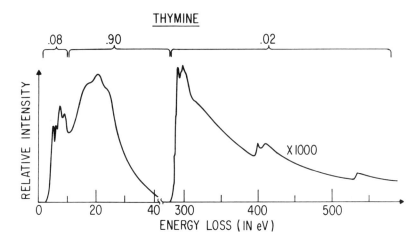

Figure 2. Characteristic electron energy loss spectrum of 25 keV
           electrons which have been transmitted through a 500Å
           thick film of the nucleic acid base thymine ($C_5N_2O_2H_6$)
           supported on a 30Å thick carbon substrate. The spectrum
           was obtained by detecting only electrons scattered
           within 6×10^{-4} radians of the incident beam direction.
           The numbers above the spectrum refer to the fraction
           of the total inelastic scattering cross-section which
           comprises each energy range. The inner shell excit-
           ation peaks corresponding to the constituent carbon,
           nitrogen and oxygen atoms are clearly seen at around
           284 eV, 398 eV and 535 eV (7, 8).

     In practice, the analysis can be obtained by placing the
spectrometer after the specimen in a conventional electron microscope. Using a "selected area" aperture to pick the region of
the sample (i.e. the image) that one wants to analyze, one allows
only electrons which have been transmitted through that area to
pass through the spectrometer. The types of spectrometers which
have been fitted on to electron microscopes are too numerous to
individually mention in this paper and I have listed some of the
articles in the literature in the references (9-15). They consist of attaching magnetic or electrostatic prisms or Wien filters
(crossed electric and magnetic fields) beneath the column of a
conventional electron microscope to inserting cylindrically
symmetric electrostatic or magnetic systems within the microscope
column (between projector lenses) to combinations of electrostatic mirror-magnetic prism or magnetic mirror-magnetic prism
devices installed beneath the objective lens of the microscope.

Another method of analysis and the one most suitable for
point by point identification can be carried out by illuminating the specimen with a small electron probe. By inserting a
spectrometer after the sample, one can obtain an energy loss
spectrum from only the area the electron beam has passed
through. One can then record spectra from different areas of
the sample by electrically moving the incident probe. Images
of the sample are then easily obtained by deflecting the probe
across the sample in a raster type fashion and using the intensity
of particular groups of transmitted electrons to modulate the
intensity of a synchronously scanned cathode ray tube. Such an
instrument is commonly called a scanning transmission electron
microscope (STEM). Detailed discussions of the principles and
operation of the STEM can be found in references 4, 16-18.

Although the technique of trying to obtain chemical information by analyzing the energy lost by electrons transmitted
through samples was first proposed and attempted by Hillier and
Baker more than three decades ago (19), the technique lay dor-
mant for over two decades and has only in the last ten years
gained a large interest due to the work of various research
groups throughout the world (e.g., 20-28). This technique has
become of interest for several reasons (29). First, the technique (ELS) detects electrons which have lost energy in traversing the sample independently of any secondary process which may
occur (e.g., X-ray emission or production of Auger electrons).
Thus, for every excitation or ionization of an atom in the
sample there is a transmitted electron which has lost energy in
producing it. Therefore, there is no fluorescent yield involved.

We can see the implication of this if we consider a simple
equation for the detectability of a given amount of material in
the absence of background. One finds that the number of atoms
N which can be detected with a counting rate S is given by

$$N = S/J\sigma\eta y \tag{1}$$

where J is the current density of the incident electron beam,
σ is the cross-section for the primary process, y is the yield
of the observed process and η is the efficiency of collecting
these events. Therefore, for ELS, y = 1, but for the detection
of emitted X-rays y = ω the fluorescent yield and for emitted
Auger electrons y = 1 - ω. So for detection of the lighter
elements (where ω << 1), ELS can be an extremely efficient means
of detecting minimal amounts of material. (We won't consider
detection of Auger electrons any further, since that is primarily a surface technique and will be covered by another paper in
this conference (30).

    In addition, the transmitted energy loss electrons tend
to be concentrated around the direction of the incident beam.
For instance, half of all electrons which have lost an energy
E are scattered into angles smaller than $5\theta_E$. For 100 keV
incident electrons this means that almost half of all electrons
which have lost less than 1 keV energy are scattered within an
angle of 10 mrad (see figure 3). Thus a spectrometer with a
small collection aperture can collect a large fraction of the
energy loss electrons.

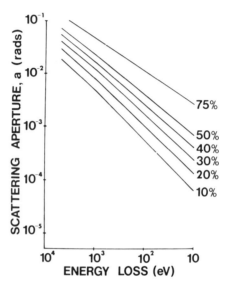

Figure 3.   Graph of the collection efficiency of a spectrometer
            with an effective acceptance half angle a, for 100
            keV electrons incident on a sample. Each line corres-
            ponds to the aperture necessary to collect a certain
            percentage of electrons which have lost a given amount
            of energy, E. That is, 0.001 radian half angle aper-
            ture will allow one to collect 30% of those electrons
            which have lost 100 eV in traversing the sample. We
            assume that the collected fraction for an energy loss
            E is given by $\ln[1 + a^2/\theta_E^2]/\ln[2/\theta_E]$ for a $< \sqrt{2}\theta_E$
            where $\theta_E = E/Pv$, P being the momentum of the incident
            electron of velocity v.

This is quite different from the detection of X-rays or Auger
electrons which to a first approximation are emitted isotrop-
ically. Thus, a state of the art energy dispersive X-ray detect-
or can collect at most a few percent of the emitted X-rays while
for most wavelength dispersive X-ray detectors, the solid angle
subtended at the sample is so small that only about a tenth of a

percent or less can be collected. Therefore, η for ELS can be
orders of magnitude greater than techniques which detect secondary products. To illustrate this, we show figure 4 which shows
the collection efficiency for detecting K shell ionizations
using transmitted energy loss electrons as a function of atomic
number and the collection efficiency for an energy dispersive
X-ray detector (assuming it to be given only by $1/4\pi$ and neglecting window absorption effects). It is easily seen that for
detection of the lighter elements (which may be of importance
in environmental pollutant analysis), in principle, ELS is
certainly a more efficient technique.

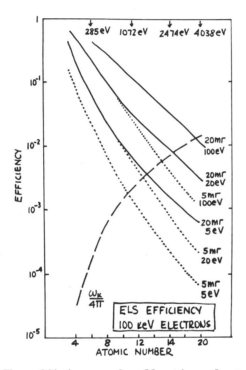

Figure 4.   The efficiency of collecting electrons which have lost
            energy in producing a K shell excitation or ionization.
            The curves have been obtained using equations (8) and
            (9) from reference 29 and the numbers next to each
            curve correspond to the acceptance half angle of the
            spectrometer and the effective energy window of the
            spectrometer (i.e., if a slit is used to record a
            spectrum it is the slit width). The long-dashed line
            is the efficiency for detecting X-rays assuming the
            X-ray detector counts all X-rays incident upon it and
            that it subtends $1/4\pi$ steradians at the specimen,
            ω_K is the fluorescent yield for K-shell excitation.

There are, however, problems which will ultimately limit the
sensitivity of ELS (for instance, we have said nothing yet about
background or what concentrations could be detectable) and we
will discuss this later. Some of the problems are instrumental
and we may be able to correct them. Others are more fundamental
and we will have to live with them. For a more complete report
on some of these problems, the reader is referred to reference 31.

Before we continue, we should first consider the types of
energy loss which an incident electron can undergo in passing
through a sample and what a typical energy loss spectrum looks
like. Electron energy loss spectra for thin films of most
materials generally show considerable structure in the region
between 0 and 50 eV energy loss, (see figure 2, for instance).
For substances in which there is a large fraction of relatively
free valence electrons (such as some metals) these electrons can
be excited in unison by the incident electron beam resulting in
a collective or "plasma" oscillation of the valence electrons
(e.g., 32-33). These plasma peaks are most dramatic in the case
of the 15 eV loss in aluminum (see figure 5).

For most materials, the valence electrons are not nearly as
free and the structure in this region is due mainly to bound
electron excitations and ionizations. Most of this structure can
be correlated with peaks in optical absorption spectra (e.g. 34).
For organic materials, the low lying structure in the energy loss
spectra is due mainly to molecular orbital excitations (in the
region less than 10-20 eV) and valence shell excitations and
ionizations (above about 10-20 eV) (7, 35). There is a broad peak
around 20 eV in all the organic materials we have studied (e.g.,
see figure 2) and it is a matter of debate whether or not collective excitations contribute to it. The main point is that since
a broad peak near 20 eV is common to most organic substances, it
can't really be used as a chemical identification marker.

The region less than 15 eV can be interesting in organic
materials as can be seen from figure 6 where we have shown the
energy loss spectra of 25 keV electrons transmitted through thin
films of various components associated with biological membranes
(36). Each component exhibits a different spectral character.
However, such structure generally gets destroyed (if the sample
is at room temperature) at incident beam doses around 60 electrons/
$\AA^2$ (37,38). Since this is less than the dose needed for analysis
at spatial resolution less than $1000\AA$, this structure may not be
useful for analysis of small areas. However, there is the possibility of reducing this destruction by keeping the sample at cryogenic temperatures and we are exploring this avenue.

Above 50 eV energy loss, the most striking feature of electron
energy loss spectra is the appearance of sharp edges corresponding

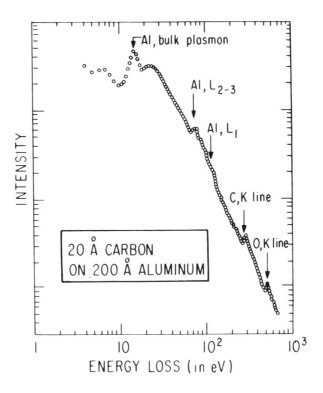

Figure 5. Characteristic electron energy loss spectrum of 25 keV
electrons transmitted through a sample consisting of
a 20Å thick carbon film deposited on a 200Å thick
aluminum film. The effective spectrometer acceptance
half angle was 12.5 mrad, the slit width was about
5 eV and the spectrum obtained in 300 seconds.

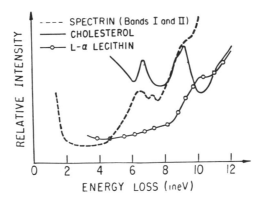

Figure 6.   Characteristic electron energy loss spectra (in the
            region less than 12 eV energy loss) of 25 keV electrons
            transmitted through thin films of various components
            associated with biological membranes. The spectra
            were obtained with total incident doses less than
            .12 electrons/Å^2. For clarity, the spectra have been
            displaced in the vertical direction (36).

to the excitation of electrons in the inner atomic shells. These
edges correspond to the classical energy necessary to ionize
the atom and are directly related to the atomic number. Close
inspection of these edges reveals a wealth of fine structure
(peaks separated by 1 eV or less) as shown in figure 7 (8, 35,
22). This fine structure represents the excitation of an inner
shell electron to a bound excited state and gives us a detailed
picture of the chemical bonding of the constituent atoms. For
example, in our studies of nucleic acid bases we have found that the
positions of these fine structure peaks near the carbon K shell
excitation edge (at 285 eV) are correlated with the calculated
charge on the constituent carbon atoms (figure 8). While this is
only preliminary, such structure might possibly be of use in
identifications of bonding states of C, N, O, P and S which would
be of interest insofar as environmental pollutants were concerned.

    In addition to the classical ionization edges and the fine
structure near the edges, there is often oscillatory structure
far away from the edges (extended fine structure) and this is
illustrated schematically in figure 9. This structure is directly related to X-ray absorption extended fine structure observed
in photoabsorption experiments (e.g., 39) and it gives us information about the immediate surroundings of the constituent atoms
(which give rise to the edge). That is, this structure gives us
some information about the positions of the atomic neighbors.

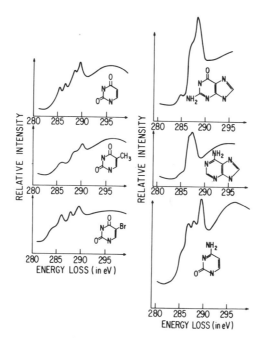

Figure 7. Electron energy loss spectra in the region of carbon
K shell excitation showing the fine structure near
the edge of six types of nucleic acid bases (after
ref. 29). Note that an energy resolution of at least
0.75 eV is necessary to distinguish these peaks and
that with a resolution of 5 to 10 eV we would just
observe a classical K shell ionization edge.

An example of this is shown in figure 10 for amorphous carbon and
graphite where we see that the regular nature of graphite gives
rise to oscillatory structure away from the classical edge, whereas in the amorphous carbon spectrum there is none.

From the point of view of microanalysis then, the characteristic inner shell edges give us elemental information whereas
the fine structure at the edges or away from the edges give us
chemical and structural information. As a practical note, one
needs only about 10-20 eV energy resolution to detect an "elemental"
edge, 1-3 eV to detect the extended fine structure away from the
edge and less than 1 eV to detect the fine structure at the edge.
This is important in determining what type of spectrometer one
wants for particular applications. (In general, the better energy

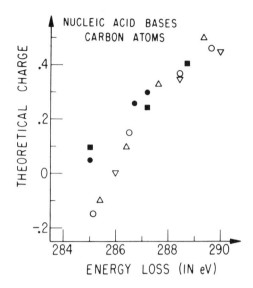

Figure 8.   A plot of the theoretical charge on the constituent
            carbon atoms of five of the nucleic acid bases shown
            in figure 7 as a function of the positions of the
            fine structure peaks observed in the energy loss
            spectra near the carbon K edge. (from ref. 29)
            ● adenine, ■ guanine, △ cytosine, ▽ thymine,
            ○ uracil.

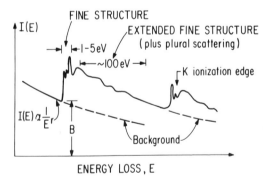

Figure 9.   A schematic diagram of an ELS spectrum in the region
            of inner shell excitation. The background beneath
            the ionization edge falls off as E^{-r} where r is
            generally between 3 and 5 for most conditions.

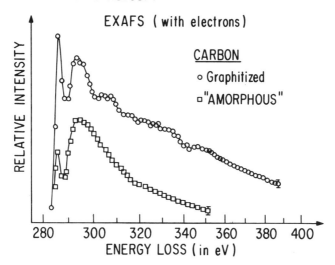

Figure 10. ELS spectra in the region of carbon K shell excit-
            ations for two different forms of carbon (amorphous
            carbon and pyrolitic graphite). The two spectra
            are plotted on a log-log scale to illustrate the
            E^{-r} dependence of intensity away from the edge. Both
            samples were 20-30Å thick and the spectra obtained
            with 25 keV incident electrons. The spectra have
            been vertically displaced for clarity.

resolution required, the more expensive will be the spectrometer!)
One other point worth noting is that the energy resolution of a
wavelength dispersive detector of X-rays is only 10-20 eV, while
for an energy dispersive detector it is only around 130 eV. Thus,
some of these fine structure effects which we are considering here
for ELS cannot be noticed by detecting emitted X-rays.

III. Microanalysis

    The ultimate goal is to analyze microscopic areas for
their chemical composition. From equation (1), one might
be led to believe that one could detect as small amount of
material as one would like. In fact, theoretical calculations
indicate that under very favorable background conditions, single
atoms may be detected and identified. However, in practical
cases, the presence of background in the ELS spectra precludes
this. One can show that one can expect to be able to detect
concentrations less than 100 PPM under certain conditions using
ELS but not orders of magnitude better than that (29). And
as the size of the area illuminated by the incident probe decreases beyond a certain value, the detectable concentration
must increase (e.g. 29, 40). Thus, although 100 PPM of sulphur
in a carbon like matrix may be possible with a 1000Å diameter

electron probe illuminating a 200Å thick sample, it may not be
possible if we reduced the probe diameter to 100Å since then
there would be only about 10^5 atoms irradiated by the probe
and detection of 100 PPM of sulphur would mean detection of
only 10 sulphur atoms!

However, even though one may not be able to detect such
small concentrations from small illuminated areas, one can
utilize the fact that one can make electron beams as small
as 2.5Å in diameter (16) and obtain energy loss spectra from
areas only 10 times as large (or smaller). For instance, if
one had a sample in which there was only 10 PPM of sulphur in
a carbon like matrix of every micron of material, but that the
sulphur might be concentrated in 100Å wide regions, this would
correspond to a 10% concentration in those areas and this would
easily be detectable.

An example similar to this (admittedly a very favorable one)
is shown in figure 11. Here we show a micrograph of a field
of ferritin molecules supported on a thin carbon substrate
(figure 11A). Ferritin is an iron-core storage protein which
consists of a protein shell about 120Å outside diameter and
75Å inside diameter. Inside this shell can be stored up to
5000 iron atoms in the form of an iron oxide hydrate (41). The
concentration of iron of the total mass of the protein is about
25% if the core is completely filled. Although this is a high
concentration, the concentration of iron due to all the ferritin
molecules on the field of view shown is only 5%. And if we had
only one ferritin molecule in the entire field of view, the
concentration of iron would be only 0.31% over the field and
would be easily detectable. Thus, even though the local concentration seems high, the relative concentration can be low.
If we stop an electron beam of 30Å diameter on the molecule
indicated by the area and record an energy loss spectrum of the
transmitted electrons, we obtain figure 11B. The peak near
53 eV corresponds to the excitation of the $M_{2,3}$ level electrons
in the iron and that near 535 eV corresponds to K shell excitation of oxygen (which is in the protein shell and in the core).
The detection of the iron peak corresponds to a mass less than
5×10^{-19} grams. Although the experimental arrangement was not
optimized it seems apparent that it may be possible to detect
extremely small amounts of material using the ELS technique.
The ability to obtain spectra from small regions of material
can also be seen in figure 12 where the spectra shown were
taken from regions about 600Å across (the regions are shown in
Figure 12). The point to emphasize is that the ELS technique
can allow us to detect local composition of objects visible
in the microscope image which may be smaller than 30Å in size.

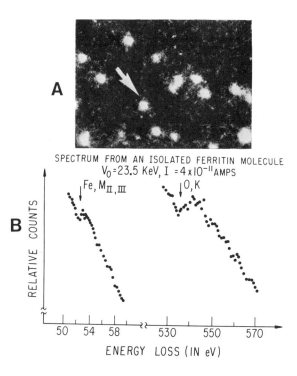

Figure 11. Microanalysis of ferritin molecules on a 30Å thick
carbon substrate. Figure 11A shows a field of view
of ferritin molecules. The full horizontal scale
is 2000Å and the micrograph was taken with a scanning
transmission microscope under conditions described
in ref. 16, 18. The arrow indicates the molecule
on which the illuminating beam was stopped and defocused to 30Å diameter (3). Figure 11B shows the
resultant energy loss spectrum obtained from electrons
transmitted through that molecule (after ref. 36).

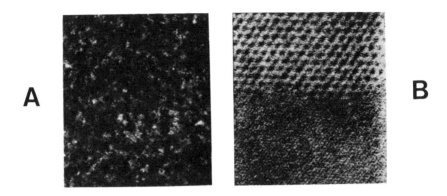

Figure 12. Dark field scanning transmission electron micrograph
            of samples consisting of 20-30Å thick: A) amorphous
            carbon film and B) graphite film showing the differ-
            ent spatial structure. The full horizontal scale
            in both cases was 600Å. The micrographs were obtained
            by forming an image consisting of elastically scatter-
            ed electrons (see ref. 16, 5).

It is really the background from the matrix material which
will hamper detectability limits in ELS. This has been discussed in more detail in the literature (see e.g. 31) and I will
mention here only several points, (see figure 9). The background preceding a classical K edge falls off as an inverse power
of the energy loss. The exact power depends upon the atomic
number of the matrix, the mass thickness of the matrix and the
angular acceptance of the spectrometer. This background is due
to both the high energy tail of lower lying losses and the
multiple valence shell excitation losses. If the sample is
more than one mean free path (for total inelastic scattering)
thick, the multiple lower lying losses dominate the background
and if the sample gets very much thicker than that these losses
swamp out the inner shell edge (i.e., the peak to background
ratio at the edge gets very small). One can show that the
optimum sample thickness for maximum peak to background at
the edge is about one mean free path for total inelastic scattering (see reference 29). Thus, if one is interested in looking
for sulphur in a carbon-like material of density 1.5 gm/cm^3,
the optimum thickness for ELS analysis would be one mean free
path for inelastic scattering in the carbon-like material which
would be about 750Å if we used 100 keV incident electrons.

It is the fact that the background is material-dependent
which will set the fundamental detectability limits. There is
still uncertainty as to the exact nature of this background
and exactly how the various parameters I have mentioned come into
play. At present there is no theoretical model which can
exactly predict the shape of the background under a given set
of conditions. When such a model is developed, then we will be
able to state with more certainty what the ultimate detectability
limits really are, and how far we can push this technique.

IV. Conclusion

Even though electron energy loss spectroscopy as a means
of microchemical analysis is being pursued throughout the world,
it is still in an extremely primitive state. At present, it
certainly can't be thought of as a quantitative technique.
However, since it can be easily coupled to microscopy at the
10Å spatial resolution level (and in some cases to single atom
microscopy), it has the potential of becoming a powerful microanalytical tool (particularly for analysis of the lighter
elements). There are several problems, at present, which will be
solved in the near future. These are problems common to any new
technique: 1) one needs to design more optimum spectrometer
microscope systems; 2) one has to develop more software for
reducing the data (background substraction, peak fitting, etc.)
and more hardware for recording the data and 3) one wants to
understand the physics of the background in order to optimize
all instrumental (and specimen) parameters (and this requires
more data).

Most certainly ELS can be thought of as a complementary
technique to X-ray emission analysis (and, in fact, there have
been suggestions to reduce the background by detecting energy
loss electrons in coincidence with emitted X-rays) (42) both
with regard to elemental detection and the fact that an electron
spectrometer neatly fits beneath the specimen whereas the X-ray
detectors are placed above it.

As more people become involved in ELS, more data will become
available which will presumably lead us to a better understanding
of the spectra and the background. And this in turn should lead
us further on the path of quantitation so that within the next
few years we should see ELS on its way to becoming a viable
microanalytical technique.

Acknowledgements: This work was supported by the U.S. Energy
Research and Development Administration. I wish to thank the
Alfred P. Sloan Foundation for the award of a faculty fellowship.
I would also like to thank Profs J. Coleman and T. Toribara for
inviting me to present this paper.

REFERENCES

1. SCOTT, W.T.: Rev. Mod. Phys. <u>35</u> (1965) 231.

2. LANGMORE, J., WALL, J. AND ISAACSON, M. Optik <u>38</u> (1973) 385.

3. ISAACSON, M. Proc. Sixth European Congress on Electron Microscopy, Jerusalem. <u>Vol 1</u> (1976).

4. ISAACSON, M. Proc. of First Workshop on Analytical Electron microscopy, Ithaca (1976). Cornell University Materials science Center Report #2763. 81.

5. CREWE, A.V., LANGMORE, J.P. AND ISAACSON, M. in: <u>Techniques in Electron Microscopy and Microprobe Analysis</u> (B. Siegel and D. Beaman, eds.) John Wiley and Sons, New York (1975) 47.

6. INOKUTI, M. Rev. Mod. Phy. <u>43</u> (1971) 297.

7. ISAACSON, M. J. Chem. Phys. <u>56</u> (1972) 1803.

8. ISAACSON, M. J. Chem. Phys. <u>56</u> (1972) 1813.

9. METHERELL, A.J.F. Adv. in Optical and Electron Microscopy <u>4</u> (1971) 630.

10. CURTIS, G.H. and SILCOX, J. Rev. Sci. Inst. <u>42</u> (1971) 630.

11. CREWE, A.V., ISAACSON, M. and JOHNSON, D. Rev. Sci. Inst. <u>42</u> (1971) 411.

12. CASTAING, R. and HENRY, L. Comptes rendus Acad. Sci. <u>255</u> (1962) 76.

13. ROSE, H. and PLIES, E. Optik <u>40</u> (1974) 336.

14. ZANCHI, G., PEREZ, J. and SEVELY, J. Optik <u>43</u> (1975) 495.

15. WITTRY, D.B., FERRIER, R.F. and COSSLETT, V.E. Brit. J. Appl. Phys. <u>2</u> (1969) 1867.

16. WALL, J. LANGMORE, J.P., ISAACSON, M. and CREWE, A.V. Proc. Nat. Acad. Sci. (U.S.A.) <u>71</u> (1974) 1.

17. CREWE, A.V. and WALL, J. J. Mol. Biol. <u>48</u> (1970) 375.

18. ISAACSON, M., LANGMORE, J. and WALL, J. in <u>Scanning Electron Microscopy/1974</u> (O. Johari and I. Corvin, eds.) IITRI Chicago, 19.

19. HILLIER J. and BAKER, R.F. J. Appl. Phys. 15 (1944) 663.

20. COLLIEX C. and JOUFFREY, B. Phil. Mag. 25 (1972) 471.

21. EGERTON, R.F. and WHELAN, M.J. Proc. Eighth Int. Congress on Electron Microscopy, I. Canberra. (1974) 384.

22. ISAACSON, M. and CREWE, A.V. Ann. Dev. Biophys. and Bioeng. 4 (1975) 165.

23. KOKUBO, Y., KOIKE, H. and SOMEYA, T. Proc. Eighth Int. Congress on Electron Microscopy I Canberra (1974) 374.

24. COLLIEX, C., COSSLETT, V.E., LEAPMAN, R.D. and TREBBIA, P. Ultramicroscopy 1 (1976) 301.

25. JOY, D.C. and MAHER, D.M. Proc. 35th Ann. EMSA Meeting, Boston (1977) 244.

26. VAN ZUYLEN, P. Proc. Sixth European Congress on Electron Microscopy, Jerusalem, Vol. 1 (1976) 434.

27. JOUFFREY, B. Proc. 35th Ann. EMSA Meeting, Boston (1977) 692.

28. NOMURA, S., TODOKORO, H. and KOMODA, T. Proc. 34th Ann. EMSA Meeting, Miami (1976) 524.

29. ISAACSON, M. and JOHNSON, D. Ultramicroscopy 1 (1975) 33.

30. LINDFORS, P.A. and HERSLAND, C.T. (this proceedings).

31. ISAACSON, M. and SILCOX, J. Ultramicroscopy 2 (1976) 39.

32. RAETHER, H. Springer Tracts in Modern Physics 38 (1965) 85.

33. MARTON, L., LEDER, L. and MENLORVITZ, H. Adv. Electronics and Elec. Physics 7 (1955) 183.

34. DANIELS, J., VON FESTENBURG, C., RAETHER, H. and ZEPPENFELD, K. Springer Tracts in Modern Physics 54 (1970) 77.

35. JOHNSON, D. Rad. Res. 49 (1972) 63.

36. HAINFELD, J. and ISAACSON, J. Ultramicroscopy 2 (in press).

37. ISAACSON, M. JOHNSON, D. and CREWE, A.V. Red. Res. 55 (1973) 205.

38. ISAACSON, M. "Specimen Damage in the Electron Microscope".
     Principles and Techniques of Electron Microscopy. Vol. 7.
     (M.A. Hayat, ed) Van-Nostrand Reinhold, New York (1977) 1.

39. DONIACH, S., HODGSON, K., EISENBERGER, P. and KINCAID, B.
     Proc. Nat. Acad. Sci. (U.S.A.) 72 (1975) 111.

40. COLLIEX, C. Proc. N.I.H. Workshop on Biological Microprobe
     Analysis, Boston (1977) in press.

41. HAGGIS, G.H. J. Mol. Biol. 14 (1965) 598.

42. WITTRY, D.B. Ultramicroscopy 1 (1976) 297.

Discussion

BEAMAN - You pointed out the advantages of ELS for light element analysis, would you comment on the use of ELS for heavier elements.

ISAACSON - Figure 4 shows that there are crossovers in the X-ray yield and the ELS collection efficiency in the regions of $Z = 10$ to $Z = 20$. There are two major considerations in the analysis of heavier elements. First, as the energy loss gets larger the angular distributions of the electrons which have lost this energy increases. Thus a larger aperture is required to collect the same number of electrons. As the aperture angle increases so do aberrations. This is not a fundamental limitation and several workers at Chicago and elsewhere are working to minimize these aberrations. Secondly, since the scattering cross-section for the K shell excitation is inversely proportional to Z^2, it falls off rapidly as Z increases. To circumvent this, one can look at the L shell excitation and try to stay within the energy range from zero to a few kilovolts.

HEINRICH - What energy electrons were used to form the images you showed in Figures 11 and 12?

ISAACSON - Our acceleration voltages were between 25 and 50 kV. With electrons of this energy the mean free path in carbon is about 500 $\overset{\circ}{A}$.

HEINRICH - Do you have trouble with the monochromaticity of these electrons?

ISAACSON - We use a field emission source so that the energy spread of the source electrons is very small.

HUNTZICKER - Are the spectral characteristics attributed to chemical effects amenable to theoretical interpretation?

ISAACSON - There is ample reason to believe that it is. The major reason that more information is not available about this portion of the spectrum, is that few workers have examined it. However, there are some workers, for example at Xerox here in Rochester, who have access to electron accelerators that use 300 kV electrons; they are using them with broad, i.e. 3-4 mm, beams to investigate these spectral characteristics.

SECONDARY ION MASS SPECTROMETRY FOR PARTICULATE ANALYSIS

Dale E. Newbury

Analytical Chemistry Division
National Bureau of Standards
Washington, DC 20234

Introduction

The technique of analysis by exciting a sample with a focused
beam of ions was first proposed by von Ardenne [1] and advanced
through the efforts of many workers, including Slodzian and
Castaing [2], who developed the ion microscope, and Liebl and
Herzog [3], principal developers of the ion microprobe. Emission
products of the interaction of ions with solids include electrons,
photons, and ions as well as backscattering of the primary ions.
If the primary ion beam is energetic enough, nuclear reactions can
result. Techniques of analysis based on all of these interaction
products have been developed. In this paper, we shall be concerned with analysis by secondary ion mass spectrometry, that is,
the spectrometry of the charged sample atoms emitted by a solid
when bombarded by ions. The principles of the technique are
illustrated schematically in Figure 1. A beam of energetic primary
ions, generally species such as $^{16}O_2^+$, $^{16}O^-$, $^{14}N^+$, or $^{40}Ar^+$ at an
energy from 1 to 35 keV, interacts with a solid target held in a
high vacuum (10^{-6} Pa), dislodging atoms lying near the surface in
the process known as sputtering. A small fraction, 10^{-4} to 10^{-2},
of the sputtered atoms are emitted in a charged state, the so-
called secondary ions. It is important to note that, as shown in
Figure 1, the primary ions are incorporated into the specimen.
Although the implanted species can be influential in controlling
the secondary ion emission as well as being sputtered subsequently,
the ions which are emitted are mainly those of the sample.
Secondary ion mass spectrometry (SIMS) consists of generating the
secondary ions by primary ion bombardment, collecting them from
the sample region through the use of electric fields, and analyz-

317

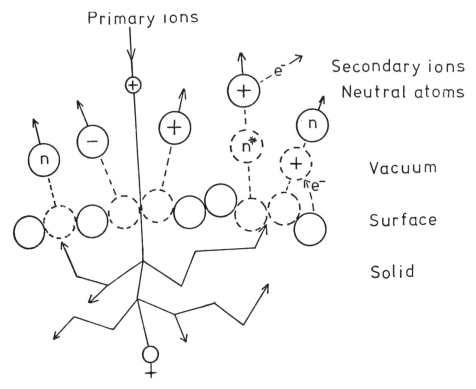

1 Interaction of primary ions with the sample causing the emission
   of neutral atoms, secondary ions, and excited neutral atoms.

ing the collected ions in a mass spectrometer according to the
mass-to-charge ratio. Except for the technique of generating the
ions, SIMS makes use of conventional mass spectrometry techniques,
and is subject to all of the strengths and weaknesses of mass
spectrometry. Qualitative SIMS analysis is performed by assigning
appropriate elemental or molecular species to the peaks identified
in the spectrum. Quantitative analysis involves the reduction of
the observed secondary ion intensities to compositional values
through the use of empirical procedures or techniques based on
physical models of the secondary ion emission process.

     An overview of the analytical information available in SIMS
can be obtained by examining a typical secondary ion mass spectrum,
in this case a positive ion spectrum from a single airborne particulate, Figure 2. Signals from all of the elements, including
hydrogen, can be observed if present. The capability for the
analysis of hydrogen, especially from small selected areas, is

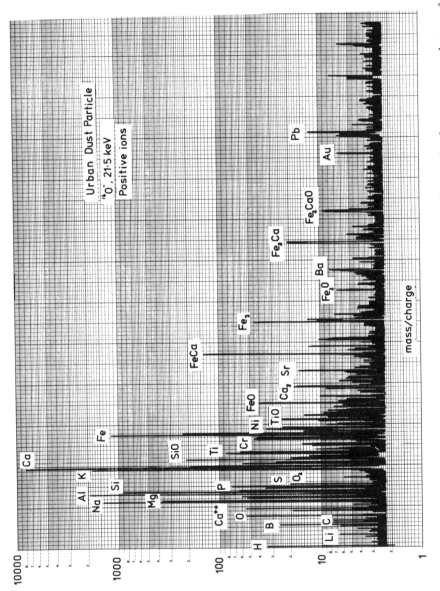

2 Positive secondary ion mass spectrum from an environmental particulate; note that only
the principal peak for each element is indicated. (Ordinate is relative intensity.)

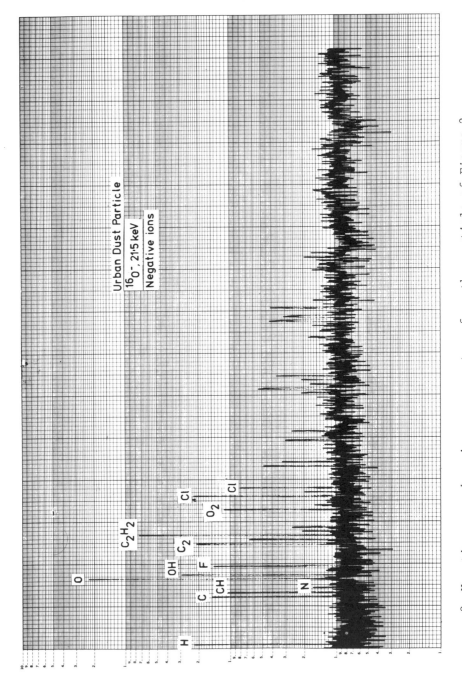

3   Negative secondary ion mass spectrum from the particle of Figure 2.
(Ordinate is relative intensity.)

available only with SIMS and the nuclear microprobe [4] (In the
nuclear microprobe high energy (MeV) ions are focused to a micrometer sized beam and used to induce nuclear reactions). For
multi-isotopic elements, e.g., silicon in Figure 2, signals for
all of the isotopes can be measured, giving the possibility of
isotopic ratio measurements for diffusion studies, age-dating,
etc. The spectrum of Figure 2 consists of positive secondary
ions. By adjusting the extraction voltage polarity, negative
secondary ions can be detected, Figure 3. As a general rule
elements with a strong electropositive character, such as the
alkali and alkaline metals, produce relatively strong signals in
the positive ion spectrum and weak or insignificant signals in
the negative spectrum, while the reverse is true for strongly
electronegative elements such as the halogens. Elements of
intermediate character often produce signals in both positive and
negative spectra; oxygen is an example in Figures 2 and 3.

     In addition to the elemental ions, signals are also observed
for molecular species, including dimers (e.g., Si_2^+), oxides
(e.g., SiO^+), other compound ions (e.g., $SiCa^+$), and multiply-
charged elemental ions (e.g., Si^{++}). A portion of the positive
secondary ion spectrum from high purity silicon is shown in
Figure 4, demonstrating the complicated nature of the SIMS spectrum. For light elements, the molecular species have an intensity
which is generally less than 10 percent of the parent peak, while
for heavy elements, such as tantalum or lead, the monoxide peak
can be as much as a factor of 10 higher than the elemental peak,
and dioxide and multiple oxide peaks also produce significant
intensities. Indeed, Benninghoven [5] has demonstrated that the
molecular signals can be used to study the nature of chemical
bonding in a sample. Benninghoven [6] has also reported that
information on complicated organic molecules can be obtained,
principally in the negative secondary ion spectrum.

     The secondary ions are emitted with a low energy, generally
1-50 eV, and they lose energy rapidly as they propagate through a
solid. Secondary ions have a significant escape probability only
when they are created within a shallow layer near the surface,
approximately 1 nm deep. SIMS is thus effectively a surface
analysis technique comparable to Auger electron spectroscopy
(AES), x-ray photoelectron spectroscopy (XPS or electron spectroscopy for chemcial analysis (ESCA)), and ion scattering spectrometry
(ISS). Space does not permit an exhaustive comparison of SIMS to
these other surface analysis techniques; the reader is referred
to several review articles in the literature [7,8,9]. SIMS is
noted primarily for high sensitivity, specificity, capability for
detection of all elements, and the detection of molecules.

     It is important to recognize the nature of the interaction
of the ions with the solid target in SIMS. SIMS is a destructive

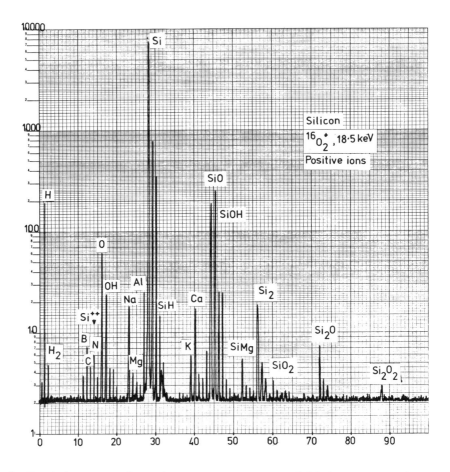

4   Positive secondary ion mass spectrum of high purity silicon.
    (Ordinate is relative intensity.)

technique, requiring that a portion of the sample be consumed to
produce an analytical signal. While the amount of sample consumed
is small, as little as 10^{-19} grams in a favorable case, it should
be noted that the exact analysis cannot be repeated. This situation is particularly significant for surface analysis since most
SIMS instruments detect only one species at a time. Hence, a
sequence of measurements of different ions cannot be made from
exactly the same region of the sample, since material is constantly eroded. Relative to this point, Benninghoven [10] has
made an important distinction between conditions of "static" and
"dynamic" SIMS. In static SIMS, the primary ion current density
and energy are maintained at a sufficiently low level so that
there is a very small probability that a surface site from which
a secondary ion is emitted has been previously hit by an incident

primary ion. The surface site is thus expected to be in a condition which is representative of the initial state of the sample.
In dynamic SIMS, the current density and beam energy are
increased to the point that there is a high probability that
surface sites from which secondary ions are emitted have been
previously altered by incident primary ions. The possibilities
of beam damage artifacts such as short range atomic mixing,
fragmentation of molecules, and the formation of molecules not
representative of the specimen must be considered. In the ion
microprobe, the typical current densities are such that a condition of dynamic SIMS is obtained. The direct spectral information
is thus expected to be principally useful for elemental analysis.

Instrumentation

The instrumentation for SIMS falls into three general
classes: the ion microscope, the ion microprobe, and broad-beam
SIMS. The general characteristics of each class will be briefly
described; a reference to a more complete description is given.*

(a) Ion Microscope [11]

A schematic diagram of the ion microscope is shown in Figure 5.
The sample is bombarded with an energetic (15 keV) ion beam
derived from a duoplasmatron ion source [12], to produce secondary
ions from a relatively large region of the sample, 20 to 100
micrometers diameter. The secondary ions are collected by an
electrostatic immersion lens which extracts the ions, maintaining
the geometrical relationship of the ray paths from the emission
volume and forming an image of the bombarded sample region at the
entrance slit of the mass spectrometer. The mass spectrometer
has a double pass configuration such that image aberrations which
result from the mass dispersion process in the first sector are
compensated in the second sector. The mass-dispersed image is
then magnified by projector lenses onto an image converter where
the ion signal is converted to an electron signal. The electrons
are used to excite a fluorescent screen which is viewed by an
optical microscope. By adjusting the magnet strength, images of
ions with various mass-to-charge ratios can be obtained sequentially. It is important to note that this imaging system is a
true microscope, with ray paths connecting each point in the
sample region to an equivalent point in the image plane. Each

---

* In order to describe materials and experimental procedures adequately, it is occasionally necessary to identify the sources of
commercial products by the manufacturer's name. In no instance
does such identification imply endorsement by the National
Bureau of Standards nor does it imply that the particular
product is necessarily the best available for that purpose.

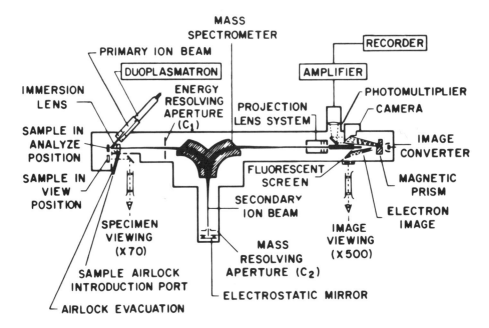

5   Schematic diagram of the ion microscope (courtesy R. Hessler,
    Cameca Instruments).

point in the image is thus continuously illuminated. The limiting
resolution of the imaging mode is approximately 0.5 micrometers
due to lens aberrations. Microanalysis can be accomplished with
this system by placing an aperture in the image plane and detecting
those ions which pass through. The practical lower limit on the
dimension of the area analyzed is about 5 micrometers diameter.
A typical ion microscope image of a copper-2% beryllium alloy is
shown in Figure 6, revealing the beryllium distribution in a
polycrystalline region. Bright regions are indicative of high
beryllium concentrations. The beryllium is localized in the
grain boundaries.

(b) Ion Microprobe [13]

A schematic diagram of an ion microprobe mass analyzer is shown
in Figure 7. The primary ion beam is again produced in a duoplasmatron souce and accelerated to an energy of 5 to 20 keV. The
energetic beam is passed through a mass spectrometer which separates the numerous ion species (atomic and molecular ions of the
source gases, and compound ions) produced by the source. The
resulting beam is then focused by two electrostatic condenser

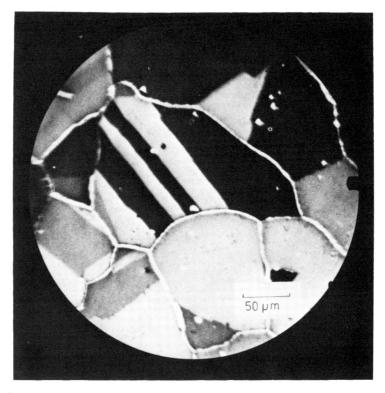

6   $^9Be^+$ image showing the beryllium distribution in a copper-
   beryllium alloy; ion microscope image (courtesy of R. Hessler,
   Cameca Instruments).

lenses to form a probe at the specimen plane. The probe diameter
is variable, with a limit of performance of about 1 micrometer
diameter in current instruments. The secondary ions produced by
the beam are extracted from the sample region without subsequent
focusing and passed through a mass spectrometer having electrostatic and magnetic sectors. The mass dispersed beam is detected
with an ion-to-electron converter and a high-gain photomultiplier.
Imaging with the ion microprobe is accomplished by scanning the
primary ion beam in a raster pattern on the sample. A signal
derived from the secondary mass spectrometer for a particular ion
is used to intensity-modulate a cathode ray tube (CRT) which is
scanned in synchronism with the raster pattern on the specimen.
The image is thus created as an intensity map which transfers
information from the specimen coordinates to CRT coordinates. To
a first approximation, the resolution is limited by the diameter
of the probe which strikes the sample and is of the order of 1

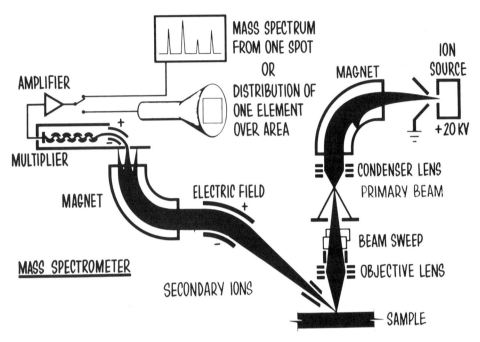

7   Schematic diagram of the ion microprobe mass analyzer
    (courtesy J. Hinthorne, Applied Research Laboratories).

micrometer. A typical ion microprobe image showing sodium-rich
particles (bright regions in the micrograph) on a silicon substrate is shown in Figure 8. Microanalysis is accomplished in
the ion microprobe by fixing the position of the probe on the
sample and detecting secondary ions excited in the probe area.
The lower limit on the size of the region sampled is about one
micrometer.

Slodizan [14] has discussed considerations of the signal-to-
noise ratio (S/N) in images observed with the ion microscope and
the ion microprobe. Usually, for a given primary ion beam current
density and picture frame time, the ion microscope has a superior
S/N ratio and hence, better image quality. This is a direct
result of the fact that every picture element of an ion microprobe
image is illuminated only for the period of time during which the
probe is resident. Consider an image of a field of particles on
an inert background. When an ion image of a constituent in the

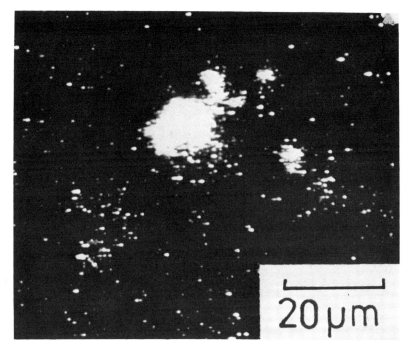

8   Ion microprobe image of sodium-rich particles on a silicon
    wafer. Primary ion beam $^{16}O_2^+$, focused to one micrometer.

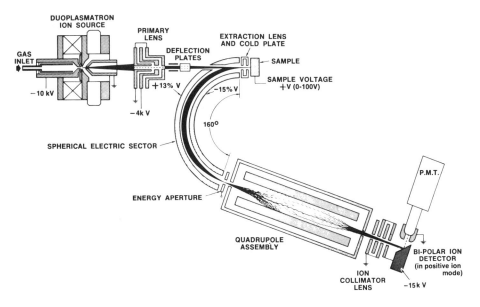

9   Schematic diagram of a broad-beam SIMS instrument (courtesy
    Applied Research Laboratories).

particles is formed with the ion microscope, every collected ion
leaving the bombarded region is brought to a corresponding point
in the recording medium since all points are continuously exposed.
However, in the ion microprobe, ions can only be collected when
the beam strikes a particle. Equal time is spent by the beam on
each point in the image, including those where there is no
particle. Advanced scanning systems which would increase the
beam dwell time as a function of signal would improve the S/N
ratio of ion microprobe images and therefore increase the image
quality.

(c) Broad-Beam SIMS

In the third general type of SIMS instrumentation, the high
spatial resolution of the ion microscope and ion microprobe is
sacrificed to obtain a much simpler instrument, Figure 9. The
probe-forming and imaging lenses are discarded. The specimen is
illuminated with a broad beam (100 micrometers to several mm)
derived from a duoplasmatron or a thermal ionization source. The
sputtered secondary ions are electrically attracted into a quadrupole mass spectrometer. Besides the obvious advantages of
simplicity of operation and lower cost, the use of the quadrupole
spectrometer allows rapid mass scanning and/or peak switching
which greatly improves data collection and gives a close approach
to simultaneous measurement. Broad-beam quadrupole SIMS systems
are frequently combined in hybrid vacuum chambers with other
surface analysis techniques such as AES, XPS or ISS.

A major limitation in the first generation of SIMS instruments has been the relatively low mass resolution available,
typically $M/\Delta M = 300$ (where M is the mass) at 10 percent peak
intensity. Because of the complicated nature of the spectra,
e.g., Figure 4, mass interferences are frequently encountered
which cannot be resolved with a low resolution spectrometer.
While it is often possible to find an alternate isotopic peak
which is unaffected by a particular interference or to calculate
the peak intensities from known isotopic ratios, the problem of
spectral artifacts is one of the greatest limitations in SIMS.
The problem is especially severe when trace levels of an element
are sought in a complicated specimen where the possibilities of
molecular ions are most numerous.

Quantitative Analysis

Quantitative analysis by SIMS is complicated by the strong
matrix effects which are observed. Figure 10 shows the variation
in secondary ion yields (normalized with the primary ion current)
observed from pure elements by Andersen and Hinthorne [16,17]. A
variation of more than five orders of magnitude is observed

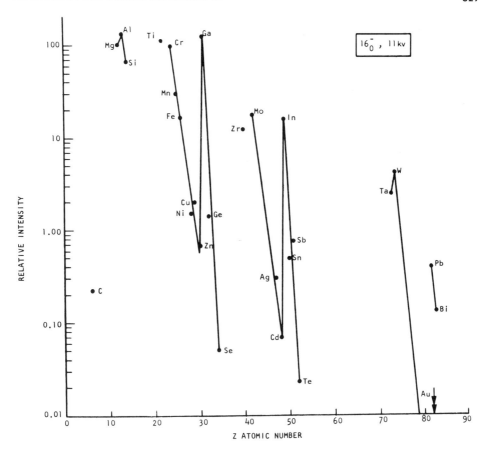

10   Normalized secondary ion yield from pure elements (courtesy
      C. A. Andersen and J. Hinthorne, Applied Research
      Laboratories).

across the periodic table. Chemically active elements such as
sodium show the highest signals while chemically inactive elements
such as gold produce low signals when the primary ion is electronegative in character, e.g., oxygen or nitrogen. For bombardment
with electropositive species such as cesium, the sensitivity
behavior is inverted. Thus, selection of the primary ion can
influence the sensitivity of SIMS.

When various elements are dispersed into a matrix of one or
two elements, the variation in sensitivity is greatly reduced.
As an illustration of this effect, the relative sensitivities of
a number of elements dispersed in a silicate glass are shown in

Figure 11. A relative elemental sensitivity factor, S, has been
calculated according to the equation:

$$S_{X/M} = (i_X/C_X f_X)/(i_M/C_M f_M) \qquad (1)$$

where i is the observed secondary ion intensity, C is the atomic
concentration, f is the isotopic fraction for the line measured,
and X and M denote two elements. In the ideal case, X is a
solute element and M is a matrix element. The values of $S_{X/Si}$ in
Figure 11 show a significantly reduced variation as compared to

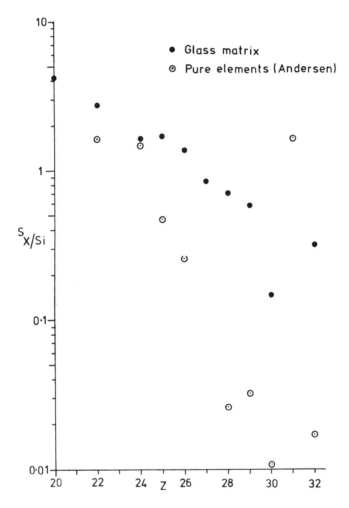

11 Relative elemental sensitivity factors measured in a glass
    matrix and calculated from the pure element yield of
    Figure 10.

values calculated from the pure element sensitivities from Figure
10. It is clear that the matrix element controls the secondary
ion emission. That is, upon dispersal in a matrix, a solute atom
behaves very much as a matrix atom with regard to secondary ion
emission rather than as it behaves in its pure element state.

Quantitative analysis can be carried out by either of two
general approaches. In the first approach, a physical model of
the process of secondary ion emission is used to predict the ion
intensities. The second approach is empirical, relying on the
use of relative elemental sensitivity factors, equation (1),
derived from standards similar to the unknown to reduce the
observed ion intensities to compositional values. To assess the
accuracy of quantitation which can be achieved by SIMS, a study
has been carried out on a series of multi-element glasses of
known composition [18]. The results of this study are pertinent
to the question of quantitative analysis of environmental particles, since the fully oxidized condition of the elements in the
glasses closely resembles the chemical state of many reaction
products of environmental interest, e.g., fly ash particles.

Of the several physical models available to describe the
process of secondary ion emission [19], the local thermal equilibrium (LTE) model of Andersen and Hinthorne [17] is the most
directly applicable to multi-element specimens such as glasses or
environmental particles. The LTE model assumes that the secondary
ions are formed in a dense plasma which is at internal thermal
equilibrium. With this assumption, the ratio of ions to neutral
atoms in the plasma for each element can be described by a form
of the Saha-Eggert ionization equation:

$$\log(n_{M+}/n_{Mo}) =$$
$$15.38 + \log 2(B_{M+}/B_{Mo}) + 1.5 \log T - \frac{5040(I_p - \Delta E)}{T} - \log n_{e-} \qquad (2)$$

where n designates the concentration of particles of element M, B
is the partition function which gives the distribution of population of possible energy states, T is the absolute temperature, I_p
is the ionization potential, ΔE is the potential depression due
to the field interactions of the charged particles, and n_{e-} is
the number of electrons per unit volume. The values of the
partition functions, ionization potential, and potential depression can be obtained from independent experiments or calculations.
For a particular ion bombardment condition, the temperature and
electron density should characterize the plasma. The LTE model
proceeds with an analysis in the following way. The analyst must
know the concentration values of at least two elements in the
specimen, the so-called "internal standards". From the measured
values of the secondary ion intensities for these elements, two

values of (n_{M+}/n_{MO}) can be calculated. Equation (2) can then be
solved as a set of simultaneous equations in the unknowns T and
n_{e-}. The values of T and n_{e-} determined in this way are then
used to solve for $n_{M}o$ for the remaining elements in the sample
for which the secondary ion intensities n_{M+} have been measured.
Since $n_{M}o$ in the plasma is expected to be directly proportional
to the number of atoms of M in the sample, the concentrations of
all elements can be calculated. The LTE model can also be used
to make corrections for the fraction of ions of an element which
appear as molecular ions.

A histogram of the errors observed in the analysis of the
SIMS spectra of the glasses with the LTE model is shown in Figure
12. An error factor has been defined to judge the accuracy of
the analysis:

$$F = C(true)/C(calculated) \qquad (3)$$

where C is the atomic concentration. In Figure 12, the error
factor or its reciprocal, for F <1, has been plotted. Various
combinations of internal standards and corrections for the molecular oxide component have been utilized in preparing the histogram.
In general, approximately 50 percent of the analyses fall within
a factor of two of the accepted compositional value and 80 percent
within a factor of five. In addition, this study revealed that
the LTE approach was most successful for elements with atomic
numbers less than 60. The LTE model tended to underestimate the
concentration of heavy elements such as tantalum and lead in these
samples.

A considerable improvement in the accuracy of analysis can
be achieved through the use of empirically determined relative
elemental sensitivity factors. The SIMS spectra of the glasses
were analyzed using average values of the sensitivity factors
relative to silicon derived from the entire set of glasses. The
resulting error histogram is shown in Figure 13. More than 80
percent of the analyses fall within a factor of two of the
accepted value and 99 percent within a factor of five. The error
factors have been replotted with greater resolution for the range
$0.5 < F < 2$ in Figure 14. More than 50 percent of the analyses
fall within the range of $0.83 \leq F \leq 1.2$. Because the glasses
cover a wide range of compositions, it seems reasonable to expect
that this level of accuracy could be obtained for materials in a
similar state, i.e., highly oxidized, through the use of sensitivity factors derived from the glasses.

A topic related to quantitative analysis is that of detection
limits. As mentioned previously, when dealing with low signals,
the problem of spectral artifacts must be carefully considered.

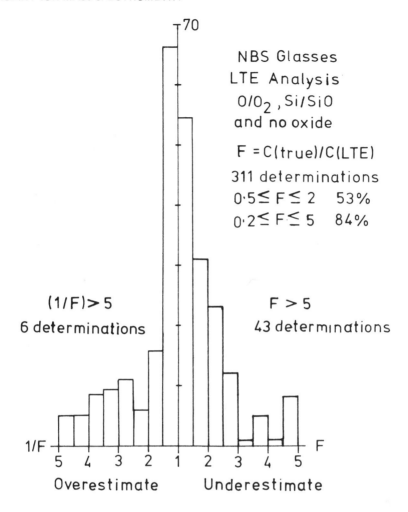

12 Histogram of error factors observed for quantitative analysis of SIMS spectra of glasses with the local thermal equilibrium model.

Provided that no interference exists, which is not usually the case in complicated samples, the following discussion is appropriate. The minimum detectable limit depends on the number of incident particles, i.e., the primary beam current. In other microanalysis techniques, such as electron probe microanalysis, the primary current can be varied with impunity over a large dynamic range (typically six orders of magnitude) without altering the specimen condition appreciably. In SIMS, the process of

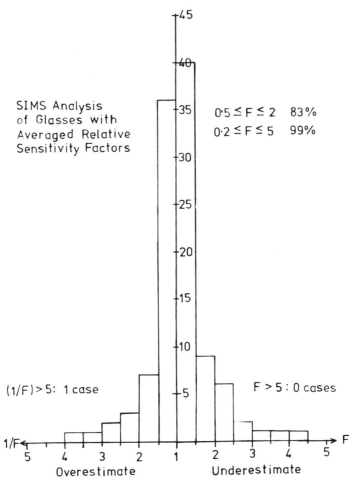

13 Histogram of error factors observed for quantitative analysis
    of SIMS spectra of glasses with average relative sensitivity
    factors.

sputtering to produce secondary ions necessarily leads to sample
destruction. Although the SIMS sensitivity can also be increased
by increasing the beam current, this increase requires increased
sample destruction. It is thus important to consider not only
sensitivity expressed in counts per second per unit current but
also counts per second per unit sputtered volume. The counting
time obviously has a strong influence on the volume of sample
consumed. Since most SIMS instruments are single channel devices
utilizing electrical detection, it is necessary to analyze a

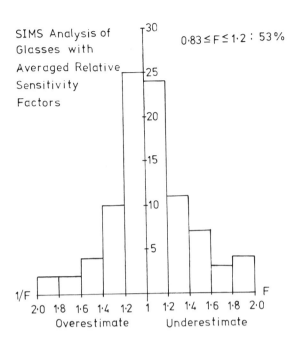

14   Histogram of Figure 13 replotted for the range $0.5 \leq F \leq 2$.

sample sequentially. It is, therefore, desirable to use short
counting times to minimize the effects of the sample variation
due to sputtering. For most microanalysis conditions in the ion
microprobe, it is best to use a low beam current (1 nA) and a
short counting time (1–10 s).

Values of the limiting sensitivity for a number of elements
in a silicon–lead–oxygen glass are given in table 1, for the conditions indicated. The limiting sensitivity has been estimated by
linearly extrapolating the working curve observed for minor constituents at the 0.1 atomic percent level to a count rate equal to
three times the background, approximately 1 c/s. The sensitivity
limit varies from element to element, but it is generally in the
parts per million (ppm) range. For the bombardment conditions
used, the sputtering rate is approximately 1×10^3 nm/s/A/cm^2 in
silicon. For a 1 s integration time, a beam of diameter 1 micrometer carrying an ion current of 1 nA of $^{16}O^-$ ions a 21.5 keV
sputters a crater to a depth of about 10 nm (100 Å) in silicon.
For an oxidic system such as a silicate, the depth would be
reduced by at least a factor of five. If the 1 micrometer beam

Table I

Detection Limits in Silicon-Lead-Oxygen Glass
(NBS K493 and K523 Glasses)

Element[a]	c/s/nA/Atom Percent[b]	1 nA detection limit[c] (1 sec. integration) ppm
Li	7.22×10^4	0.42
B	4.69×10^3	6.4
Mg	4.66×10^4	0.64
Al	6.32×10^4	0.47
P	1.32×10^3	23
Ti	5.37×10^4	0.56
Cr	3.01×10^4	1
Fe	1.77×10^4	1.7
Ni	1.13×10^4	2.7
Ge	7.45×10^3	4
Zr	4.20×10^4	0.71
Ba	7.41×10^4	4.0
Ce	1.85×10^4	1.6
Eu	6.45×10^4	4.6
Ta	2.51×10^3	12
Th	1.38×10^4	2.2
U	1.42×10^4	2.1

[a] All elements as dilute solutes in the indicated matrix.

[b] Corrected for mass abundance.

[c] Concentration which is expected to produce three times the background count rate (1 c/s).

was raster-scanned in an area of 10 x 10 micrometers, the current
density would be reduced and the erosion rate would decrease to
0.1 nm/s. It should thus be possible to analyze a surface layer
10 nm deep on a 10 micrometer diameter particle for 10 constituents, allowing a 5 second counting time and 5 seconds for spectrometer setting. Such data collection would be possible with computer
control of a sequential measurement.

Depth Profiling

Since the target is progressively sputtered away in a controlled fashion, it is possible to obtain the distribution in
depth of an elemental or molecular species by monitoring the
secondary ion signal as a function of erosion time. The time of
erosion can be converted into an equivalent depth by measuring
the depth of a crater generated under a known current density in
a material similar to the one under study. The sputtering rate
is generally inversely related to the strength of chemical
bonding, and it is, therefore, desirable to calibrate the sputtering rate for each material to be studied.

The cross sectional profile of a crater generated by ion
bombardment is directly related to the intensity distribution of
the primary ion beam. For a focused beam, the current density
follows a Gaussian distribution along a diameter, and such a beam
would produce a Gaussian profile across the erosion crater.
Since secondary ions would escape from all points on the crater
surface, the signal would be integrated over a range of depths.
This defect can be avoided by two technqiues. First, the beam
can be defocused, which leads to a uniform current density except
near the edge of the beam. Second, the focused beam can be
scanned in a raster pattern, which gives a uniform current density
across the scanned area except at the edge where the scan termi-
nates. A sputtered crater produced by raster scanning a focused
beam on silicon is shown in an optical interference micrograph in
Figure 15. The initial surface of the silicon wafer was extremely
flat, as evidenced by the straight interference lines, and this
flatness is preserved in the crater bottom. The walls have a
sloped profile indicated by the curvature of the interference
lines. Through the use of signal gating as a function of scan
position, the signal acceptance can be confined to the flat bottom
of the crater. This procedure can produce depth profiles with
depth resolutions of the order of 10 nm. The depth resolution is
limited by several interaction effects, including atom displacements which causes atomic mixing and differential etching effects
[20]. An example of a depth profile for indium implanted in
lead-tin-telluride covering a dynamic signal range of four orders
of magnitude is given in Figure 16.

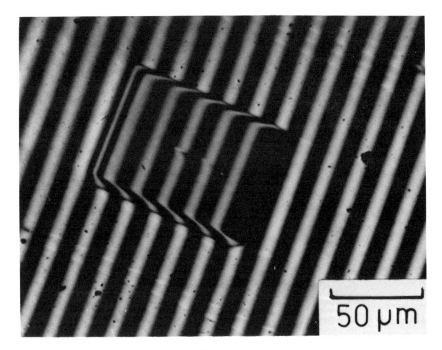

15   Crater eroded into a silicon wafer by a raster-scanned,
     focused ion beam.   Optical interference micrograph.

Application to Particulate Studies

     SIMS offers several interesting capabilities for the study
of particulates: (a) quantitative analysis of all elements with
good sensitivity in most cases, including light elements; (b) the
capability for surface studies; (c) the determination of elemental
depth profiles; and (4) the observation of molecular information.

(a) Quantitative analysis

     McHugh and Stevens [21] demonstrated that the ion microprobe
could be successfully applied to the determination of the composition of environmental particulates.   They have discussed mounting
techniques and instrumental operation for optimum results.   For
the present work, samples have been mounted on a gold background
layer deposited by thermal evaporation and overcoated with a thin
carbon layer to improve stability during charged particle bombardment.

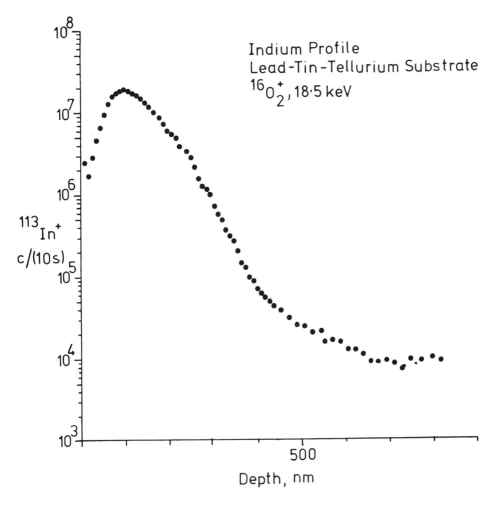

16   Depth profile of indium layer applied to a lead-tin telluride
     wafer.

     SIMS spectra such as those in Figures 2 and 3 can readily be
obtained from particles as small as 1 micrometer diameter. For
particles with diameters less than about 10 micrometers, it is possible to make use of positive primary beams to excite the secondary
ions despite the fact that they are insulators. Operation with a
positive primary beam is a great advantage since the duoplasmatron
ion source is approximately 20 times brighter for $^{16}O_2^+$ than for
$^{16}O^-$, allowing a much higher beam current for a given probe size in
the positive mode. For particles above 10 micrometers in diameter,

a negative primary ion beam is necessary in order to provide charge compensation for insulating specimens [16].

In order to convert the observed ion intensities into compositional information, it appears that the use of empirical sensitivity factors is to be preferred. The glasses [21] described previously provide an excellent source of materials chemically similar to environmental particles from which to derive the necessary sensitivity factors. In addition, the glasses have been prepared in the form of micrometer-sized fibers and particles. These particulate forms are useful in assessing the effects of particle geometry on SIMS signals and in serving as direct standards of the same size as the particles of interest. An example of a SIMS spectrum obtained from a 10 micrometer diameter fiber is shown in Figure 17. The intensities relative to silicon observed for the fiber are virtually identical to those observed from the bulk sample of the same glass, implying that size effects are not significant, at least for small particles.

Using sensitivity factors relative to silicon derived from these glasses, the major elements observed in the positive secondary ion spectrum of the environmental particulate of Figure 2 have been converted to compositional values in Table II. The sum of the metallic and metalloid elements was assumed to total 40 atomic percent, an approximate average for fully oxidized samples.

The analysis of Table II conceivably could have been carried out by a combination of electron probe microanalysis (EPMA) and Auger microprobe analysis. However the EPMA would have integrated a major portion of the particle volume, while the Auger microprobe would have obtained only surface information from 1 nm deep. SIMS can be used to analyze over a range of depths, and furthermore, is the only surface technique which can easily analyze insulators. It is questionable if the detection sensitivity of the EPMA and the Auger techniques would have been capable of detecting many of the elements observed in the SIMS spectrum.

(b) Surface Studies

As an example of the capability for the detection of surface deposits, consider the following illustration. In the course of heat treating ground glass of the composition used for the fibers of Figure 17 in order to produce spherical particles, a number of the particles were recovered after they had made contact with the hot wall of the furnace. SIMS examination produced the spectrum of Figure 18. Comparison of Figures 17 and 18 reveals that the heat treated particle was contaminated by contact with the furnace walls, principally by lithium, sodium, potassium, and copper, and by numerous minor elements. Subsequent ion erosion reduced the

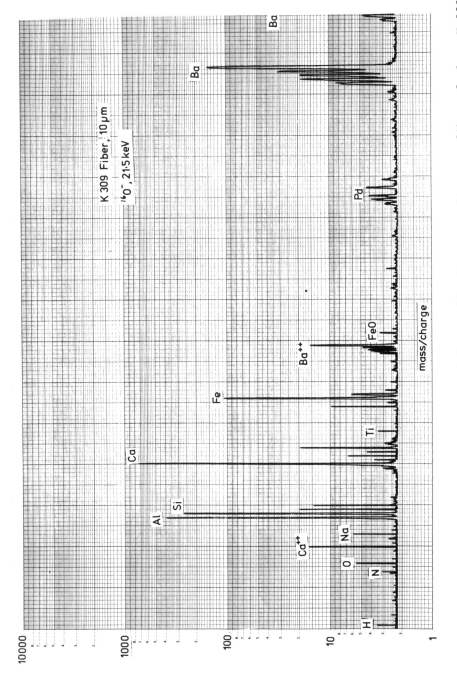

17 Positive secondary ion mass spectrum obtained from a 10 micrometer diameter fiber of glass K-309. (Ordinate is relative intensity.)

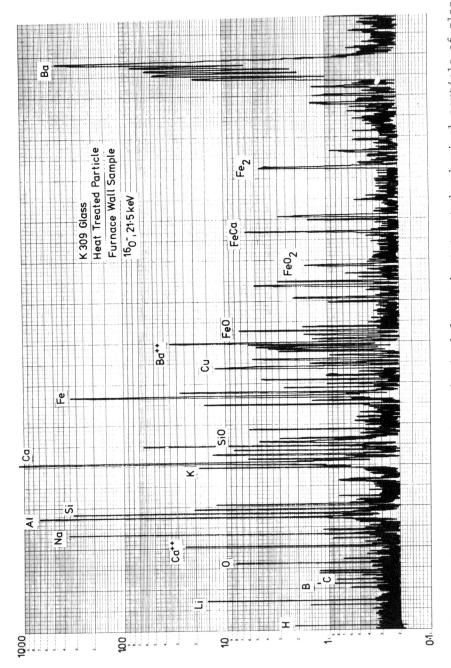

18   Positive secondary ion mass spectrum obtained from a heat-treated spherical particle of glass
      K-309. The particle was collected from the wall of the furnace. (Ordinate is relative
      intensity.)

Table II

Analysis of the Surface Layer of an Urban Environmental
Particulate with Relative Sensitivity Factors Derived
NBS Glasses

Element	Concentration (atomic percent)
Li	0.0067
B	0.62
Na	3.3
Mg	1.3
Al	4.0
Si	6.1
P	2.1
K	1.7
Ca	12.8
Ti	0.25
Cr	0.21
Fe	6.8
Ni	0.42
Ba	0.012
Pb	0.28

The sum of all the metallic and metalloid elements was
assumed to be 40 atomic percent; balance - oxygen.

signals from the extraneous elements, leaving the basic constituents of the glass as the principal signals, and thus demonstrating that the contamination resided mainly on the surface.

(c) Depth Profiles

Linton et al. [23] have applied SIMS to the study of depth
distributions in individual large (50-100 micrometer) environmental particulates. They compared the surface and interior
concentrations of elements such as lead and reported a surface
predominance of several species.

The depth profiling of particles is subject to geometrical constraints which considerably reduce the quality of the depth profile as compared to the "ideal" case illustrated in Figure 15 and 16, i.e., a flat sample of large dimensions. The smallest practical "window" for signal gating during raster scanning is of the order of 20 micrometers square, and over this area, even a large particle of 100 micrometers diameter will have considerable curvature. The sputtering rate varies sharply with the angle of beam incidence [24], so that the curvature effect will lead to differential sputtering rates. For particles in the size range of particular interest, 1-10 micrometers, it is not possible with existing instruments to avoid uneven erosion.

Within these limitations, experiments can be performed on small particles which give at least a qualitative impression of the depth behavior of elements. An example of such a profile is shown in Figure 19, obtained from a particle of approximately 10 micrometers diameter. The beam was rastered over an area larger than the particle to provide a uniform current density and mass spectra were periodically recorded. The erosion rate was estimated from that observed on a glass with similar constituents and was assumed constant over the whole particle. The signals for the elements studied were normalized with the silicon intensity to compensate for possible changes in the absolute intensities due to changes in the chemical state [20]. The intensity of silicon, a matrix element, is expected to reflect any overall changes due to bonding differences. The elements studied show a marked decrease as a function of depth, except for lithium. The behavior of lead appears to follow that noted for larger particles by Linton et al. [23]. After a depth of 1.5 micrometers was passed, the signals increased, which suggests that the particle is not uniform in composition. The erosion of 2 micrometers of oxidized material with a scanning beam required approximately 90 minutes.

(d) Molecular information

Although it is possible to identify a number of molecular peaks in the spectra of Figures 2 and 3, we cannot unambiguously relate these molecules to the chemcial state of the sample. At the current density and beam energy used to record the spectra, a condition of dynamic SIMS existed, raising the possibility of molecular fragmentation and formation of artifact compounds. In order to gain sensible information on the state of chemical bonding in particles by SIMS, it will probably be necessary to use either of two approaches. First, a condition of static SIMS could be achieved by reducing the current density and beam energy, although a great decrease in the sensitivity of analysis from small particles will necessarily occur. An alternative would be to continue to use dynamic SIMS, but to make use of ion fragmentation patterns

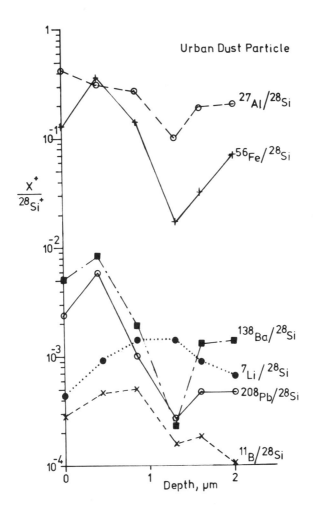

19 Depth profile obtained from a single urban environmental particle.

derived from known compounds to relate the molecular ions observed in the particle spectra to the original state.

Summary

SIMS offers a number of interesting and potentially powerful capabilities for the analysis of individual particles. All elements can be detected, with sensitivities frequently in the ppm

range from micrometer sized areas. Quantitative analyses with a
relative error less than a factor of two are feasible using relative elemental sensitivity factors. Analysis can frequently be
confined to a near surface region, even in the microanalysis mode,
and the determination of distributions in depth are possible.
Molecular information is available, but the interpretation of the
bonding state is ambiguous in the microanalysis mode. Images
showing the lateral distribution of elements can be obtained.

Acknowledgements

The author wishes to thank Mr. D. Blackburn of the Inorganic
Glass Section for the preparation of the glasses and
Dr. J. Small of the Microanalysis Section of NBS for the particles
used in this study.

References

1   VON ARDENNE, M.:   Z. Tech Phys 20 (1939) 344.

2   CASTAING, R. and G. SLODZIAN: J Microscope 1 (1962) 395.

3   LIEBL, H. J. and R. F. K. HERZOG: 12th Annual Conference
    Mass Spectrom and Allied Topics, ASTM E-14 (1964) p 393-397.

4   COOKSON, J. A., A. T. G. FERGUSON, and F. D. PILLING: J
    Radianal Chem 12 (1972) 39.

5   BENNINGHOVEN, A.:   in Book of Abstracts, 3rd Annual Meeting
    Federation of Analytical Chemistry and Spectroscopy Societies,
    (Nov. 15-19, 1976), Philadelphia, paper 303.

6   BENNINGHOVEN, A., D. JASPERS, and W. SICHTERMANN: Appl Phys
    11 (1976) 1.

7   EVANS, C. A., JR.: Anal Chem 47 (1975) 818A.

8   CZANDERNA, A. W., ed.: Methods of Surface Analysis (Amster-
    dam, Elsevier, 1975).

9   MORABITO, J. M.: in Secondary Ion Mass Spectrometry, ed.
    K. F. J. Heinrich and D. E. Newbury, National Bureau of
    Standards Special Publication 427, US Department of Commerce
    (1975) 121.

10 BENNINGHOVEN, A.,: Surf Sci 53 (1975) 596.

11 MORRISON, G. H. and G. SLODZIAN: Anal Chem 47 (1975) 932A.

12   VON ARDENNE, M.:   Tab. d. Electronenphysik, Ionenphysik, and
     Ubermikroskopie, vol 1, VEB (Berlag d. Wissensch., Berlin,
     1956) 544.

13   LIEBL, H.:   in Secondary Ion Mass Spectrometry, ibid, 1.

14   SLODZIAN, G.:   in Secondary Ion Mass Spectrometry, ibid, 33.

15   FRALICK, R. D., H. J. RODEN, and J. R. HINTHORNE: in Surface
     Analysis Techinques for Metallurgial Applications, ASTM STP
     596 (Philadelphia, 1976) 126.

16   ANDERSEN, C. A.: in Microprobe Analysis, ed. C. A. Andersen
     (Wiley, New York, 1973) 531.

17   ANDERSEN, C. A. and J. R. HINTHORNE, Anal Chem 45 (1973) 1421.

18   NEWBURY, D. E.:   in Quantitative Surface Analysis, ASTM
     Special Publication, Proceedings of a Symposium on Quantita-
     tive Surface Analysis, Cleveland-Pittsburgh Meeting on
     Analytical Chemistry (1977).

19   SCHROEER, J. M.:   in Secondary Ion Mass Spectrometry, ibid.,
     121.

20   McHUGH, J. A.:   in Secondary Ion Mass Spectrometry, ibid, 179.

21   McHUGH, J. A. and J. F. STEVENS: Anal Chem (1972) 2187.

22   Research Materials 30 (bulk glasses) and 31 (fibers), avail-
     able from the Office of Standard Reference Materials,
     National Bureau of Standards, Washington, DC 20234.

23   LINTON, R., LOH, A., NATUSCH, D. F. S., EVANS, C. A., and
     WILLIAMS, P.:   Science 191 (1976) 852.

24   CARTER, G. and J. S. COLLIGON, Ion Bombardment of Solids
     (Elsevier, New York, 1968) 313.

Discussion

LINDFORS – When you compare dynamic and static SIMS, how much sensitivity do you lose in static SIMS?

NEWBURY – I would estimate three or four orders of magnitude in sensitivity would be lost in static SIMS, especially if a small beam is used. If a large diameter beam, 1 mm or larger, can be employed, then the low current density required for static SIMS can be achieved along with high current and sensitivity. Obviously such a beam would not be adequate for studying individual particles. Such a beam could be used to study molecular species adsorbed on large targets or aggregates of particles.

It is difficult to predict the required conditions for static SIMS on an arbitrary specimen since the ion damage cross-sections are not well known. If one sees complicated molecules in the SIMS spectrum, then a condition of static SIMS has probably been achieved. Benninghoven has demonstrated this by studying thin layers of nucleic acids adsorbed on metal substrates. When the current density is large, only elemental ions and dimmer molecular ions are observed, while at low current densities, the parent molecular ion as well as many fragment ions are observed.

BEAMAN – Can you mention how sensitive SIMS is to orientation (crystal) effects?

NEWBURY – Orientation dependent secondary ion yields have been observed, particularly with noble gas primary ions. Slodzian has produced excellent crystallographic contrast in ion microscope images of polycrystalline samples. With oxygen primary ions, crystallographic effects are considerably suppressed. We have observed the creation of an amorphous surface layer on a silicon crystal during oxygen primary ion bombardment.

SURFACE CHEMICAL ANALYSIS OF PARTICLES BY AUGER ELECTRON SPECTROSCOPY AND ESCA

Paul A. Lindfors and Claire T. Hovland

Physical Electronics Industries, Inc.
6509 Flying Cloud Drive
Eden Prairie, Minnesota, 55343

ABSTRACT

The basic phenomena of Auger electron spectroscopy (AES) and ESCA (electron spectroscopy for chemical analysis, also called XPS; X-ray photoelectron spectroscopy) are described. These surface sensitive (a few atomic layers) analytical methods were used to study flyash samples from the inlet and outlet of an electrostatic precipitator. AES results from the inlet particles indicated uniformity and a surface layer ($\sim$20Å thick) rich in Na and K. AES results from the outlet sample indicated two types of particles. One was similar to the input particles. A second was rich in Ca and C relative to the other particles. ESCA results from both inlet and outlet samples indicated a surface layer ($\sim$10A thick) rich in Na, K and C.

AUGER ELECTRON SPECTROSCOPY (AES)

Auger electron spectroscopy is a method of analysis yielding identification of elements in the top few atomic layers on a sample surface,[1] and estimates as to the concentration of the elements.[2] The emission of an Auger electron by an atom is shown pictorially in Figure 1. If one bombards a target in an ultrahigh vacuum chamber with electrons, (represented by the arrow coming in from the right) and if these excitation electrons have sufficient energy, one can ionize a core level of a target atom. This ionization is represented

349

by the "B" electron. The core level vacancy can be
filled by transition of an electron from a higher
lying electron energy level in the atom. This is
represented by the electron moving down from "C" to "A".
Conservation of energy now requires the atom to emit
either an x-ray photon or an Auger electron with an
energy equal to the energy difference involved in the
transition of the electron to fill the core level
vacancy. Because electron energy levels in all elements are discrete, emitted Auger electrons have energy
information characteristic of their parent atoms.
Because two electron energy levels in an atom are required for Auger electron emission, H and He cannot be
detected using AES.

     Bombarding a sample with electrons of sufficiently
high energy to ionize core levels (typically 2 to 10
keV) and an electron energy analyzer to detect the
energies of emitted electrons (typically 50 to 2000 eV)
allows one to do AES. The strength of the Auger signal
is proportional to the concentration of the emitting
species and quantitative calculations can be made from
the data. The minimum concentration that can be
detected is typically 1 part per 1000. Of perhaps more
importance is the surface sensitivity of AES. Only
those Auger electrons emitted from atoms in the first
few atomic layers of a sample will not lose sufficient
energy before they are emitted into the vacuum such that
one can recognize their parent atoms. Figure 2 shows
the escape depth of electrons with 50 to 2000 eV
energies. Note that it is 4 to 20A, or perhaps 1 to
5 atomic layers. The surface sensitivity of AES allows
lower thresholds of detection for elements concentrated
in particle surface layers. Consider the situation
shown in Fig. 3. If the electron beam diameter is equal
to the particle diameter and the particle is 1 μm in
diameter, then the limit of detectability for an element
concentrated in the outermost layer of atoms on the particle would be $\sim$5 parts per million. For a particle
10μm in diameter and an electron beam 1μm in diameter
the limit of detectability would be $\sim$1 part in 10^8.
Fig. 4 demonstrates the kind of resolution that can be
obtained using commercially available scanning AES
equipment with an electron beam diameter $\sim$0.2μm.
Review articles describing AES and details of analytical
apparatus are available in the literature.[3,4,5]

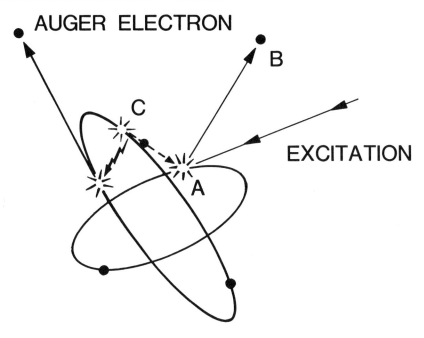

Fig. 1. - Pictorial representation of Auger electron emission and
x-ray photoelectron emission.

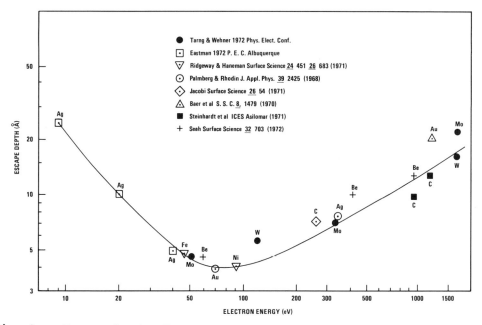

Fig. 2. - Escape depth of electrons in various materials vs. electron
energy.

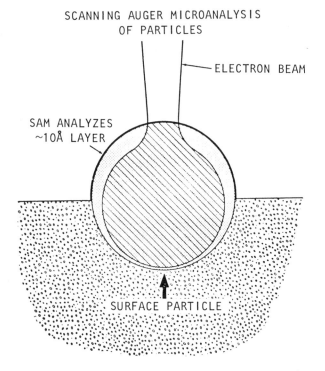

Fig. 3. - Pictorial representation of scanning Auger
analysis of particles.

Fig. 4 - Secondary electron image of flyash from the precipitator inlet obtained using scanning Auger microprobe.

ESCA (XPS)

ESCA is another surface sensitive analytical method yielding indentification of elements[6] and estimates of their concentrations.[7] ESCA stands for Electron Spectroscopy for Chemical Analysis and is often called X-ray Photoelectron Spectroscopy or XPS. The latter title is a more accurate description of the process. The phenomena of this analytical method are also represented in Figure 1. In ESCA, X-ray photons bombard the target atom and result in ejection of photoelectrons. In Figure 1 the excitation arrow represents an x-ray photon, and the electron at "B" represents a photoelectron. If one uses a monochromatic x-ray source, then the electrons emitted by the atom due to absorption of an x-ray photon will have an energy equal to the x-ray energy minus the binding energy of the electron. The binding energies of electrons in atoms are discrete so the parent species can be identified. ESCA requires a monochromatic x-ray source and an electron energy analyzer. As in AES the strength of the ESCA signal is proportional to the relative population of a species, and one can make quantitative calculations.[8] Moreover, the typical range of photoelectron energies in ESCA is similar to that of typical Auger electron energies, and thus ESCA is equally surface sensitive. The minimum concentration ESCA can detect is also 1 part per 1000. Detection of H, and to a lesser extent He, is difficult due to interference by background signals.

One very important aspect of ESCA has been omitted thus far. Using high resolution energy analysis one can detect shifts in the energies of the emitted photoelectrons which give information regarding the bonding of target atoms.[11] Because none of the results from the flyash analysis involves energy shift considerations, this topic will not be discussed. Review articles describing ESCA and analytical apparatus are available in the literature.[9,10]

EXPERIMENTAL RESULTS

At the request of the source the origin of the flyash samples must remain confidential. They were taken from the input and output of an electrostatic precipitator in a pulverized coal burning electrical generating plant. Further, they were obtained using the Environmental Protection Agency Method 5, regarding the sampling of particles in a gas stream.

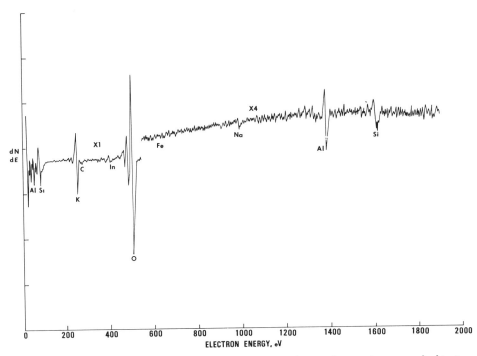

Fig. 5. - Auger spectrum of flyash particle from the precipitator
inlet, As received condition.

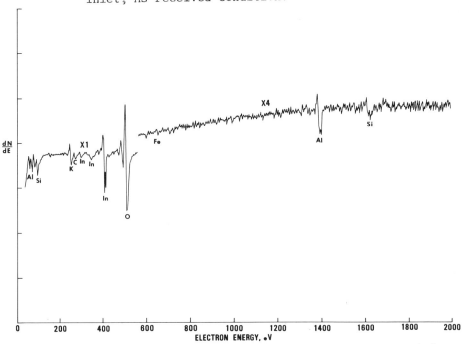

Fig. 6. - Auger spectrum of flyash particle from the precipitator
inlet, After ∿20Å removed.

AES RESULTS

Most of the inlet particles are electrical insulators which can lead to sample charging during Auger analysis. For this reason, sample preparation was a key consideration. The Auger spectrum in Fig. 5 was obtained from flyash pressed into In foil. The incident electron beam was approximately 2.5 micrometers in diameter. The In peaks in Figure 5 indicate some overlap of the electron beam beyond the particle edges. Two methods were used to distinguish Auger peaks due to the particle and not due to the overlap area. First, smaller and smaller areas containing a particle were analyzed. Each decrease in analysis area meant the particle was a larger fraction of the total area and one could observe which peaks increased and which decreased under such conditions. The peaks which increased were assigned to the particle. Secondly, Auger spectra were obtained from adjacent areas not including a particle to note what peaks were present in analysis of such regions. Both methods of peak assignment indicated Si, Al, K, Na and Fe were associated with the particles only. Another Auger spectrum obtained after approximately 20A of material was removed from the analysis area by ion etching with Ar^+ ions is shown in Fig. 6. Consideration of the changes from Fig. 5 to Fig. 6 suggests Na and K were concentrated in the outer 20A of the particle while Fe, Al and Si were more bulk constituents. All particles from the inlet sample gave similar results to those shown in Fig. 5 and 6. In addition, the entire inlet sample was of a uniform tan color.

Visually, some outlet particles were dark though the bulk of the sample was the same tan color as the inlet sample. There was a distinct difference in the Auger spectra from the dark and light particles as shown by consideration of Fig. 7 and 8. In Fig. 7, from a tan particle, the peaks are similar to those from the upstream sample particles. For the dark particle the spectrum in Fig. 8 shows more C, some Ca and N, and less Al, Si, K and Na than for the tan particle. Both the particles resulting in Fig. 7 and 8 were pressed into the same piece of In foil. The dark particles in the outlet sample were better electrical conductors than the tan particles.

Fig. 9 shows an Auger spectrum from a second dark particle prepared for analysis in a different manner. This sample was mounted on a thin Cu wire, wet with Ag

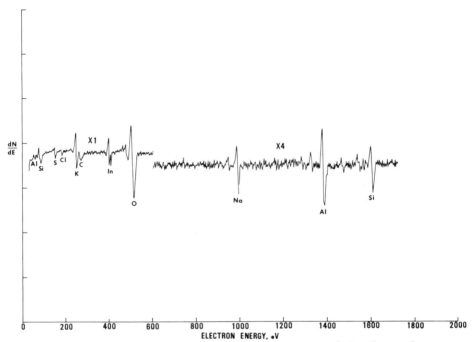

Fig. 7. - Auger spectrum of a tan flyash particle from the
        precipitator outlet.

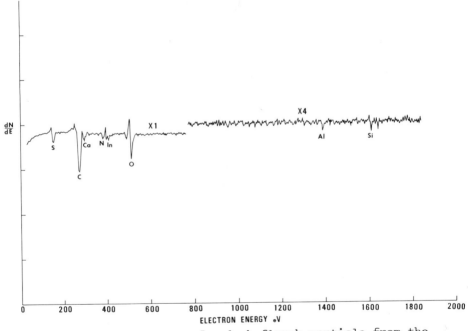

Fig. 8. - Auger spectrum of a dark flyash particle from the
        precipitator outlet.

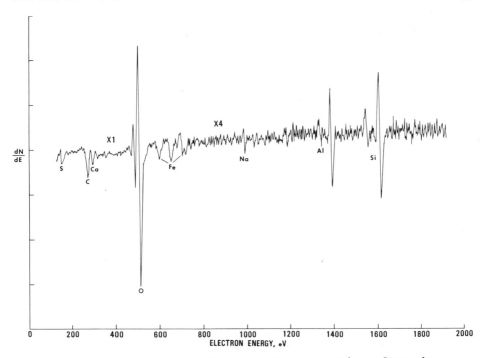

Fig. 9. – Auger spectrum of a dark flyash particle from the precipitator outlet.

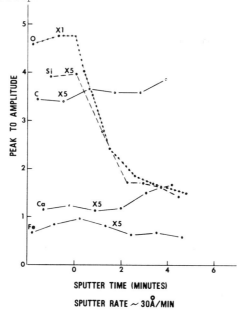

Fig. 10. – Depth profile of Auger peak amplitudes as functions of sputter etching time, same particle as in Fig. 9.

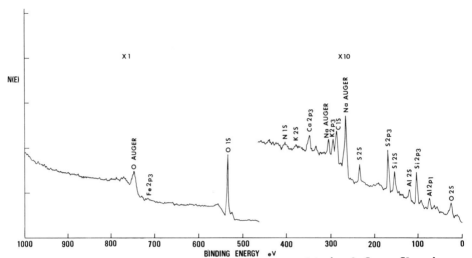

Fig. 11. - Typical ESCA survey spectrum obtained from flyash
          samples.

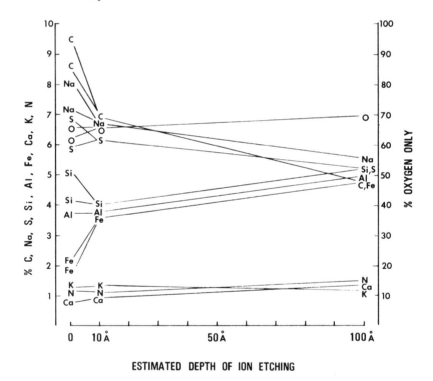

ESTIMATED DEPTH OF ION ETCHING

Fig. 12. - Estimates of constituent concentrations (atomic
          percent) vs. depth of ion etching for material collec-
          ted at the precipitator inlet.

paint, and the dark particle stuck to the end while
observation was made under a microscope. In addition
to the spectrum in Fig. 9 an Auger profile was obtained
from the sample. An Auger profile is a plot of the
amplitudes of Auger peaks as a function of ion etching
time into a sample and is shown in Fig. 10. Note O and
Si decrease while C and Ca increase as a function of
depth into the particle. The spectra and profile from
the dark particles are in contrast to the spectra and
changes with sputtering found for the inlet particles
and the tan colored outlet particles. This indicates
the electrostatic precipitator modified particle character in addition to acting as a filter.

ESCA RESULTS

Fig. 11 above shows a typical ESCA spectrum that
was obtained. This spectrum was taken from an inlet
sample before any ion etching. The area analyzed in
the ESCA measurements was about 3 millimeters in diameter, so the results are an average over a great many
particles. Sample charging was not a problem and a
thick layer of sample material was analyzed. Therefore,
all the peaks in a spectrum, such as Fig. 11, could
definitely be assigned to the particles. Thus the
number of definite constituents is increased from Si,
Al, Na, K and Fe to also include C, S, O, N and Ca.
Some of the peaks are labeled as Auger peaks, not photoelectron peaks because the original core level vacancy
starting the process of Auger electron emission could
also result from creation of a photoelectron. Changes
in the abundance of the elements with ion etching to
depths of approximately 10A and 100A are shown in Fig.
12. These data indicate a surface layer rich in C, Na,
K and S with O, Si, Al, Fe, Na and Ca more abundant
further into the particle. This is in good agreement
with the Auger results from the inlet sample particles.

A similar summary of ESCA data vs. ion etching
for the outlet sample is shown in Fig. 13. The changes
in Fig. 13 are similar to those in Fig. 12 and are
compared in the Table A, Fig. 14. The dark outlet
particles were only a small fraction of the total, and
did not appreciably influence the "averaged" ESCA results. Most of the constituent concentrations and
changes in concentrations are similar for both the
inlet and outlet samples except for the concentrations
of O, Na and Fe. It is possible that the reduction of

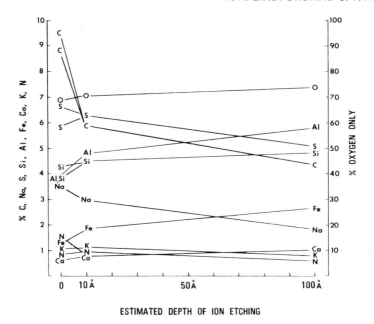

Fig. 13. - Estimates of constituent concentrations (atomic percent)
vs. depth of ion etching for material collected at the
precipitator outlet.

	ESTIMATED CONCENTRATION %					
	INLET			OUTLET		
ELEMENT	0Å	100Å		0Å	100Å	
O	65	70	*	70	75	
C	9	5		9	4.5	
Na	7.5	5	*	3.5	2	
S	6.5	5		6.5	5	
Si	4.5	5		4.5	5	
Al	3.5	5		4	6	
Fe	2	5	*	1	3	
K	1	1		1	1	
N	1	1		1	1	
Ca	0.8	1.3		0.5	1	

*Concentrations differ inlet to outlet

Fig. 14. - Table of changes in estimates of constituent concentrations shown in Figs. 12 and 13.

Na and Fe concentrations from inlet to outlet reflects more effective removal of particles with large Na and Fe concentrations. The increased O concentration from inlet to outlet could reflect less effective removal of oxides.

SUMMARY

These preliminary AES (from individual particles) and ESCA (averaged over many particles) studies indicate surface layers of a few atomic layers thickness and different from the bulk are present on flyash particles. The sample preparation technique of stick-small individual particles to a Cu wire, wet with Ag paint appears to be a promising method to minimize charging effects during Auger analysis. The AES results indicate uniformity in the inlet particles, and some variety in the outlet particles.

REFERENCES

1. LANDER, J. J.: Phys. Rev. 91 (1953) 1382.

2. DAVIS, L. E., MAC DONALD, N. C., PALMBERG, P. W., RIACH, G. E., and WEBER, R. E.: Handbook of Auger Electron Spectroscopy, 2nd Edition, Physical Electronics Industries, Inc., Eden Prairie, Minn. 1976.

3. JOSHI, A., DAVIS, L. E., and PALMBERG, P. W.: Auger Electron Spectroscopy, Ch. 5, Methods of Surface Analysis, (Czanderna, A. W., Ed). Elsevier Scientific Publishing Co. New York (1975)

4. STEIN, D. F., WEBER, R. E., and PALMBERG, P. W.: J. Metals, 23 (1971) 39.

5. WEBER, R. E.: J. Cryst. Growth, 17 (1972) 31.

6. SIEGBAHN, K.: Alpha, Beta and Gamma Spectroscopy, North-Holland, Amsterdam (1965).

7. WAGNER, C. D.: Anal. Chem, 44 (1972) 1050.

8. SCOFIELD, J. H.: Lawrence Livermore Laboratory, Rep UCRL-51326, (1973).

9.   RIGGS, W. M, and PARKER, M. J.: Surface Analysis by
     X-Ray Photoelectron Spectroscopy, Ch. 4, Methods of
     Surface Analysis, (Czanderna, A. W., Ed.). Elsevier
     Scientific Publishing Co. New York (1975).

10. RIGGS, W. M., and SWINGLE, R. S.: Critical Reviews
     in Anal. Chem 2, (1975) 267.

11. NORDLING, C., SOKOLOWSKI, E., and SIEGBAHN, K.:
     Ark. Fys., 13 (1968) 483.

Discussion

BEAMAN - You have some numbers on there (Fig. 14) of 7.5, 1, 2, and 8. What kind of accuracy would ascribe to those numbers?

LINDFORS - I believe them to be within a factor of two.

BEAMAN - You show an oxygen difference between 65% and 70%. Can you really be sure that's a different number?

LINDFORS - Because oxygen was the dominant species in the specimens, I tend to believe the difference is real. If oxygen was a less constituent and the difference was ten versus twelve percent, I wouldn't make that statement. Certainly additional analyses would be required before I'd go to court and try to defend the numbers.

COLEMAN - In the analysis of fly ash precipitate you didn't show any heavier elements. Are any missing?

LINDFORS - I didn't find evidence of heavier elements. In conversations here, some of the other participants have said lead, selenium, and/or cadmium might precipitate out on or condense on the outer layer of fly ash. If the heavier elements were present in concentrations of even a few percent, I would have seen it. I didn't squeeze to get the ultimate sensitivity from my analyzer.

LODGE - You probably pumped the heavier elements off.

LINDFORS - It's possible, it was a minimum of a half hour of pumping before I started analysis.

BEAMAN - Can you give me some indication of how to handle an electron backscattering problem. My problem is that I'm looking at a layer that's twenty angstroms thick. I've got a primary beam exciting that, but I've also got additional excitation of the layer on my surface due to backscattered electrons. I'd like to know how to factor that into a quantitative correction. For example, in gold 50% of the primary beam will be backscattered so I've got an additional 50% excitation.

LINDFORS - Taung and Wehner (J. Appl. Phys., 44: 1534 (1973)) found about 20% increase in Mo signals when deposited on a W substrate. They ascribed the increased Mo signal to the high electron backscattering of the W. Not much work has been done in this area and an experimental calibration may be required.

NEWBURY - These backscattering problems were discussed at the Miami EMSA. Ted Hall uses a band of continuum to normalize the data.

SESSION V:

PHYSICAL ANALYTICAL METHODS

SIZE MEASUREMENT OF AIRBORNE PARTICULATES BY

TIME-OF-FLIGHT SPECTROSCOPY

Barton Dahneke

Department of Radiation Biology and Biophysics
University of Rochester
Rochester, New York 14642

ABSTRACT

The use of time-of-flight spectroscopy for measuring the
aerodynamic size distribution of airborne particulates expanded
into an aerosol beam in a vacuum environment is described. Results
obtained using two different prototype models are presented. Both
of these instruments use scattered light signals generated by the
passage of a sample particle through one or two focused laser beams
to detect and measure the particles. Although convenient to use
for larger particles, the intensity of the scattered light signals
depends very strongly on particle size so that particles smaller
than about 0.1 μm diameter cannot be detected and measured. The
detection of sample particles by electron beam scattering is considered in detail. By use of this detection method, it seems
possible to extend the range of measurable particle size to an
order of magnitude smaller.

Introduction

The nature of small suspended particulates in the atmosphere
is a topic pertinent to a broad variety of scientific and industrial interests. Meteorologists are interested in the atmospheric
aerosol because of its influence on the condensation process and
its possible effect on climate. Engineers are attempting to minimize the contributions of combustion and industrial processes to
the atmospheric aerosol. The general public and particularly

specialists in health related fields are concerned about the influence of atmospheric aerosols on health. Aerosol physicists and
chemists are attempting to discover the origins, extent, dynamics
and fate of atmospheric aerosols and to develop theoretical and
experimental techniques that properly characterize these aerosols.

For all of the above examples, characterization of the aerosol
by both size (or volume or mass) and chemical distribution is
highly desirable. The ability of a nucleus to act as a condensation center depends on both its size and composition. Likewise,
the influence of atmospheric aerosols on climate and public health
is related to both the size and chemical composition of the particulates. Measurement of both size and chemical distributions
seems essential to discovering the exact origins, extent, dynamics
and fate of atmospheric aerosols. Thus, full characterization of
an atmospheric aerosol for each of these purposes and others must
include both the size and chemical composition distributions.

The atmospheric aerosol contains large numbers of particles,
at least 10^4 particles/cc in a typical urban atmosphere. Thus,
millions of particles are contained in a small sample volume.
Meaningful measurement of the particles therefore requires instruments which can quickly process large numbers of particulates and
provide the distribution of their size and composition. Microprobe techniques wherein single particles are examined individually
are not efficient for obtaining such statistical information because they require manual operation and are too slow. Moreover,
for an adequate number of measurements, the number of complete
spectra of individual particles is long incomprehensible unless it
is reduced to distribution form. Routine analysis of airborne particulates is therefore best performed by techniques that measure
the desired distribution functions directly.

No single method currently used in the measurement of aerosols
is able to provide all the desired information. Combination of
methods such as elemental analysis of impactor samples by X-ray
fluorescence can provide much of the desired information, but not
all. Consider the measurement of particle size distribution alone.
No single method is presently capable of on-line measurement of
aerosol size distribution over the "respirable size range", i.e.,
the size range between about .01 μm and 5 μm particle diameter.
Two or more methods can be combined to obtain size data over this
range, but such combinations introduce complexities and additional
expense. More important, present size measurement methods generally either bias or preclude later composition measurement of
the sample or restrict the type of composition measurement that
can be made. Filter and impactor samples, for example, may lose
water and other volatile components such as light organics before
they can be analyzed. Composition analysis in a parallel system

limits correlation of size and composition data. Clearly, improved
techniques for aerosol characterization are needed.

This paper describes one method of measuring the size of individual particles, time of flight (TOF) spectroscopy. The method
appears capable of on-line particle size measurement. Moreover,
the size measurement is nondestructive of the sample particulates
and does not preclude rapid composition measurement of the individual particles.

Time of Flight Aerosol Beam Spectroscopy

Several articles (1,2,3) and a U.S. Patent (4) describe the
details of this measurement technique and the performance of prototype models of time of flight aerosol beam spectrometers (TOFABS).
The essentials of the technique comprise:

(a)   Expansion of the aerosol sample through a suitable
nozzle or capillary into a vacuum chamber where the
particles form an aerosol beam analogous to a molecular beam.

(b)   Measurement of the TOF of each beam particle as it
traverses a fixed path length in the vacuum chamber.

(c)   Identification of the particle's "aerodynamic size"
from the measured TOF. Since larger particles lag
the gas velocity in the highly accelerating jet flow
more than the smaller particles, the measured TOF
uniquely identifies the particle aerodynamic size.
Calibration data relating particle mass, shape,
orientation and TOF can be obtained either analyt-
ically or experimentally.

(d)   Storage and display of the size distribution data.

To measure particle TOF across a fixed path length, various
methods have been used. The original TOFABS (1) used a pair of
focused laser beams to define the fixed path length. A photomultiplier tube (PMT) detected each pair of scattered light signals
defining the particle TOF. The corresponding pair of electrical
signals from the PMT was used to turn a high speed counter on and
off causing the particle TOF to be counted and printed. This
spectrometer is shown schematically in Fig. 1 and example signals
are shown in Fig. 2.

A later model instrument (2) is shown schematically in Fig. 3
with example PMT signals in Fig. 4. In this model the particle is
sensed over the full duration of its flight across a focused laser
beam. In this case the passage of a particle through the focused
laser beam, which is Gaussian in its spatial intensity distribution,
causes the scattered light signal to be Gaussian in its temporal

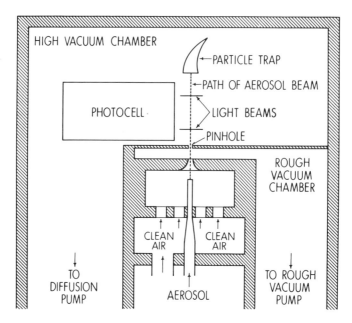

Figure 1.   Schematic diagram of a TOF aerosol
beam spectrometer.

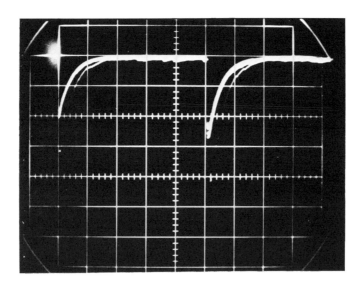

Figure 2.   Example PMT signals obtained from the
TOFABS of Fig. 1.

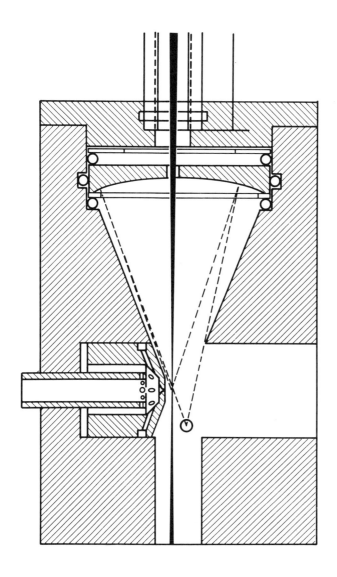

Figure 3. Schematic diagram of a second TOFABS using a single light beam.

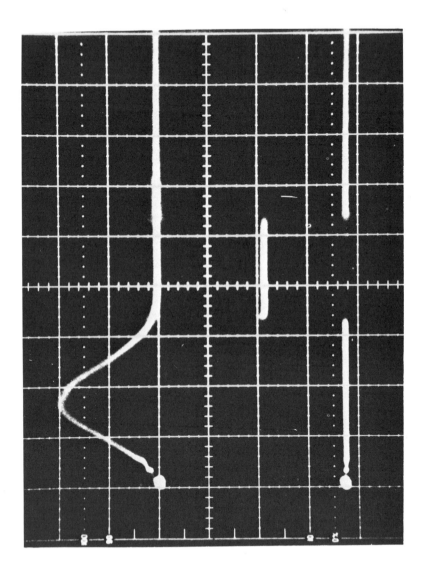

Figure 4.   Example PMT signals (upper trace) and logic signals of width
equal to the particle TOF (lower trace) for the TOFABS of Fig. 3.

distribution, as shown in the upper trace of Fig. 4. Thus, instantaneous particle location in the beam is given by the instantaneous intensity of the scattered light signal. The duration of the signal for which its instantaneous value exceeds 1/3 of the maximum signal amplitude, shown as the logic signal trace of Fig. 4 delayed 0.5 μsec from the PMT trace, is the particle TOF across the fixed "width" of the laser beam.

Other methods of measuring particle size are also, in effect, TOF methods. One example is the passage of a particle through an interference pattern generated by combining two laser beams at small angle. The frequency of the scattered light signal caused when the particle passes through this field at constant velocity is a measure of the particle velocity or TOF.

Velocity selectors, long used in molecular beam research, can also be used to measure (select) particles having a certain TOF. In this case the filter passes a fixed fraction of particles having TOF in a selected range. By scanning the TOF ranges and counting the particles passed by the selector in each range, the particle size distribution can be obtained. The size resolution of this method is poor compared to the above methods, but still fully adequate to determine the "respirability" of atmospheric particulates.

In each of these systems the smallest detectable particle size is limited by the particle detector capability. For systems using scattered light signals to detect the particles, the smallest detectable particles are about 0.1 μm diameter.

Thus, although the prototype systems have demonstrated the utility, speed and accuracy of particle size measurement by TOF spectroscopy, application of the method to aerosol size measurements over a broad size range, such as the respirable size range, must await improved detection systems. Although this limitation may not apply in the use of a velocity selector to pass only a selected size sub-range which can be subsequently measured, other limitations do. For instance, only a single particle size can be measured at one time.

Instrument Calibration

Calibration of TOFABS instruments can be accomplished experimentally or theoretically. In this section the theoretical calibration of an instrument is described and example results are compared to experimental data.

The motion of an aerosol particle in a gas jet is described by the particle equation of motion in the axial direction

$$m \frac{dv_p}{dt} = f(v_f - v_p)$$

where m is the particle mass, v_p the axial particle velocity, t the time, f the particle friction coefficient and v_f the axial gas velocity. For a spherical particle of radius a and mass density ρ_p

$$f/m = 9 \ \mu\kappa/2a^2\rho_p C_s$$

where μ is the local gas viscosity, κ the dynamic shape factor that corrects for non-negligible particle Reynolds number and non-spherical particle shape (3,5), and C_s the local slip correction factor (6,7).

Use of the theory of isentropic flow of a perfect gas to obtain v_f in the nozzle (8) and the solution of Ashkenas and Sherman (9) for the flow in the expanding free jet in the vacuum chamber provides the fluid velocity near the axis throughout the nozzle - free jet system.

The particle velocity is obtained by solving the above equation of motion

$$v_p(t) = e^{-\int_o^t \alpha d\phi} \left\{ v_f(o) + \int_o^t e^{\int_o^\theta \alpha d\phi} v_f(\theta)\alpha(\theta)d\theta \right\}$$

where $\alpha = f/m$ and ϕ and θ are dummy variables. Axial particle location at time t is given by

$$x(t) = x(o) + \int_o^t v_p(t)dt.$$

Thus, an iterative approach based on these expressions is used to obtain axial particle velocity vs. axial position for various particle sizes, shapes, orientations and masses. The particle TOF across a specified path length in the vacuum chamber is easily calculated from these data. Example results for unit mass density spheres are compared with experimental data in Fig. 5. Several example calculations of this sort are described by Dahneke (3).

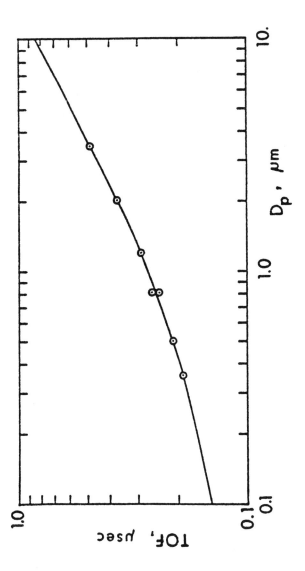

Figure 5. Comparison of experimental and analytical calibrations
for aerosols of latex spheres and the TOFABS of Fig. 3.

Particle Detection by Electron Beam Scattering

TOF spectroscopy of aerosol particles has been limited to the particle size range above 0.1 μm diameter because present methods of detecting the passage of the high-speed beam particles cannot detect smaller particles. Recent calculations (3) suggest the measurable size range can be extended to include particles about an order of magnitude smaller by detecting the passage of charged particles through a focused electron beam of energy less than 10 eV. As the previously charged beam particles pass through the electron beam, they deflect the electron trajectories by coulombic interaction causing scattered electron signals which can be detected by use of an electron multiplier.

The amplitude of the signal obtained when a particle of N elementary charge units (positive or negative) passes through a sheet electron beam of thickness 2y and current density J_o is given by (3)

$$S = (\pi/4)(bL/y)^2 J_o$$

where $b = 2Ne^2/(mu^2)$, e is the magnitude of the electronic charge, $mu^2/2$ is the kinetic energy of the beam electrons and L is the axial distance from the scattering particle to the plane of the forward scattered electron detector. This expression for S assumes a parallel electron beam (i.e., y = constant) as well as other assumptions (3). Nevertheless, it provides a simple expression which has adequate accuracy to estimate the signal levels obtainable.

Background noise in such a system is due to poorly focused electrons and to electron scattering from beam collimators and from background gas molecules in the vacuum chamber. Noise due to the former cause can be minimized by proper design. Noise due to scattering from background gas molecules is sufficiently low if the chamber pressure is sufficiently low. The ratio of signal to noise due to electron scattering from background gas molecules is

$$S/N_{bg} = \frac{(\pi/4)(bL/y)^2 J_o}{2yw L_o n\sigma\eta_{fs} J_o}$$

where $2ywL_o$ is the volume of the electron beam in view of the scattered electron detector, n the number density of the background gas molecules

$$n = 3.2 \times 10^{16} \text{ p molecules/cm}^3 \text{ torr}$$

for air at pressure p and normal temperature, $\sigma \approx 10^{-15}$ cm^2 the scattering cross-section of the gas molecule and η_{fs} the average coefficient of forward scattering onto the detector.

For example, consider the case of a .01 μm radius water droplet charged to its Rayleigh limit N = 200. Assume L/y = 100,
$2ywL_o = 10^{-2}$ cm^3 and η_{fs} = 0.3. These values give

$$S/N_{bg} \approx 10^{-5} \text{ torr/p.}$$

Thus, a pressure of $p = 10^{-6}$ torr gives a signal to noise ratio of
10. Since the assumed parameter values are reasonable and vacuum
chamber pressures of 10^{-6} torr are not difficult to obtain, the
detection of water droplets and other particles of ~ .01 μm radius
and larger by low energy electron beam scattering is feasible if
one final condition is met.

Since it is necessary to detect each passage of a particle
through the electron beam, at least several scattered electrons
must be contained in each signal. That is, in addition to an adequate signal to noise ratio, each signal must possess an adequate
number of electrons. For the above example of a particle with
N = 200 and for a transit time of 0.1 μsec through an electron
beam of 3 eV energy and 30 μamp/cm^2 current density, the electrons
scattered into the detector exceed 30 per signal.

Thus, TOF spectroscopy for the sizing of small particles by
low energy electron beam scattering seems feasible. This technique
may, for the first time, provide the basis of a single instrument
capable of measuring the distribution of aerosol particles in size
over the full respirable size range.

Because intermediate and high energy electron beams are more
conveniently generated and focused, the use of these beams to detect small particles has also been investigated. For such beams,
particle detection by forward scattering of electrons does not
seem promising because the electron trajectories are deflected so
slightly for these higher energy beams. Therefore, the signal due
to backscattered electrons and secondary electron emission was investigated. The investigation of particle detection by electron
backscattering is summarized in the Appendix. This investigation
shows that the use of intermediate energy electron beams to detect
small particles is not practical because in order to obtain adequate signal levels the electron beam energy and current are so
high the particles would be vaporized. For low energy electron
beams, the heating of particles is not significant.

Composition Analysis of Aerosol Beam Particles

Combination of TOF spectroscopy and a suitable method of

measuring particle composition in an on-line instrument would provide the capability of fully characterizing the aerosol particles
with respect to their size and chemical distributions.

Two methods of obtaining particle composition information
seem readily compatible with TOF spectroscopy. These are surface
ionization mass spectroscopy (SIMS) and thermal analysis.

Davis (10) and Myers and Fite (11) were the first to demonstrate the use of SIMS in the composition measurement of airborne
particulates. They showed that SIMS of airborne particulates is
capable of measuring many compounds present in the particles but
cannot resolve the signals due to individual particles present at
high concentrations because of the relatively long time required
for particle vaporization on the heated filament. Thus, to obtain
size - composition distributions for concentrated aerosols, a "continuous" SIMS composition analysis scanning the various compounds
present could be made sequentially for several size sub-ranges.
These size subranges could be selected by use of a velocity selector, by aerodynamic deflection of the aerosol beam so that only a
desired aerodynamic size range obtains a certain trajectory or by
the coupling of TOF spectroscopy with a suitable gating device
that passes only particles within a selected TOF range. We are
currently assembling in our laboratory a device of the first type
using a velocity selector to size sort the aerosol and SIMS to
sequentially analyze the various size sub-ranges.

Husar (12) has demonstrated the application of thermal analysis to the measurement of aerosol composition. He showed the
light scattering coefficient of laboratory aerosol samples varies
with temperature of the sample pre-heater chamber in characteristic
ways for various aerosol materials. In measurements of atmospheric
aerosols he was able to obtain qualitative information about the
aerosol composition. He concludes the method is promising but requires fundamental studies to provide more quantitative results.

Improved techniques for on-line analysis of aerosol particles
are needed.

References

1. Dahneke, B.: Nature Phys. Sci. 244 (1973) 54.
2. Dahneke, B.: Time of flight aerosol beam spectrometry: a new
    technique for measuring airborne particulates, Atmospheric
    Pollution (M. Benarie, Ed.) Elsevier, Amsterdam (1976).
3. Dahneke, B.: Aerosol beams, Recent Developments in Aerosol
    Science (D. Shaw, Ed.) Wiley-Interscience, New York (in press).
4. U.S. Patent 3,854,321.

5. Dahneke, B.: Aerosol Sci. 4 (1973) 139.
6. Dahneke, B.: Aerosol Sci. 4 (1973) 147.
7. Dahneke, B.: Aerosol Sci. 4 (1973) 163.
8. Shapiro, A.H.: The Dynamics and Thermodynamics of Compressible Fluid Flow, Vol. 1, Ronald Press, New York (1953).
9. Ashkenas, H. and Sherman, F.: The structure and utilization of supersonic free jets in low density wind tunnels, Fourth International Symposium on Rarefied Gas Dynamics, Vol. 2, (J.H. deLeeuw, Ed.) Academic, New York (1966).
10. Davis, W.D.: Continuous mass spectrometric analysis of particulates using surface ionization, Rpt. N. 76CRD069, General Electric Corporate Research and Development, Schenectady (1976). See also article by W.D. Davis in this volume.
11. Myers, R. and Fite, W.: Environmental Sci. and Technology 9 (1975) 334.
12. Husar, R.B.: J. Thermal Analysis 10 (1976) 183.
13. Pierce, J.R.: Theory and Design of Electron Beams, Van Nostrand, Princeton (1954).
14. Oatley, C.W.: The Scanning Electron Microscope, Cambridge (1972).
15. Dahneke, B., Flachsbart, H., Mönig, F.J. and Schwarzer, N.: Cooling of particles in aerosol beams, in Rarefied Gas Dynamics (K. Karamcheti, Ed.) Academic, New York (1974).

APPENDIX

Particle Detection by Scattering of Electron Beams

Detection of aerosol beam particles by scattering of low energy electron beams seems to be feasible. However, low energy beams (< 10 eV) are difficult to generate and focus. Intermediate energy beams (100 to 5000 eV) are more easily generated and focused and would therefore allow the system and associated theory to be simpler. This investigation considers the possible use of intermediate energy electron beams to detect small aerosol beam particles.

A schematic diagram of a possible detector system is shown in Fig. 6. A horizontal aerosol beam passing through the vacuum chamber is intersected by vertical sheet electron beams at K and K'. These originate from the heated tungsten ribbon filaments near the cathode A, are accelerated by the potential V_A between the cathode and the grounded collimating plate B, are focused by the electrostatic slit lenses CDE, pass through the slits in the shield G and are absorbed in the trap JIH.

When a particle passes through the electron beams, first at K and then K', electrons are backscattered by the particles through

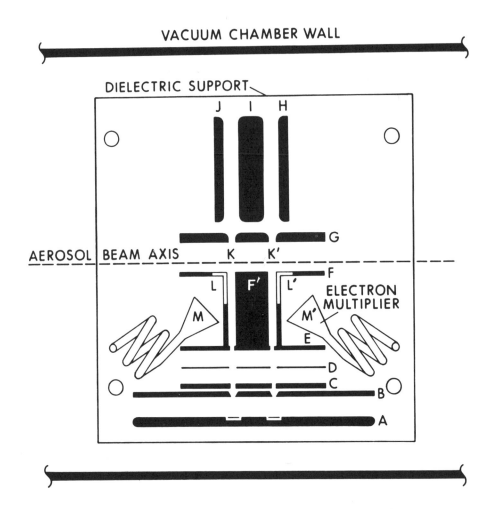

Figure 6.   Schematic diagram of an electron beam
            system for measuring the TOF of aerosol
            beam particles.

the apertures L and L' and collected in the electron multipliers M
and M'. The passage of each particle through the electron beams
can be detected in this way provided the magnitude of the scattered
electron signal is sufficiently high.

The calculations summarized in the following sections each
pertain to the calculation of the signal and noise levels to determine the feasibility of the detection of aerosol beam particles by
backscattering of an electron beam.

Electron Beam Generation

The electron beams are generated by thermal emission of electrons from the resistance heated filaments. Upon emission the
electrons are accelerated by the negative cathode potential V_A towards the grounded collimating plate B containing two collimating
slits .05 cm wide by 1 cm long.

The beam currents are controlled by space charge limited emission. That is, the beam currents are limited by the cathode potential V_A and not by the filament temperatures. This mode is desirable for several reasons including convenient control of the beam
currents and minimal effect of local temperature variation of the
filament surface.

For space charge limited current between two parallel plane
electrodes, the current density is given by the Child-Langmuir law

$$J(x) = \alpha V^{3/2}/x^2 \qquad\qquad [1]$$

where J is the current density (amps/cm^2), x the distance from the
cathode plane (cm), V(x) the local potential referenced to the
cathode potential (volts) and α the proportionality constant
2.335×10^{-6}. Table I shows current density and current vs. cathode
potential as calculated with the Child-Langmuir law for the system
of Fig. 6 with anode-collimator spacings of x = 0.2 and 0.5 cm.

Electrostatic Focusing of Sheet Electron Beams

Consider the electrostatic lens consisting of the three slit
apertures in C, D and E in Fig. 6. The lens for one of the electron beams is also shown in Fig. 7. The potentials V_C, V_D and V_E
are referenced to the cathode potential. Thus, although V_C and
V_E are both grounded, they have the positive values $V_1 = -V_A$ while
$V_2 = V_D - V_A$.

For the two-dimensional case of the present problem the par-
axial electron equation of motion is (13)

$$2\,V\,y'' + V'y' + V''\,y = 0 \qquad\qquad [2]$$

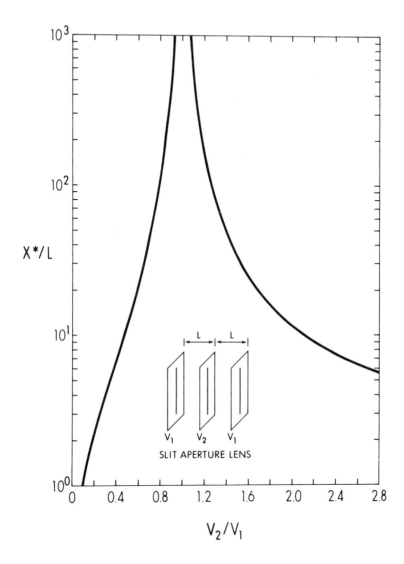

Figure 7.   Location of zero thickness of a
            focused sheet electron beam.

Table I. Calculated current densities at B and beam currents vs.
          cathode potential V_A.

V_A, volts	A-B spacing = 0.2 cm		A-B spacing = 0.5 cm	
	J_B, mamp/cm^2	I, mamp	J_B, mamp/cm^2	I, mamp
-100	58.4	2.92	9.34	0.468
-200	165.	8.26	26.4	1.32
-500	653.	32.6	104.4	5.22
-1000	1846.	92.4	295.4	14.78
-2000	5221.	261.0	835.4	41.8
-5000	20639.	1032.0	3302.	165.2

where y is the distance from the beam axis plane, $V = V(x)$ the
local potential on the axis ($y = o$), x the axial distance and the
primes denote differentiation with respect to x.

Application of [2] to the lens of Fig. 7 gives for the focal
length f

$$-y_o'/y_o = 1/f = (3/8L)(V_2/V_1-1)[4 -(V_2/V_1)^{\frac{1}{2}}- 3(V_1/V_2)^{\frac{1}{2}}] \qquad [3]$$

where y_o' and y_o are the slope and location of the beam envelope
at the lens exit and L is the axial separation of the lens
elements. A parallel beam was assumed at the lens entrance.

Influence of Space Charge on Electron Beam Geometry

The influence of space charge on electron beam geometry has
been described by Pierce (13). For two-dimensional (sheet) electron beams the expression for the beam half-width y(x) is

$$y(x) = y_o + y_o' x + \frac{\eta J_o y_o}{2\varepsilon u^3} x^2 \qquad [4]$$

where y_o, y_o' and J_o are, respectively, the half-width, slope and
current density of the beam at $x = 0$, η is the electron charge to
mass ratio, ε the permittivity of free space and u the axial
electron velocity.

Accordingly, the axial location of minimum beam thickness is

$$x^* = -(y_o'/y_o)(\varepsilon u^3/\eta J_o). \qquad [5]$$

If the electron lens conditions are controlled so that

$$-y_o'/y_o = (2\eta J_o/\varepsilon u^3)^{\frac{1}{2}} \tag{6}$$

then a monoenergetic "parallel" sheet electron beam reaches zero
thickness at x^*. Of course, because of variations in thermal
energy of the electrons and lens distortions, beams of zero thickness are never attainable. The actual electron beam thickness is
discussed below.

Combination of expressions [3], [5] and [6] gives the relation
between the spacing and potentials of the lens elements and the
location of zero beam thickness

$$x^*/L = (16/3)(V_2/V_1 - 1)^{-1}$$
$$[4 - (V_2/V_1)^{\frac{1}{2}} - 3(V_1/V_2)^{\frac{1}{2}}]^{-1}. \tag{7}$$

A plot of this function is shown in Fig. 7 and Table II lists
values of V_1 and V_2 for the system of Fig. 6 for which $x^*/L = 5$.

Electron Beam Thickness

The thickness of the electron beam is necessary to the calculation
of both signal and noise levels since the beam current density is
required in these calculations. Because of lens distortions and
variation in thermal energy, electron beams always possess a finite minimum thickness. A complete analysis of the beam thickness
is beyond the scope of this study. Rather, we shall only consider
the minimum beam thickness due to spread in the axial energy of
the beam electrons. This thickness provides a correct order of
magnitude value for the estimation of the signal and noise levels.

In the field free region beyond the lens, the electrons have nominal axial velocity

$$u = (2\eta V_1)^{\frac{1}{2}} \tag{8}$$

Table II. Lens element potentials for various cathode potentials.

V_A, volts	V_C, volts	$V_1 = V_C - V_A$, volts	$V_2 = V_D - V_A$, volts	V_D, volts
-100	0	100	35	-65
-200	0	200	70	-130
-500	0	500	175	-325
-1000	0	1000	350	-650
-2000	0	2000	700	-1300
-5000	0	5000	1750	-3250

Variations in electron thermal velocity in the axial direction
cause a spread in u which can be equivalently regarded as a spread
in V_1.

According to [5] and [6] the minimum half-thickness of the
electron beam of energy V_1 is

$$y_{min} = y_o - \varepsilon u^3 y_o'^2 / (2\eta J_o y_o). \qquad [9]$$

If the lens is adjusted so that a monoenergetic beam of electrons
having energy equal to the maximum electron energy present in the
beam would obtain zero thickness at the prescribed axial location
x^*, electrons of lower energy will not obtain zero beam thickness
but rather the half thickness given by

$$y_{min} = \frac{dy_{min}}{dV_1} \Delta V$$

where $\Delta V < 0$. The quantity $|\Delta V|$ corresponds to the spread in
thermal energy of the beam electrons.

By use of [3], [6], [8] and [9] we obtain

$$\frac{dy_{min}}{dV_1} = \frac{\partial y_{min}}{\partial y_o'} \frac{dy_o'}{dV_1} + \frac{\partial y_{min}}{\partial u} \frac{du}{dV_1} \cong - (11/2) \; y_o/V_1.$$

Thus, the minimum beam thickness is

$$2y_{min} = - 11 \; y_o \Delta V/V_1.$$

Table III shows calculated values of $2y_{min}$ when $y_o = .05$ cm,

Table III. Estimated values of beam thickness, current density
and particle transit time through the beam near the
beam focus.

V_A volts	$2y_{min}$ um	J_{focus}, A-B spacing = 0.2 cm amp/cm^2	J_{focus}, A-B spacing = 0.5 cm amp/cm^2	τ, sec.
-100	55.0	0.53	0.085	1.8×10^{-7}
-200	27.5	3.01	0.480	9.2×10^{-8}
-500	11.0	29.7	4.75	3.6×10^{-8}
-1000	5.5	168.	26.9	1.8×10^{-8}
-2000	2.75	949.	152.	9.2×10^{-9}
-5000	1.10	9380.	1500.	3.6×10^{-9}

$x^*/L = 5$ and $\Delta V = -1$ volt. Calculated current densities and particle transit times through the electron beam for an assumed particle velocity of 3×10^4 cm/sec are also shown in Table III.

These estimates of electron beam thickness are larger than the
minimum value obtainable when the lens potentials are adjusted to
minimize the electron beam thickness. When this is done the envelope defined by the electrons of maximum axial velocity which
cross the axis can be made to coincide at x^* with the envelope defined by the electrons of minimum axial velocity which do not cross
the axis. The estimates of Table III are, however, adequate for
the present calculations.

Signal to Noise Ratio

The backscattered electron signal levels can be estimated by
use of the data of Tables I and III and the other pertinent parameters, namely, the particle cross-section πr^2, the backscattering
coefficient for the particle η_{bp} and the fraction of the total
solid angle occupied by the electron multiplier detector δ. The
electron current backscattered into the electron multiplier is

$$S \cong \delta \eta_{bp} \pi r^2 J_{focus} \qquad [10]$$

where J_{focus} is the current density in the electron beam at the
beam focus. Table IV shows values of S for spherical particles of
radius r, $\delta = .08$ and $\eta_{bp} = 0.1$ (14).

Since we wish to detect each event in which a particle passes
through the electron beam, we require a minimum of several electrons
per signal. For example, when the particle radius is .01 μm, this
condition is minimally met for $V_A = -5000$ volts with an A-B spacing
of 0.2 cm. In other words, to detect .01 μm radius particles by
backscattered electrons, high energy and high current beams are
required and even then the signal is marginal. Larger particles
can be detected with lower energy, lower current beams.

To estimate the noise level, consider only noise due to electrons backscattered from the background gas molecules in the vacuum
chamber. Backscattered electrons from forward scattered electrons
striking the shield G may also contribute to the noise, perhaps in
a comparable amount. The noise current scattered by background
gas into the electron multiplier is

$$N = \delta_g \eta_{bg} n' \sigma J_{focus} \qquad [11]$$

where n' is the number of background gas molecules in the electron
beam volume "seen" by the detector at average solid angle fraction
δ_g, η_{bg} the backscattering coefficient and σ the scattering

Table IV. Estimated signal levels.

V_A, volts	r, μm	A-B spacing = 0.2 cm		A-B spacing = 0.5 cm	
		S, amp	τS, electrons	S, amp	τS, electrons
-100	.01	1.4×10^{-14}	1.6×10^{-2}	2.2×10^{-15}	2.5×10^{-3}
	0.1	1.4×10^{-12}	1.6	2.2×10^{-13}	2.5×10^{-1}
	1.0	1.4×10^{-10}	1.6×10^{2}	2.2×10^{-11}	2.5×10^{1}
	10.0	1.4×10^{-8}	1.6×10^{4}	2.2×10^{-9}	2.5×10^{3}
-200	.01	7.5×10^{-14}	4.3×10^{-2}	1.2×10^{-14}	6.9×10^{-3}
	0.1	7.5×10^{-12}	4.3	1.2×10^{-12}	6.9×10^{-1}
	1.0	7.5×10^{-10}	4.3×10^{2}	1.2×10^{-10}	6.9×10^{1}
	10.0	7.5×10^{-8}	4.3×10^{4}	1.2×10^{-8}	6.9×10^{3}
-500	.01	7.5×10^{-13}	1.7×10^{-1}	1.2×10^{-13}	2.7×10^{-2}
	0.1	7.5×10^{-11}	1.7×10^{1}	1.2×10^{-11}	2.7
	1.0	7.5×10^{-9}	1.7×10^{3}	1.2×10^{-9}	2.7×10^{2}
	10.0	7.5×10^{-7}	1.7×10^{5}	1.2×10^{-7}	2.7×10^{4}
-1000	.01	4.2×10^{-12}	4.8×10^{-1}	6.5×10^{-13}	7.5×10^{-2}
	0.1	4.2×10^{-10}	4.8×10^{1}	6.5×10^{-11}	7.5
	1.0	4.2×10^{-8}	4.8×10^{3}	6.5×10^{-9}	7.5×10^{2}
	10.0	4.2×10^{-6}	4.8×10^{5}	6.5×10^{-7}	7.5×10^{4}
-2000	.01	2.4×10^{-11}	1.4	3.8×10^{-12}	2.2×10^{-1}
	0.1	2.4×10^{-9}	1.4×10^{2}	3.8×10^{-10}	2.2×10^{1}
	1.0	2.4×10^{-7}	1.4×10^{4}	3.8×10^{-8}	2.2×10^{3}
	10.1	2.4×10^{-5}	1.4×10^{6}	3.8×10^{-6}	2.2×10^{5}
-5000	.01	2.3×10^{-10}	5.3	3.7×10^{-11}	8.4×10^{-1}
	0.1	2.3×10^{-8}	5.3×10^{2}	3.7×10^{-9}	8.4×10^{1}
	1.0	2.3×10^{-6}	5.3×10^{4}	3.7×10^{-7}	8.4×10^{3}
	10.0	2.3×10^{-4}	5.3×10^{6}	3.7×10^{-5}	8.4×10^{5}

cross-section for the background gas molecules. For air at normal temperature, use of the perfect gas law gives

$$n' = 3.2 \times 10^{16} \frac{\text{molecules}}{\text{cm torr}} \, y_{min} \, p$$

for an electron beam region of 1 cm axial length by 0.5 cm wide.

The signal to noise ratio S/N is given by the ratio of [10] to [11].

$$S/N = \delta \, n_{bp} \, \pi r^2 / \delta_g n_{bg} \, n' \sigma \qquad [12]$$

If we assume $\delta \cong \delta_g$ and $n_{bp} \cong n_{bg}$ we obtain for the background gas pressure

$$p = (\pi/3.2) \, \frac{10^{-16} \, r^2}{y_{min}} \, \sigma \, S/N$$

where p is in torr, y_{min} in cm and r^2 and σ both in cm^2. As an
example, if $\sigma \approx 10^{-16} cm^2$, $y_{min} \approx 10^{-4}$ cm and we require $S/N \geq 10$ when
$r = .01$ μm we obtain the condition $p \lesssim 10^{-9}$ torr. Operation at
such low pressure is not convenient.

Heating of Particles in the Beam

Since relatively high energy, high current electron beams are
required to detect small particles, these particles may be substantially heated and quickly vaporized. Significant loss of particles
by this mechanism would preclude the detection of small particles
by backscattering of electron beams. It is therefore essential to
consider the heating and vaporization of particles in the electron
beam.

The heating process can be regarded most simply as the absorption of high energy electrons by the particle followed by
emission of much lower energy electrons. As the simplest approximation of the particle heating rate, we accordingly assume the
total energy of each electron striking the particle is converted to
thermal energy of the particle. Thus, the particle heating rate
is given by

$$Q = \pi r^2 \, \varepsilon \, J_{focus}$$

where ε is the electron kinetic energy.

Because the heating of the particle is highly non-uniform, all
of the thermal energy being absorbed by a thin surface layer on
the face of the particle exposed to the electron beam, and because
the particle is in a vacuum environment, local evaporation of molecules will occur. If L_m is the molecular latent heat of vaporization, the rate of vaporization of the particle (molecules/sec) is

$$E \cong \pi r^2 \, \varepsilon \, J_{focus}/L_m.$$

The number of molecules in a particle is $N = 4\pi r^3 n_p /3$, where n_p is
the molecular number concentration of the bulk particle material,
so that

$$E = - \frac{dN}{dt} = - 4\pi r^2 n_p \frac{dr}{dt} = \pi r^2 \, \varepsilon \, J_{focus}/L_m.$$

Thus, the particle radius is given by

$$r(t) = r_o - \varepsilon\, J_{focus}\, t/(4n_p L_m)$$

where r_o is the initial particle radius at the instant the particle enters the electron beam ($t = 0$). For particle transit time τ through the beam the fractional decrease in particle radius is

$$[r_o - r(\tau)]/r_o = \varepsilon\, J_{focus}\, \tau/(4n_p L_m r_o).$$

The Table III values of J_{focus} and τ were used to estimate fractional decrease in the radius of water droplets at 0°C for which $n_p = 3.34\times10^{22}$ molecules/cm^3 and $L_m = 4.64\times10^{-4}$ eV/molecule. The results are shown in Table V. Fractional decrease values greater than unity mean the droplet was entirely evaporated.

Table V. Fractional decrease in droplet radius during transit through the electron beam.

V_A, volts	J_{focus} amp/cm^2	r_o, μm .01	0.1	1.0	10.
-100	0	6.3×10^{-3}	6.3×10^{-4}	6.3×10^{-5}	6.3×10^{-6}
	0.085	1.6×10^{-1}	1.6×10^{-2}	1.6×10^{-3}	1.6×10^{-4}
	0.53	9.8×10^{-1}	9.8×10^{-2}	9.8×10^{-3}	9.8×10^{-4}
-200	0	3.2×10^{-3}	3.2×10^{-4}	3.2×10^{-5}	3.2×10^{-6}
	0.48	8.9×10^{-1}	8.9×10^{-2}	8.9×10^{-3}	8.9×10^{-4}
	3.01	5.6	5.6×10^{-1}	5.6×10^{-2}	5.6×10^{-3}
-500	0	1.2×10^{-3}	1.2×10^{-4}	1.2×10^{-5}	1.2×10^{-6}
	4.75	8.6	8.6×10^{-1}	8.6×10^{-2}	8.6×10^{-3}
	2.97×10^1	5.4×10^1	5.4	5.4×10^{-1}	5.4×10^{-2}
-1000	0	6.3×10^{-4}	6.3×10^{-5}	6.3×10^{-6}	6.3×10^{-7}
	2.69×10^1	4.98×10^1	4.98	4.98×10^{-1}	4.98×10^{-2}
	1.68×10^2	3.1×10^2	3.1×10^1	3.1	3.1×10^{-1}
-2000	0	3.2×10^{-4}	3.2×10^{-5}	3.2×10^{-6}	3.2×10^{-7}
	1.52×10^2	2.8×10^2	2.8×10^1	2.8	2.8×10^{-1}
	9.49×10^2	1.8×10^3	1.8×10^2	1.8×10^1	1.8
-5000	0	1.2×10^{-4}	1.2×10^{-5}	1.2×10^{-6}	1.2×10^{-7}
	1.50×10^3	2.7×10^3	2.7×10^2	2.7×10^1	2.7
	9.38×10^3	1.7×10^4	1.7×10^3	1.7×10^2	1.7×10^1

Also shown in Table V for comparison is the fractional decrease in water droplet radius due to evaporation of the droplet at a constant uniform temperature of 0°C denoted as the case $J_{focus} = 0$. Because of cooling, the evaporation rate will actually be less (15). In any case, the values of Table V demonstrate the evaporation rate of the unheated droplets is negligible compared to those heated in the electron beam.

For the higher energy electron beams, the small droplets are completely vaporized in passing through the beam. High energy electron beams can therefore not be used for TOF spectroscopy of small droplets.

Conclusions

The detection of small particles by backscattering of electrons as the particles pass through an electron beam does not seem practical. In order to obtain adequate signal to detect the small particles, relatively high energy, high current electron beams are required. Small particles are quickly vaporized in such beams, precluding the use of such beams to detect them.

Acknowledgement: The author expresses his thanks for grant support to the National Institute of Environmental Health Sciences, to the Environmental Protection Agency and to the Energy Research and Development Administration. This paper has been assigned Report No. UR-3490-1194.

Discussion

HUNTZICKER - How precisely can you predict or determine the charge on an aerosol particle? You mentioned a hundred or two hundred charges per particle. Is the scattered electron signal sensitive to the number of charges?

DAHNEKE - It's proportional to the number of charges squared, so the signal is rather sensitive to the number of charges. Hall and Beeman [J. Appl. Phys. 47: 5222 (1976)] measured the charge obtained by polystyrene latex particles in an aerosol beam passed through an electron beam of about a hundred volts energy. They found the saturation charge for the case of half micron particles to be about 960 elementary charges and for three tenths micron particles it was about six hundred elementary charges. From these values they calculated the saturation surface potential for the polystyrene spheres to be five volts. When a particle's surface potential exceeded this characteristic limiting value, electrons were ejected so the particles reached a saturation charge level, 5 volts for the latex spheres. If one knows the saturation surface potentials for various compounds of interest, one could calculate the limiting particle charge. I think this would be an appropriate and necessary study investigating the detection and measurement of aerosol beam particles by electron beam scattering, and we hope to make such a study.

LODGE - It looked to me as though you had some fine structure on your quadruplet peak obtained with the two beam instrument [Nature Phys. Sci. 244: 54 (1973)]. Is that due to formation of the various possible four-particle structures, viz., linear chain tetrahedra, etc.?

DAHNEKE - I think it could be due to formation of these various structures. At the same time, since there weren't enough quadruplet particles to fill out the distribution there must be some noise because of the statistics.

LODGE - Didn't Stöber and Flachsbart [Environ. Sci. Tech. 3: 1280 (1969)], among others, feel they saw the separation of the different structures; that is, the linear, the planar and others?

DAHNEKE - Yes they did. I don't know if we can get the kind of resolution they obtained in their instrument . We can certainly distinguish between the singlets and the doublets because we saw sharp, distinct peaks for these two cases representing a 19% change in aerodynamic size. We could resolve size differences down to five percent. That's more than adequate for classifying a particle in most cases.

LODGE - I noticed that in the triplet peak you seem to be getting some broadening compared to the singlet and doublet peaks that could be interpreted as looking at the difference between linear and triangular configurations which would have somewhat different aerodynamic sizes.

DAHNEKE - Yes, it could be. On the other hand, I wouldn't make any firm conclusions about the occurrence of structural differences because these monodisperse aerosols are not really monodisperse. There is a certain finite width to their size and when you start putting three together you might compound rather small size differences into broader ones by adding together particles at the same extreme in the size distribution.

BEAMAN - If you had fibers of varying diameters and lengths, what would be the effect of their orientation?

DAHNEKE - We've never tried fibers in an aerosol beam and I think there is a very genuine question as to whether they would come out straight because of the high orientation dependence of their aerodynamic properties. An inclined particle could experience a large lift force in addition to the drag force in the high relative velocity between the particle and the gas. If a fiber was inclined it could be lifted right out of the beam. Cylindircal fiber particles tend to orient parallel to the flow axis in hydrodynamic flow but there is a statistical distribution. I think about 75% of them tend to be oriented parallel and the others tend to be oriented in the perpendicular direction.

ROBILLARD - If you were to use the electron beam assay for studying a sample that had particles of different size and composition and thus obtained different charging potentials, would you be able to distinguish composition differences from differences in particle size?

DAHNEKE - Such an approach seems possible and may be useful, since additional methods would be used to determine particle size. It would be rather nonspecific, I'm afraid, because I suspect variations in the charging of a given size particle would be small for a lot of compositions and I therefore don't think it would be possible to precisely determine what the composition was. Of course, one could see gross differences, but particles of the same charge may have significantly different composition. I probably should say I don't know enough to really answer that question.

SMITH - Since your laser beam has a wavelength comparable to the particle size, did you attempt any diffraction pattern measurements? Or instead of just scattered signal intensity at one angle, did you attempt moving the angle?

DAHNEKE - Well, first of all, I think you could indeed get information of an optical nature using our instruments. But our interests in characterizing particles relates to such things as particle penetration into the lung so that aerodynamic sizing of the particles is more pertinent. That, coupled with the fact that I don't know very much about optics, leads me to look in the other directions.

SMITH - Has anybody tried hooking a mass spectrometer onto the end of your jet?

DAHNEKE - Yes, and that question serves as a good introduction to our next speaker. I first heard Bill Davis talk on the topic of mass spectrometry of aerosol beam particles in 1973 in San Francisco at the American Society for Mass Spectrometry and Allied Topics Meeting and his talk was very well received at that time and has been a frequently cited and highly useful work.

CONTINUOUS MASS SPECTROMETRIC ANALYSIS OF

ENVIRONMENTAL POLLUTANTS USING SURFACE IONIZATION

William D. Davis

General Electric Company
Corporate Research and Development
Schenectady, New York 12301

ABSTRACT

An instrument for real-time analysis of trace impurities in
air, including particulates was constructed and evaluated. The
sampling system impinges a beam of particles and air on a heated
Re ribbon and the resulting ions analyzed in a 7.6 cm radius magnetic sector mass spectrometer. Oxygen in the air raises the
work function of the Re to about 7.2 eV at 1,000°K allowing analysis of elements and compounds with ionization potentials as high
as 8 eV. For some low ionization potential elements, particles
containing 1,000 atoms, particle concentrations of less than 1
per cm^3 and average concentrations of 10^{-12}-10^{-13} g/m^3 can be
detected. Organic compounds can have ionization efficiencies as
high as 20% and can be detected in concentrations as low as
0.001 ppb. Water is analyzed by forming an aerosol. Alkali metals and other easily ionized metals can be detected at concentrations of 10^{-13} g/cm^3 of water.

INTRODUCTION

Positive ions and neutrals are emitted from a heated surface
according to the Saha-Langmuir equation:

$$\frac{n_+}{n_o} = \frac{g_+}{g_o} \exp \left(\frac{\phi - IP}{RT} \right)$$

where n_+/n_o is the ratio of positive ions to neutral atoms, ϕ is
the work function of the surface and IP is the ionization poten-

tial of the evaporating atom or molecule. g_+/g_0 is the ratio of
the statistical weights of the ion and atom and has a value of
the order of unity. Mass spectroscopists often use this method
of ionization for the analysis of solids. Particulates are usually analyzed by drawing air through a filter or other separator
and then transferring the sample to the ionizing filament. However, by impinging the air stream directly against a heated ionizing filament, real-time analysis is obtained and transfer problems are eliminated. At the same time, the added oxygen adsorbs
on the filament and increases its work function. The ordinary
gaseous components of the air have ionization potentials much too
high to be ionized and hence do not present any interference problems. At sufficiently high temperatures, each particle as it hits
the filament produces a burst of ions. The number of ions in the
burst gives a measure of the amount of that element in the particle and the number of bursts per second is proportional to the
number of particles per cm^3 in the air.

Gaseous impurities in the air with low ionization potentials,
primarily organic compounds, can be detected at very low concentrations because of the high ion currents obtainable with surface
ionization and the lack of interfering "air" ions. Impurities in
water can also be determined by atomization, drying and then analyzing the resulting aerosol.

This report describes the apparatus used to study these techniques and some of the results obtained[1]. An independent investigation along the same lines was conducted by Myers and Fite[2].

APPARATUS

The basic ion source and mass spectrometer used in these studies was of conventional design. The sampling system and ion
source used for particulates is shown in Fig. 1. The air sample
first passes through a needle valve which serves to adjust the
air flow to the optimum value. The air then travels down about
25 cm of 1.6 mm ID stainless steel tubing and impinges on a small
hole of about .05-.07 mm diameter. Most of the data reported
here was obtained with a .071 mm diameter hole produced by a
laser in a 0.76 mm thick sheet of stainless steel. Most of the
air is removed by a 160 liter/min mechanical pump attached to the
small chamber shown in the figure. Some of the sample particles
and air pass through the hole and impinge on a 3.2 mm X 19 mm
metal ribbon heated directly by an electric current to 600-2,000°C.
The ions produced as a particle hits the surface of the ribbon are
accelerated and focused by a simple system of three split-plate
electrodes and then mass analyzed in the usual manner. The first
pair of electrodes also served as an electron bombardment ion
source for analyzing the neutral gaseous components present in
this region. The doughnut-shaped liquid N_2 trap above the ribbon

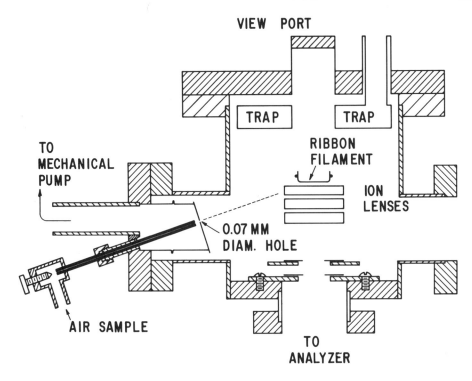

Figure 1. Diagram of sample inlet system and ion source chamber.

was occasionally used to reduce the partial pressure of interfering organic impurities in the vacuum.

    The mass spectrometer (Fig. 2) was a 7.62 cm (3 inch) radius
-90° magnetic sector type instrument with entrance and exit slits
of 0.254 mm and 0.381 mm respectively. At the normal acceleration voltage of 500V, the resolution was sufficient to separate
the mass peaks up to about mass 115. Numerous measurements both
with ions from the surface ionization filament and from the electron bombardment source showed that the overall transmission of
ions from source to detector was about 4%. The background count
rate of the electron multiplier detector was usually less than 1
count/sec. The system was evacuated by a 4 inch oil diffusion
pump and liquid nitrogen trap backed by a mechanical pump and was
normally baked at 300°C for several hours in order to reduce the
organic background.

    Only two metals, W and Re, have been used in this study. The
Re ribbon was fabricated from electron-beam zone refined material.

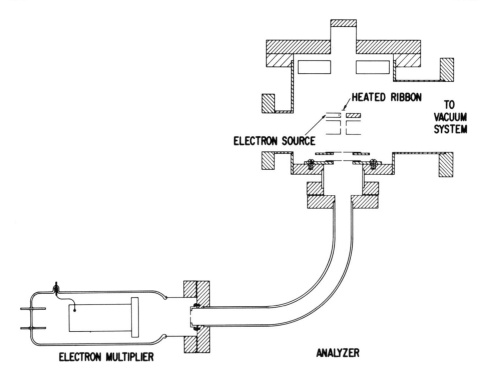

Figure 2. Diagram of complete instrument including the mass ana-
          lyzer.

The tungsten ribbon was ordinary type 218 tungsten ribbon. The
thickness of the ribbons was 0.030 mm for the Re and 0.015 mm for
the W. Ribbon temperature was measured with an optical pyrometer.
The initial studies were made with W but Re proved to be greatly
superior to W. Most of the data presented here is with the Re
ribbon.

EXPERIMENTAL RESULTS

     The amount of oxygen adsorbed on the ribbon, and hence the
increase in the work function, increases with oxygen pressure and
decreases with temperature. The pressure of air at the metal surface was limited to the range 1-5 X 10^{-5} torr. The temperature
used depended on several factors but was usually a compromise
between efficiency of ionization and rate of evaporation of the
element from the surface. The background of undesired ions also
changes with temperature and must be considered when choosing the
optimum temperature. For the elements investigated, the optimum
temperatures were in the range 800-1,500K.

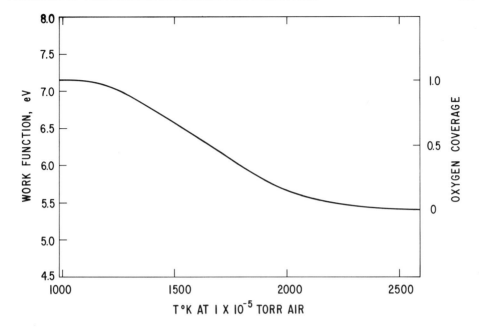

Figure 3.   Work function of Re ribbon in 10^{-5} torr air.

Fig. 3 shows the result of experimental measurements of the
work function of the Re ribbon.   The value of 5.4 eV for the clean
surface was obtained from measurements of the electron emission
of the ribbon in high vacuum.   This high value indicates that the
surface of the ribbon consists predominantly of 0001 planes[3,4,5].
A similar determination of the W ribbon yielded the accepted value
of 4.6 eV.

The high temperature results for oxidized Re were obtained by
measuring the drop in electron emission on admitting air.   The
lower temperature results were obtained by comparing the output
ion yield of Pb^+ or Bi^+ at several temperatures for an approximately constant dose of dust particles of Pb_3O_4 or $BiCO_3$ onto the
ribbon.   Another approach used was to determine the number of ions
produced at the ribbon by particles of Cr_2O_3, CuO and Pb_3O_4 of
known size.   The size was estimated by applying Stokes law for
settling velocities to a recently shaken container of the powdered
compound.   The number of ions was calculated from the measured
multiplier gain and analyzer transmission.

Fig. 4 shows the ionization efficiency $n_+/(n_o + n_+)$ calculated from the Saha-Langmuir equation using $g_+/g_o = 1$ and
the work functions indicated by the curve of Fig. 3.   It will be

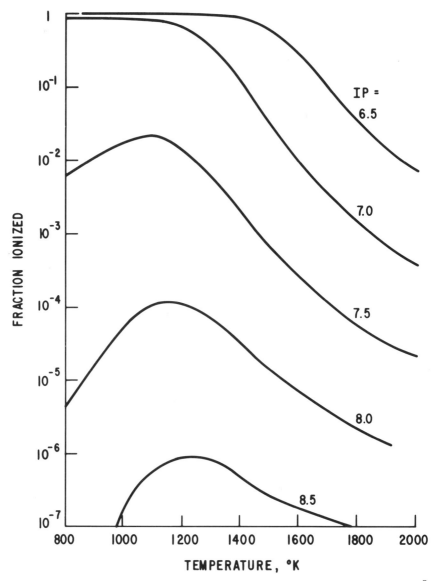

Figure 4. Calculated ionization efficiency for Re in 1X 10^{-5}
          torr of air for different ionization potentials (IP).

seen that for high ionization potential elements, a maximum in
the ionization efficiency occurs at 1,100-1,200K in agreement with
experimental observations.

The number of particles hitting the ionizing ribbon exhibited
a broad maximum at a pressure of about 3-4 X 10^{-5} torr in the ion-

ization chamber. At this point, the flow of atmospheric air at
the inlet was about 24 cm^3/sec. The particle yield was estimated
by measuring the number of particles per cm^3 of air with a calibrated condensation nuclei counter and at the same time counting
the number of particles hitting the ribbon and producing an ion
burst at the detector. For best accuracy, synthetic aerosols of
an alkali metal salt were used. The maximum particle yield measured was 0.3% but 0.2% (1 particle detected in 500 sampled) is a
more typical value. This means that a particle density in the air
of 20/cm^3 will give 1 ion burst per second at the detector. The
ion current background of the instrument normally consisted of a
random emission of individual ions rather than bursts of ions so
that detection of particle densities much less than 1 per cm^3
should be possible for many of the elements.

For most particles, the emission of metal ions showed an
abrupt rise as the particle hit the ribbon followed by an approximately exponential decay with time. This is the behavior one
would expect if the particle completely evaporated or diffused
onto the surface of the ribbon in a time short compared to the
residence time of the resulting atoms on the surface. Even at
temperatures where the decay time was of the order of tens of
microseconds, the rise to the maximum ion current was usually only
about 10 microseconds (for example $CsNO_3$ on W at 950°C).

A study of $SrCO_3$ and $SrCl_2$ particles on W showed that the
decay time of the Sr^+ burst was the same for both compounds and
was in reasonable agreement with published values for Sr on oxidized W[6]. This again indicates that the atoms of the particle
are dispersed over the surface before the ions are emitted. A
more thorough study of the dynamics of ion emission would be needed however to determine exactly how the various kinds of particles
interact with the hot surface.

Inorganic Particulates

The minimum amount of element in a particle that can be relia-
bly detected is shown in Table I for some representative elements
listed in order of increasing ionization potential. In all cases,
the limitation is due to background ions arriving at the detector,
not detector noise. Na, K, Rb and Cs containing particles give
results similar to Li with the minimum amount of element detectable being about 1,000 atoms.

Fig. 5 shows a recording of the ion bursts from U-containing
particles in the laboratory air. Uranium is unique in that it is
the only element observed to desorb as a compound ion. The amount
of element per particle is calculated from the number of detected
ions in the burst, the transmission of the analyzer (4%) and the
ionization efficiency for that element estimated from the Saha-

TABLE I - Minimum number of atoms detectable in single particles.

Element	IP, eV	Re Temperature °C	% Ions	Atoms
Li	5.4	830	100	800
Sr	5.7	1730	50	10^4
U	6.1	1070	50	1300
Cr	6.8	1500	0.5	3×10^5
Pb	7.4	910	3	2×10^5
Cu	7.7	1130	.02	7×10^6

Langmuir equation. For Uranium, even the largest ion bursts correspond to only a few thousand uranium atoms per particle indicating a low concentration of this element in the particles.

For a more abundant element such as K, Na, Li or Pb, the average ion burst is far above the background. For example, pulse height analysis of K^+ ion bursts from laboratory air particles showed the average number of K atoms per particle was 2×10^5.

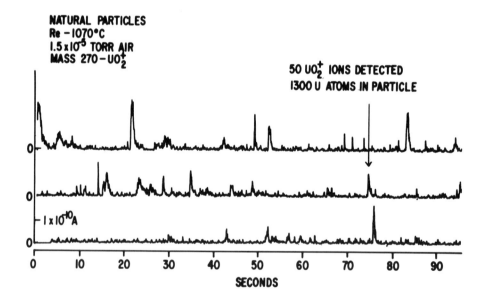

Figure 5. Bursts of UO_2^+ ions from individual particles.

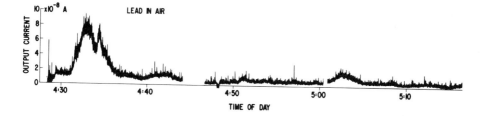

Figure 6. Continuous recording of Pb^+ signal from laboratory air.

The average concentration of an element may be estimated from the average ion current and the particle efficiency (0.2%) of the sampling system. In most cases, the ribbon temperature is lowered so as to produce a more or less steady signal rather than individual bursts of ions. The lower temperature also raises the work function of the surface and in some cases, reduces the background signal. Fig. 6 shows a recording of the Pb concentration in the laboratory using a Re temperature of 920°C. The large peak corresponding to about 0.5 µg Pb/m³ was caused by laboratory employees starting their automobiles after quitting work at 4:30. A smaller peak caused by employees quitting at 5:00 can also be observed.

TABLE II - Instrument background levels and typical concentrations of particulates in laboratory air.

Element	IP, eV	Re Temperature, °C	% Ionized	Background ng/m³	Laboratory Air, ng/m³
Cs	3.9	600	100	.002	2
Rb	4.2	600	100	.0008	.05
K	4.3	600	100	.0004	.5-200
Na	5.1	600	100	.0004	.4-40
Li	5.4	840	100	.0001	.1-6
Sr	5.7	1500	100	.02	10
U	6.1	1070	50	.003	< .003-.7
Cr	6.8	1130	50	.005	.05
Pb	7.4	910	3	.6	60-600
Cu	7.7	1000	0.1	1	< 1-15

Table II lists some representative results using this technique. The background levels shown are obtained by filtering the incoming air of all particles and thus represent instrumental background only.

The background levels for high mass elements such as Pb and U were caused primarily by organic compounds in the vacuum system and hence could be improved by using more care to obtain a clean system. The backgrounds for Na and K were caused primarily by the elements themselves present in the filament material. The Ca, Ba, Sr and Cr backgrounds appeared to be caused primarily by evaporation of compounds of these elements from nearby surfaces. The surfaces were undoubtedly contaminated by the large amounts of pure compounds of these elements used during the preliminary phases of this study and the backgrounds are probably not representative of a clean ion source.

For some elements, lower concentrations can be measured by reducing the current to the ionizing ribbon, collecting particles on the cooler surface for about a minute or longer and then rapidly reheating the ribbon to obtain a burst of metal ions. This technique is especially applicable to those elements like Pb or U where the background is caused primarily by organic compounds. About 1 or 2 orders of magnitude gain in signal to noise level can be obtained, probably because interfering organic compounds are too volatile to be absorbed. Best results are obtained by lowering the temperature of the ribbon to a temperature which is sufficiently cool to retain the metal atoms but hot enough to oxidize or prevent condensation of the organic compounds. Fig. 7 for U shows that by using 600°C as the lower temperature, the organic interference can be essentially completely eliminated. In the lower trace with filtered air, the organic ions continue to evolve at the lower temperature and no burst is evident on heating to the higher temperature. From this figure, one can estimate that the background is roughly equivalent to 0.0003 ng U/m^3. It will be noticed that the burst of UO_2^+ is roughly proportional to the collection time of 1 minute and 1/2 minute.

Organic Compounds

Many organic compounds have sufficiently low ionization potentials to be efficiently ionized on oxidized Re. For the Re ribbon used in this study, the ion current resulting from 100% ionization of a gas with a molecular weight of around 100 would be approximately 20A/Torr. Electron bombardment ionization sources typically yield only 10^{-3} A/Torr for 10^{-4} A electron current and the spectrum is usually more complicated than that from surface ionization. Because of this high output current capability and the complete lack of "air ions", extremely low concentrations of organic compounds can be detected.

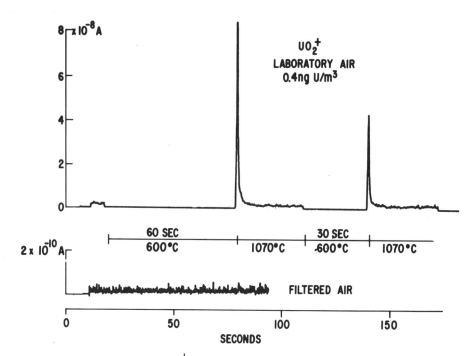

Figure 7. Bursts of UO_2^+ produced by heating to 1070°C after
           collecting natural air particles on ribbon at 600°C.

For the study of volatile organic compounds, a small 5 cm
radius magnetic sector mass spectrometer pumped by a mercury diffusion pump was used. The ion source was similar to that depicted
in Fig. 1 except that the inlet system was of conventional design.
The ion transmission from the Re ribbon to detector was about 1%.
By baking out the system at 400°C, the residual organic impurities
could be reduced to a level sufficient to detect about 10^{-12} parts
of a compound with an ionization efficiency of 10%. Two such compounds are dimethylnaphthalene and nicotine with ionization efficiencies of 13 and 20% respectively. Compounds such as mesitylene
and pyrene with ionization efficiencies of about 1% were detectable at the 10^{-11} concentration level. At the other extreme, benzene yielded no detectable ions even at high concentrations. Toluene was 0.01% ionized ($C_7H_7^+$) but a strong background peak at
mass 91 frequently interfered with measuring this compound.

Fig. 8 is a comparison of electron bombardment with surface
ionization. The concentration of dimethylnaphthalene was adjusted
to 0.02 ppm (based on electron bombardment data) at which level

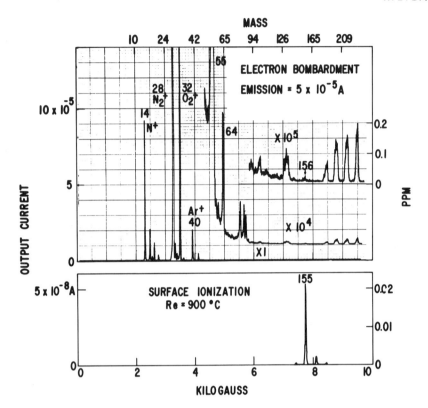

Figure 8.   Comparison of electron bombardment with surface ioniza-
            tion using dimethylnaphthalene.

the parent peak at mass 156 was barely discernible with electron
bombardment even though it was far removed from the enormous N_2^+
and O_2^+ peaks (the peaks above mass 156 are Re oxides). With surface ionization however, no interfering ions are apparent. The
main M-1 peak due to loss of one H atom is typical of hydrocarbons
when surface ionized and is more than 1000 times larger than the
electron bombardment peak. Minor peaks are at 141, 169 and 183.

    Fig. 9 shows the spectrum obtained for a volume concentration
of dimethylnaphthalene in air of 10^{-11}. At this level, the alkali
metal ions as well as residual organic ions at masses 81, 91 and
105 can be seen. The mass 155 peak of dimethylnaphthalene however
is still well above the background.

    Analysis of the normal laboratory air with the particulate
analyzer showed that the particles contained large amounts of organ-

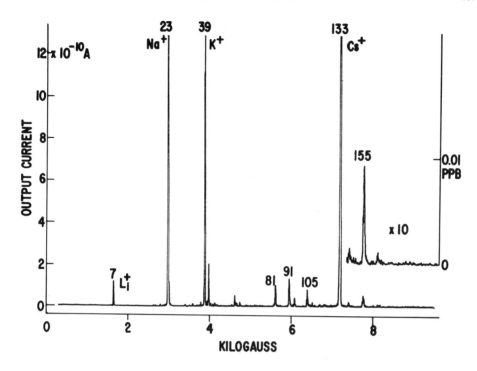

Figure 9.   Mass spectrum at 0.01 ppb dimethylnaphthalene.

ic materials but that no organic impurities could be identified
after filtering. Apparently all the ionizable vapors had adsorbed
on the particles. The spectrum from the particles was too complex
to clearly identify any one compound but large peaks at 84, 133
and 161 AMU indicated 2 ppb nicotine[1].

Analysis of Water

A little explored but potentially useful extension of this
technique is in water analysis. This is accomplished by atomizing
the water, drying the droplets with additional air and then analyzing the resulting particles in the same manner as normal air particulates. A variety of methods are available for forming the
water aerosol but a convenient method for small volumes is to pass
compressed air through a fine fritted disk moistened with the
water solution. By this means, a particle concentration of a few
million per cm^3 can be obtained. The equivalent amount of solution in the air was approximately $4X10^{-6}$ cm^3 of solution/cm^3 of air.

From alkali metal salt solutions of known concentration, out-

put ion currents of $5X10^6$ ions/sec for 1 micromole/liter were
obtained. This implies that the equivalent of $2X10^{-7}$ cm^3/sec of
solution was impinging on the ribbon. Background levels varied
from 20 ions/sec for Li and Cs to 60 ions/sec for Na and K so that
concentrations of approximately 10^{-11} moles/liter should be detectable. (For Li, $3X10^{-14}$g/cm^3). The minimum detectable levels (S/N
=1) for U and Cr were 10^{-12} g/cm^3 and for Pb, 10^{-9} g/cm^3. However, for high ionization potential metals like Pb, the accumulation of impurities in natural water on the surface of the Re ribbon apparently lowers the work function of the surface. The most
likely impurity is Ca which adheres strongly to Re and does not
evaporate rapidly at the low temperatures required to ionize high
ionization potential metals. As a consequence, determination of
these metals can be subject to serious errors in some cases. Periodic cleaning of the filament by flashing may give some improvement but was not investigated.

An example of the results is shown in Fig. 10 for a $3.9X10^{-5}$
molar solution of $LiNO_3$. A filter to exclude particles is applied
before the Li particles were sampled to illustrate the background
level.

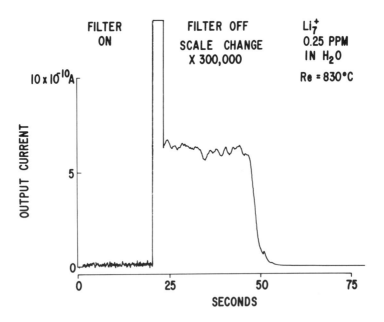

Figure 10. Analysis of particles produced from a solution of $LiNO_3$.

DISCUSSION OF RESULTS

This study has demonstrated that very low levels of impurities in air and water can be continuously analyzed using surface ionization mass spectrometry. The chief limitation is the requirement with the present ionizing surface that the ionization potential of the impurity be less than about 8 eV. The accuracy of this method has not been determined with any precision however and comparison with standard methods of analysis for environmental samples is needed. In particular, the accuracy of the ionization efficiency predicted by the Saha-Langmuir equation should be determined for a variety of elements, compounds and mixtures. For particles, the assumption that this efficiency does not vary with particle size and composition needs to be evaluated.

REFERENCES

1 DAVIS, W.D.: Environ. Sci. and Technol., 11 No. 6 (June 1977); 21st Annual Conf. Mass Spectrom. San Francisco, May 20-25, 1973, p. 329 and 343.

2 MYERS, R.L., FITE, W.L.: Environ. Sci. and Technol., 9, 334-336 (1975).

3 WICHNER, R. and PIGFORD, T.H.: Thermionic Conversion Specialist Conf., Houston, Texas, November 3, 1966, p. 405.

4 MC HUGH, J.A.: Intl. J. Mass Spectrom. Ion Phys., 3, 267-276 (1969).

5 ZANDBERG, E.Y. and TONTEGODE, A.Y.: Soviet Phys.-Tech. Phys. 11, 713-725 (1966).

6 MÜLLER, R. and WASSMUTH, H.W.: Surface Science, 34, 262 (1973).

Discussion

HEINRICH - What was the resolution of your spectrometer and did you have any interference problems?

DAVIS - With a small three inch mass spectrometer such as I used, you can only resolve up to about mass 150. Interference due to this low resolution was not a major problem for the elements studied but more serious cases undoubtedly can occur. The ubiquitous organic ions caused the most trouble and above mass 150, the background of these ions would probably be reduced if the resolution were increased. Resolution sufficient to resolve the organic ions from the inorganic ions of the same intagral mass would be a major advantage but this is difficult to achieve while still retaining a high transmission. High intensity ions such as K^+ scattered over a large mass range are a potential source of interference but for this problem, a cascade mass spectrometer would be more advantageous then higher resolution.

COLEMAN - One of your last slides showed considerable amounts of uranium and lead in the deionized water. Is that a common finding?

DAVIS - No, those species were intentionally added. I was just trying to find out how low a concentration I could detect.

DAHNEKE - One of your slides showed a signal labeled as the number of atoms in a particle. Actually, should that not more accurately be called the number of atoms in the signal, or were you able to determine otherwise that all the material in your particle was being ionized and delivered to the mass spectrometer, and that all of the material in the particle was of the mass you were looking at?

DAVIS - What I determined, of course, is the number of ions of a particular element received at the electron multiplier detector which, knowing the 4% transmission efficiency, infers the number of ions that came off the filament following the collision with a particle. The calculated or measured ionization efficiency was combined with this result to obtain the number of atoms of the particular element in the particle.

HEINRICH - Dr. Dahneke mentioned that the typical urban atmosphere has about ten to the four particless per cubic cm. and you mentioned about twenty. Is there a discrepancy or is your laboratory that clean.

DAVIS - No, what I said was if you had twenty particles per cubic cm. in the air you would get one ion burst per second in the instrument itself. If you had twenty thousand particles per cubic cm. you would get a thousand bursts per second. If the filament temperature is high enough these ion bursts last only a few seconds but, of course, the higher the temperature the lower the ionization efficiency.

STEHL - Were the organic compounds you monitored introduced as a mist or spray of organics or were they introduced as vapors?

DAVIS - Most of them were vapors except for cinchonine. Cinchonine is a compound which doesn't readily vaporize so I dissolved it in water, formed a mist from the solution and analyzed the resulting particles.

STEHL - How did you generate the mist or spray?

DAVIS - I used a sintered disk onto which I put some solution. A fine mist formed when I passed air through the sintered disk. You can get about a million particles per cubic cm. by this method.

STEHL - What was the inside diameter of the capillary you used?

DAVIS - The capillary used to introduce the particles had an inside diameter of a sixteenth of an inch.

OBSERVATION OF THE RAMAN EFFECT FROM SMALL, SINGLE PARTICLES: ITS

USE IN THE CHEMICAL IDENTIFICATION OF AIRBORNE PARTICULATES

Edgar S. Etz and Gregory J. Rosasco
Institute for Materials Research
National Burea of Standards
Washington, D.C. 20234

John J. Blaha
Department of Chemistry, Kansas State University
Manhattan, Kansas

ABSTRACT

A Raman microprobe developed at the NBS is used to observe the normal Raman effect from single microparticles of size 1 μm and larger. The resulting spectra of the Raman–scattered light are characteristic of molecular and crystal vibrations in the solid and are the basis for the qualitative identification of major molecular constituents. Particles composed of a broad range of compounds exhibit vibrational Raman spectra and thus can be identified spectroscopically by comparison with spectra obtained from known particles or by reference to Raman data available for bulk samples. The application of Raman spectroscopy to the analysis of small, single particles thereby offers the opportunity for the determination of the chemical states of major elements. This capability for the speciation of principal molecular components is not available from other microprobe techniques which furnish elemental composition data only.

The instrument and technique are applied to molecular micro-analysis of individual particles in various types of environmental particulate samples. Results are discussed on the detection and identification of major constituents of particles in the primary size fraction (>2 μm) of ambient air particulate dusts. These include the speciation of inorganic sulfur (e.g., SO_4^{2-}) and nitrogen (NO_3^-) compounds and the detection of a form of environmental carbon (e.g., soot) or residual hydrocarbon matter as a common material found with airborne particles. In micro-

413

mineralogical studies performed with the Raman microprobe, a
number of chain- and sheet silicates are investigated in the form
of fibrous and non-fibrous microparticles. For several of the
common types of asbestos minerals (e.g., amphiboles) a distinc-
ion, on the basis of the Raman spectrum, cannot be made between
the asbestos and non-asbestos (or massive) variety of a given
mineral species. A third application of the probe has involved
the molecular characterization of particulate emissions from
power plants. Results are presented from measurements on particles sampled from the stack of an oil-fired power plant. Raman
spectra observed from single microcrystals formed on the collection substrate show vanadium pentoxide (V_2O_5) as a principal
molecular constituent of the stack emissions. Finally, the
extension of the technique to the analysis of microparticulate
organic pollutants is discussed. The potential for identification of microparticulate pesticides in environmental samples is
demonstrated by Raman spectra obtained from microparticles of
selected chlorocarbon insecticides. The discussion of results
obtained in these four areas of particle microanalysis indicates
the unique information obtained through application of the Raman
microprobe. The usefulness of the technique as a complement to
the methods of electron- and ion beam microanalysis is indicated.
Prospects for future work in the application of the probe are
outlined as the more important limitations of the technique are
noted.

I. INTRODUCTION

It is not surprising that the identification of subnanogram
particles is becoming an increasingly significant problem in
pollution monitoring, health hazard evaluation, quality control
and trouble shooting.   The chemical complexity of dust and dust
particles requires detailed characterization, and many types of
analytical schemes have been utilized to accomplish this [1-7].
Most dust analyses are made on bulk samples or gross collections
of particles. Only recently has the study of individual particles been considered important. From single particle studies
more is learned about the heterogeneity of the dust .at the
microscopic level. The results may give a more precise definition
of the chemical species present, and — in favorable cases — may
permit a more conclusive identification of sources and particle
transformation processes.

Light microscopy has been used for small particle identification [8,9]. Chemical elemental analysis is performed by electron-
and ion microprobes [10-13]. The newer electron spectroscopy
techniques also permit the investigation of the surface composition [6,12,14]. However, these techniques do not yield molecular
information on the compound(s) present in the sample.

We have, in earlier work, described the application of Raman
spectroscopy as a nondestructive, highly specific and sensitive
technique of investigating molecular vibrational phenomena in
microsamples as a means of identifying the principal molecular
constituents [15,16]. These studies have involved the use of a
Raman microprobe developed at the NBS for the routine analytical
measurement of individual microparticles of size one micrometer
and larger. The vibrational Raman spectra observed from microparticles serve as a unique fingerprint of the chemical species in
such samples. In addition, such spectra often contain information
on the local molecular environment, the structural coordination
or the crystalline (or glassy) phase in which the species is
present. The Raman microprobe method of analysis represents an
important complementary technique to the methods commonly applied
to the identification of small, single particles.

In this paper we report on recent applications of the micro-
Raman spectroscopy technique to the molecular characterization of
microparticles in samples of environmental interest. This discussion is preceded by a brief review of the principles of (normal)
Raman scattering spectroscopy and an outline of the important
experimental requirements for successful Raman analysis of microparticles. This includes a description of the Raman microprobe
and of sample handling and measurement techniques. As specific
examples of application we present results on the identification
of urban air particulates, the micro-mineralogy of fibrous and
non-fibrous particulate silicates, the characterization of
particulate emissions from oil-fired power plants and the study
of microparticulate organic pollutants. The results obtained in
each of these four areas of investigation demonstrate the broad
applicability of micro-Raman spectroscopy and the unique information derived from application of the technique to environmental
analysis.

II. RAMAN SPECTROSCOPY FOR PARTICLE ANALYSIS

A. Principles of Raman Spectroscopy

The measurements performed to obtain the Raman spectra of
microparticles are based on the observation of the normal or
spontaneous Raman effect. This effect provides the basis for
Raman spectroscopy and has been extensively described [17-19].
When a monochromatic beam of visible light, usually from a laser
source, is focused into or through a sample (which may be solid,
liquid or gas), a small fraction of the scattered light will be
shifted in frequency from that of the incident light, provided
Raman activity exists. This phenomenon is known as the Raman
effect. The frequency shifts (i.e., displacements of the Raman
spectral lines from the exciting line) are identified with the

frequencies of molecular vibrations in the sample. Thus these
spectra (distributions of scattered light intensity with wavenumber) of the Raman-scattered light identify the scattering
molecules and — in the case of solid samples — contain additional
information on the molecular order of the solid phase from which
the nature and extent of crystallinity may be inferred. The
intensity of the scattered light at a particular wavenumber is
proportional to the number density of the scattering molecules in
a specific energy state.

The kind of information provided by Raman spectroscopy is
the same in essence as that of infrared spectroscopy, namely the
frequencies and intensities of spectral lines due to molecular
vibration. Thus, as is the case for infrared spectroscopy,
molecular symmetry governs the Raman activity of the so-called
"normal" vibrations. The vibrational frequencies, in turn, are
related to the masses of the vibrating atoms, the bond forces
uniting them, and their geometrical arrangement. These properties
of the sample at the molecular level determine the specific and
characteristic relationship between the molecular units in the
sample and the lines in the Raman spectrum. However, the factors
which determine the Raman scattering from molecules are quite
different from those involved in the absorption of infrared
radiation [18]. The result is that vibrations which are weak or
non-existent in the infrared are often strong in the Raman effect
and vice versa.

One drawback of Raman spectroscopy is that the normal Raman
effect is a very weak (i.e., low light level) phenomenon. Especially for microanalytical applications it becomes experimentally
difficult to detect, thereby placing difficult requirements on
the performance of a Raman spectrometer. By extending Raman
spectroscopy to the analysis of microscopic particles it is
assumed that all the physical phenomena underlying the Raman
scattering of light from samples of macroscopic dimensions equally
apply to the Raman scattering from particles approaching dimensions comparable to the wavelength of the exciting radiation.
Our earlier feasibility studies had indicated [20] that the Raman
spectra of microparticles — for a broad range of inorganic and
organic materials — are in good qualitative agreement with the
spectra of the same compounds measured in the form of bulk samples
(i.e., crystals). This correspondence among the spectra of
particles and their bulk sample counterparts permits the use of
Raman data reported in the literature as an important source of
reference for the qualitative identification of unknown particles.

Quantitative analysis of microparticles for major molecular
constituents is more complicated because particle size, shape and
refractive index influence the intensity of Raman scattering.

Attempts to quantify these various effects have been made by
other workers [21] and are presently being investigated by us
experimentally and theoretically [22].

B. Experimental Requirements

We have discussed in earlier published work [15,20] the
experimental requirements that must be met to permit the recording
of the Raman spectrum from microparticles of size down to 1 µm.
These criteria are high spectral sensitivity to extremely low
signal levels, effective rejection of optical interferences, and
appropriate choice of energy density levels (or irradiance, power
per unit area) placed in a focused laser beam that will not bring
about modification or destruction of the sample by heating or
photodecomposition.

In the following sections we describe the important features
of the NBS Raman microprobe constructed for this purpose and
present some detail on sample handling and measurement procedures.

1. The Raman Microprobe

A complete description of the instrument and its optical and
mechanical performance has been presented elsewhere [15]. The
instrument consists of (i) an excitation source, (ii) a sample
stage, (iii) the optics for the focusing of the exciting radiation
and for the collection of the scattered light, (iv) the spectrometer (a double monochromator and photomultiplier detector) and
(v) photon counting electronics coupled to a data logging system.

For the experiments the results of which we report here, we
have employed an argon ion/krypton ion laser utilizing the Ar
514.5 nm line, operating at powers in the range from 120 mW to 5
mW (measured at the sample). Non-lasing emission frequencies are
removed by the use of a Pellin-Broca prism. The laser beam is
focused onto the sample by any one of several focusing lenses
providing a lateral spatial resolution of a few micrometers
(typically 2-20 µm) determined by the diameter of the beam at its
focal point. The size of the focal spot and the incident laser
power determine the power density (watts/cm^2) placed on the
sample. Typically, irradiance levels in the range from several
megawatts/cm^2 to values of several kilowatts/cm^2 are utilized.
The Raman microprobe employs a 180° scattering geometry. In this
configuration, the total light (Mie/Rayleigh and Raman) scattered
from the particle is collected in the backward scattering direction (over a large solid angle) by an ellipsoidal mirror, and
transfer of the scattered radiation is made to a double monochromator employing concave holographic gratings (Ramanor HG-2,

Instruments SA, Inc., J-Y Optical Systems)[1]. The scattered light
at the position of the exit slit of the double monochromator is
detected by a cooled photomultiplier tube with measured dark
count of under 5 counts per second. Conventional photon counting
equipment is used for amplification and monitoring of the Raman
signal. The sample (consisting of the particle mount on a
suitable substrate) is accurately and reproducibly positioned for
measurement by a stage driven by remotely-controlled piezoelectric
translators. The microprobe is equipped with a built-in microscope to permit the viewing of the particle sample at various
magnifications and to aid in the location and precise positioning
of particles of interest. The spectrometer system is interfaced
to a minicomputer for data logging and total system control
(e.g., sample positioning, wavelength scan, slit settings, etc.).
The Raman microprobe employed in these studies is a monochannel
Raman spectrometer in which one radiation detector is used to
analyze the optical spectrum by means of a scanning monochromator.
A molecular microscope/microprobe (which also utilizes the Raman
effect)-different in design from the NBS Raman probe — has
recently been introduced as a commercial instrument and its
potential application to the microanalysis of heterogeneous
samples has been described [23-25]. This instrument, in addition
to having the capability of a monochannel spectrometer, can also
be operated in a multichannel detection mode which permits
obtaining the Raman images of samples — as well as multichannel
spectra — through television-type camera detection of characteristic Raman frequencies.

2.   Sample Preparation

In our experiments the spectroscopic measurements are performed on single microparticles (frequently of mass as little as
10^{-12} g) supported by a suitable particle substrate. High-purity
sapphire (α-Al_2O_3) and lithium fluoride of single-crystal quality
are routinely used as substrate materials in the form of small
rods (4 mm diameter, 6 mm long) held in a support collar. Sapphire is a weak Raman scatterer and does not give rise to significant spectral interferences. Its use as a substrate material is
preferred in the collection of chemically reactive particles (as
in the sampling of stack aerosols). Lithium fluoride lacks first

----

[1] Certain commercial equipment, instruments or materials are
identified in this paper in order to adequately specify the
experimental procedure. In no case does such identification
imply recommendation or endorsement by the National Bureau of
Standards, nor does it imply that the material or equipment
identified is necessarily the best available for the purpose.

order Raman activity, shows no fluorescence, and therefore serves
as an excellent substrate material.

Various techniques have come into routine use for the preparation of particle samples for micro-Raman analysis, and these
have been described with particular reference to the characterization of ambient air particulates [16]. The experiments reported
here have involved the use mainly of two methods of sample preparation. In one, particle mounts on substrates are obtained from
dispersions of bulk samples (e.g., filter-collected samples of
airborne dusts). A small fraction of the bulk dust sample is
transferred onto the substrate to achieve a low density particle
deposit devoid of extensive particle agglomeration. In some
cases, individual particles of interest that have been isolated
from a bulk sample have been transferred onto Raman substrates by
the usual particle handling techniques used by particle analysts
[8]. In many cases involving the study of airborne or stack-
suspended particles we have obtained particle samples for Raman
characterization by in situ sampling of the aerosol directly on
Raman substrates. Here our experiences have been mainly in the
sampling of urban air particulate dusts and the collection of
stack emissions from power plants. In this second method of
sample preparation we have tried to optimize the collection of
particles through the use of multi-stage cascade impactors
[26,27]. For most size-classified samplings of suspended particulates we have used a five-stage impactor of the Battelle design
[28] modified to accept on each impaction stage the Raman substrate of standard design. By using this type of sampler and
short sampling times we have sought to achieve minimum quantity
sampling in order to minimize all opportunities for changes to be
incurred in the physical and chemical properties of the collected
particles [29,30].

3.   Measurement Procedures

Full detail on measurement techniques employed in the micro-
Raman analysis of particles has been presented in earlier publications [15,16]. In a typical scheme of analysis, the particle
sample is first examined by light microscopy to single out and
locate (with the aid of a particle finder grid shadowed onto the
substrate surface) particles of interest to Raman microprobing.
The observable parameters include particle shape, size, color,
transparency, surface roughness, birefringence and extinction.
Recognition of these properties can be of value in the subsequent
Raman investigation.

If a particle shows intrinsic color, this may determine the
choice of excitation frequency and the level of laser irradiance
used. Examination in the optical microscope (supplemented by
optical micrographs taken of the sample before and after Raman

analysis, to check on possible beam damage) is then followed by
measurement in the microprobe. The spectra that are presented in
this paper have been recorded by scanning the monochromator and
plotting, on a strip chart recorder, the analog signal proportional to the number of photon counts observed in a selected
measurement time. Typically, these spectra have been obtained by
using irradiance levels ranging from a maximum of 1 megawatt/cm^2
down to values less than 5 kilowatts/cm^2. The lowest practical
power densities are often required for the probing of unknown,
environmental particles. These may be radiation absorbing and
could therefore be modified due to heating. Measurements performed using low irradiances generally require signal integration
times from 1-2 seconds at scan rates of 50 cm^{-1}/min to 20
cm^{-1}/min. A scan (with typical 3 cm^{-1} spectral resolution) which
covers the Raman spectral range from 0 to 3600 cm^{-1} (frequency
shift, ν) may therefore require from 1-3 hours of measurement
time. The spectra of essentially colorless particles can be
obtained employing moderate to high power densities without
deleterious effects on the sample. Such scans are recorded at
integration times from 0.2 to 0.5 seconds at scan rates of 200 to
50 cm^{-1}/min. Optimized instrument performance furnishes spectra
of high signal-to-noise ratio with insignificant spectral interferences from either substrate (e.g., sapphire) emissions or the
elastic scattering component. We found it necessary, in some of
our work, to apply electron- and ion-probe analysis to obtain
data on the elemental composition of particles under study. The
techniques followed in these investigations are those that have
been described by other workers in the field [13]. For particle
samples that require overcoating by a conductive film (e.g., Al,
C, Au) to permit electron beam analysis, it is necessary that
they be overcoated after Raman analysis has been performed. This
is because these films of thickness $\sim$20-40 nm are sufficiently
radiation absorbing to induce particle heating, modification or
even destruction. X-ray area scans that indicate the micron and
submicron inhomogeneity of a particle have been found to cor-
roborate the results of micro-Raman analysis. Because of the
higher elemental sensitivity, the interpretation of ion probe
mass spectra and their correlation with Raman spectra is more
complicated.

   Experiences gathered to date in the joint application of the
three types of microprobes available in our laboratory indicate
that the combination of electron-, ion- and Raman-probe techniques
results in an extremely powerful microanalytical approach.

III. APPLICATIONS OF RAMAN MICROPROBE ANALYSIS

A. Identification of Urban Air Particulates

1. Study of Ambient Air Particulate Pollution

Most of the analyses reported on air particulates have
involved the characterization of bulk samples which have furnished
extensive data on the elemental composition of air pollution
dusts [31]. Particular interest has centered on the identification
of the principal sulfur and nitrogen species, such as sulfate
(SO_4^{2-}), nitrate (NO_3^-) and ammonium (NH_4^+). The characterization
of atmospheric sulfate has received the greatest attention
[29,32-35]. An important objective of our work has been to apply
the Raman spectroscopy technique to the study of airborne particles. We have therefore — in the first year of the use of the
Raman microprobe — applied the instrument to the characterization
of microparticles in the primary or coarse size fraction (>2 μm)
of urban dusts [16]. From the study of known materials we have
shown detectability for the common oxides, sulfates, carbonates,
phosphates, nitrates and ammonium as major constituents of single
particles of size 1 μm and larger. The sharpness and frequency
position of the spectral bands for any one of the species (e.g.,
sulfates) studied allowed one to detect small spectral shifts and
therefore enabled discrimination between the various compounds
(e.g., Na_2SO_4, $CaSO_4$, $(NH_4)_2SO_4$) of a given class. From the
measurement of a large number of particles in urban dust samples,
we have made observations on the importance of unambiguous sampling
and have seen evidence of particle transformations associated
with the bulk collection of urban dusts.

2. Raman Spectra of Airborne Particles

In the following we have chosen two examples from the study
of urban air particulates to illustrate the kind of spectroscopic
information that is obtained in this field of application. This
discussion centers on the qualitative interpretation of the
spectra of two particles originating from two different samples
of urban dust. The spectra of these particles are shown in
Figure 1 and Figure 2. The frequency shifts in wavenumber units
for the Stokes-Raman scattering are displayed on the horizontal
axis. Plotted along the vertical axis is the value of the scattered light intensity in photon counts observed in a measurement
time referred to as the time constant. The zero of light intensity
is indicated by the solid horizontal baseline. All spectra have
been recorded at room temperature with an effective resolution of
approximately 3 cm^{-1} Frequency calibrations were obtained by
recording the neon and argon emission lines, providing an accuracy
of ±3 cm^{-1}. In the collection of the scattered radiation from

the particle an exit pinhole of size (i.e., diameter) 140 μm has
been utilized in all measurements [15]. Its main function is to
define and limit the effective light collection volume, thereby
optimizing the collection of scattered light from the particle
and minimizing the collection of light by the supporting substrate
(important when sapphire is used).

The particle whose spectrum is shown in Figure 1 is one
isolated from a bulk sample of St. Louis urban dust collected on
a glass fiber filter through the use of a conventional high-
volume air sampler. Other particles in this collection of dust
and in other such filter samples have furnished spectra qualitatively identical to the one discussed here. The substrate
material chosen in the preparation of this sample is LiF, so that
the information contained in the spectrum derives entirely from
the particle under study. Other information pertinent to the
spectral measurement is indicated in the figure. The probing
laser beam has not been focused down to a potentially much smaller
(on the order of 2 μm) focal point diameter so that in this
measurement only moderate power density is placed on the particle.
At the scan rate employed, 30 minutes were required to record the
spectrum which exhibits good signal-to-noise ratio on a broad,
continuous spectral level of approx. 320 counts per 0.5 sec.

Qualitative comparison of the spectrum of the particle with
reference spectra obtained from microparticles of well-characterized inorganic solids indicates the identity of the "unknown"
particle. The major constituent is calcite ($CaCO_3$), along with an
unknown quantity of anhydrite ($CaSO_4$). No attempt is made in
this analysis to quantify this second constituent. The Raman

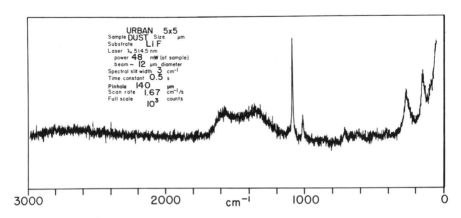

Figure 1.   Raman spectrum of a particle of calcite ($CaCO_3$) in
            urban air particulate dust.

frequencies characteristic of the normal vibrational modes of $CaCO_3$ are all present in the spectrum of Figure 1. Five Raman active modes are expected for this crystalline solid and these have been assigned frequency shifts 1432, 1088 (most characteristic totally symmetric stretching mode), 714, 283 and 156 cm^{-1} [36,37,38]. Of these, the first three Raman shifts arise from internal vibrational modes of the carbonate anion, while the latter two shifts — falling in the low frequency region of the spectrum — are identified with external (or lattice) vibrations of the solid. Four of these bands are clearly resolved. The shift at 1432 cm^{-1}, however, is obscured by the broad feature dominating the 1200-1700 cm^{-1} region. This component of the spectrum, showing two distinct maxima, is not expected in the spectrum of calcite or anhydrite. The presence of $CaSO_4$ as a minor constituent is indicated primarily by the band around 1000 cm^{-1}. We are familiar with the spectrum observed from anhydrite particles through our earlier measurements on urban dust particles [16]. As in the bulk (i.e., as single crystals or powders), microparticles of the common sulfate salts and sulfate minerals give strong and sharp Raman spectra. These have been reviewed, along with the vibrational spectra (infrared data included) of other classes of inorganic compounds and minerals, in the recent literature [39,40]. Most characteristic of the solid sulfates is the strong symmetric S-O stretch near 1000 cm^{-1}, and for $CaSO_4$ this band is reported at 1018 cm^{-1} [41,42]. Other bands due to the remaining expected normal vibrational modes are of weaker intensity and more difficult to discern against the relatively high background signal level. The bending modes of the SO_4^{2-} ion in anhydrite give rise to bands in the 400-700 cm^{-1} region. The asymmetric stretching modes produce bands with shifts around 1100 cm^{-1}, but in the spectrum of the dust particle these are obscured by the strong carbonate symmetric stretch at 1088 cm^{-1}. A number of lattice oscillations have been reported for $CaSO_4$, but these are normally quite weak.

It is noteworthy that the $CaSO_4$ present in this particle does not exist as the hemi-hydrate (plaster of Paris, $CaSO_4 \cdot \frac{1}{2}H_2O$) or the di-hydrate (gypsum, $CaSO_4 \cdot 2H_2O$). This illustrates the high specificity of the Raman effect which permits distinguishing between each of these three forms [43].

Accounting for anhydrite as the second constituent of the particle still leaves unexplained the origin of the broad, double-maximum feature in the 1200-1700 cm^{-1} region. In our study of several different samples of urban dusts collected on filters and on impactor stages, we have observed bands similar to these from many different types of particles (e.g., sulfates, oxides, nitrates). In the spectra of some particles these, in fact, were the only distinguishable features. We attribute these broad

bands — in a large number of cases — to the presence of a form of
carbon (or hydrocarbon matter) associated with these particles.
In recent work by other investigators these same features have
been observed in the Raman spectra of bulk particulate samples
collected from motor vehicle emissions and ambient air [44,45].
These workers have suggested that these bands are due to finely-
divided soot, a form of carbon chemically similar to activated
carbon and with physical structures similar to polycrystalline
graphite. Thus, those species producing the bands in the 1200-
1700 cm^{-1} region have been termed "graphitic soot" [45]. These
bands center on a line at about 1350 cm^{-1} and about 1600 cm^{-1} and
are very characteristic of elemental carbon in microcrystalline
and amorphous form as has been shown in earlier Raman studies
[46,47]. Soot as a form of environmental carbon is regarded as
profusely associated with atmospheric particles and its existence
as a component of atmospheric aerosol has been interpreted as
having important implications to atmospheric chemistry, from the
viewpoint of soot-catalyzed heterogeneous reaction of SO_2 to
H_2SO_4 [45]. The high catalytic activity of microparticulate
carbon towards SO_2 oxidation to SO_4^{2-} involving atmospheric aerosol
has been shown in recent work [48].

While soot may well be a widespread "contaminant" of environmental particulate samples (and probably more so with particles
emitted by various kinds of combustion processes), our experimental
observations also indicate that carbon species of similar form
may also result from laser-induced decomposition of organic or
biological material associated with inorganic microparticles. In
these cases, then, the presence of carbon bands in the particle
spectrum represents an artifact of the measurement. From our
experiments we can therefore suggest two explanations for the
appearance of carbon bands in the spectra of environmental
particles. By one mechanism these arise from the presence of
combustion-produced carbon (e.g., soot) in or on microparticles,
as we have shown by model experiments involving the spectroscopic
study of inorganic particle samples which we have coated (by
thermal evaporation in vacuum) with thin, nearly transparent
(approx. 50 nm thickness) carbon films. The spectra of carbon-
coated particles exhibit the two characteristic bands centered
around 1350 and 1575 cm^{-1} consistent with the spectral data
reported for polycrystalline graphite [46,47].

By a second mechanism — as we conclude from a separate
series of model experiments — these bands may be created as a
consequence of the measurement itself, from particles bearing
finite amounts of organic or biological components which decompose
to a carbonaceous residue upon exposure to intense laser irradiation. This effect was first observed in probe measurements on
microparticles of chicken eggshell, of which the major inorganic

component is calcite. The high-irradiance spectra of these particles were identical to those observed from $CaCO_3$, except for the presence of the carbon bands. We attribute their existence in these spectra to the possible decomposition of a protein (e.g., egg albumin) in the inorganic matrix of the particle. We have obtained other spectroscopic evidence for this form of sample-laser beam interaction by coating, in one type of study, various kinds of known inorganic particles with encapsulating, thin films of certain hydrocarbon liquids and recording their Raman spectra as a function of laser irradiance. The decomposition of the organic layer or contaminant was observed in these measurements by the attendant formation and evolution of the characteristic carbon bands.

Further studies along these lines are in progress and will be published. Of particular interest is the spectroscopic behavior of contaminating surface layers (on inorganic or mineral particles) of certain polycyclic organic compounds suspected to exist as pollutants of ambient air [49].

The second example chosen to illustrate the Raman analysis of particles in ambient air is the detection and identification of inorganic nitrate as a constituent of atmospheric particulate matter. Figure 2 shows the result of two successive measurements on a $\sim$3 µm particle contained in a multi-particle collection of urban dust from downtown Washington, D.C. The particle is one of many analyzed in a sample of dust collected on a sapphire substrate mounted on stage 2 of a five-stage cascade impactor (Model DCI-5, Delron Inc.)[1]. Two spectra are shown for the particle which is identified — by reference to particle spectra of known compounds and to literature data — as crystalline sodium nitrate ($NaNO_3$). The measurement parameters used in the recording of these spectra have differed, as indicated in Figure 2. Laser power and spot size determine the power density to which the particle is subjected. The upper spectrum was obtained at an irradiance of 5.1 kW/cm^2 whereas the lower spectrum was run at 0.32 MW/cm^2. The most pronounced difference in the two spectra consists of the two broad bands (i.e., carbon bands) in the 1200-1700 cm^{-1} region present in the bottom spectrum but absent in the spectrum of the first measurement. Other differences are noted in both the intensity and in the spectral definition of bands attributed to the microparticle. The predominant features in the top spectrum are lines of medium intensity with Raman shifts around 1390 and 1070 cm^{-1} — both attributed to the particle — and several bands below 800 cm^{-1} characteristic of the Raman scattering from sapphire [50]. The Raman spectrum of the substrate (recorded by moving the particle out of the focal spot of the laser) is included with the spectra of Figure 2 and appears below the trace of the high-irradiance particle spectrum. The expected sapphire

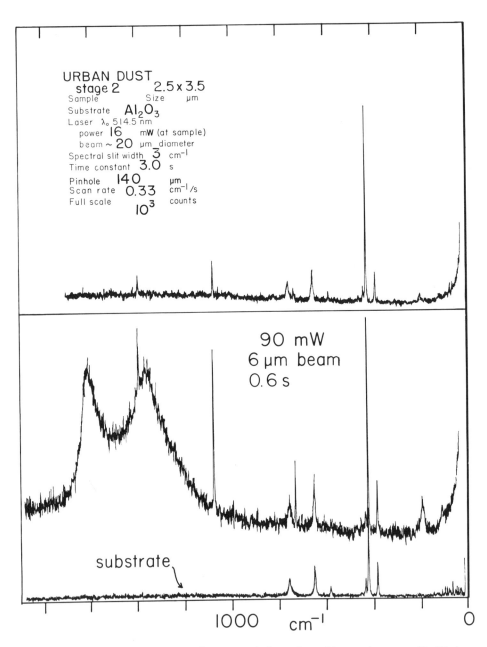

Figure 2.   Raman spectra of a particle of sodium nitrate ($NaNO_3$)
in urban air particulate dust.

modes with Raman shifts 751, 645, 578, 432, 418 and 378 cm^{-1} are seen in the spectrum of the substrate. The relative intensities of these bands are strongly dependent on the orientation of the crystal. The spectrum of $NaNO_3$ is known to us from our earlier studies of microparticles of single-crystal $NaNO_3$ [16]. The molecular solid has the same crystal structure as calcite, so that its Raman spectrum exhibits the same fundamental modes as have been discussed earlier for $CaCO_3$. Thus, three intra-ionic vibrations are expected for the NO_3^- ion and two normal modes of the lattice vibrations. The characteristic lines have been assigned frequency shifts 1385, 1068, 724, 185 and 98 cm^{-1} [37,51]. The strong nitrate symmetric stretch at 1068 cm^{-1} is well resolved and of good intensity in the upper spectrum. The other lines are of much weaker intensity but can be discerned upon a generally low signal level background. The spectrum recorded at high irradiance shows each of the five expected normal modes much more clearly. These are superimposed upon a much higher background spectral level, presumably arising from fluorescing compounds produced by the action of the focused beam or other impurities associated with the particle. Continuous background signal levels this high are not observed in the spectra of particles from pure solids (e.g., single-crystal $NaNO_3$). The carbon bands in this spectrum are well developed indicating a graphitic species possessing microcrystalline structures. The appearance of these features must clearly be interpreted as an artifact of the measurement, resulting from the laser-induced decomposition of an unknown contaminant producing a carbonaceous residue. Carbon bands of this same spectral definition have been observed in the spectra of microparticles coated with carbon films and non-volatile organic liquid layers studied in the course of modeling experiments referred to earlier.

Several other areas of concern to air pollution analysis have come under preliminary investigation. These include the study of atmospheric particles collected in the Antarctic to demonstrate the existence of stratospheric, microcrystalline sulfate species [52]. In a unique and special application of the Raman microprobe the feasibility of studying microdroplets (<10 μm diameter) of liquid sulfate aerosol has recently been demonstrated [16]. The spectra recorded in these measurements permit the qualitative characterization of the $H_2SO_4 - HSO_4^- - SO_4^{2-}$ acid sulfate system in single, liquid microdroplets. Because of the overall importance of sulfate aerosol, work has been started also on the characterization of sulfite (SO_3^{2-}) species. The vibrational spectrum of the SO_3^{2-} ion is sufficiently different from that of the SO_4^{2-} ion and other oxygenated sulfur species to permit distinguishing between crystalline sulfite and sulfate in microparticles. This can be important since the existence of stable sulfite species in ambient suspended particulates has been proposed [53].

B.   Micro-Mineralogy of Fibrous and Non-Fibrous
Particulate Silicates

1.   Analysis of Asbestos Minerals

Studies indicate that fibrous (asbestos) minerals in the
form of air- or water-suspended microparticles are a common part
of the environment [54,55]. It is known that many of these
minerals pose a potentially serious health hazard and that certain
malignancies are related to exposure to these substances [56-60].

Asbestos is a generic term that applies to a number of
naturally occurring hydrated mineral silicates which are separable
into fibers. These fibrous minerals consist of approximately 40-
60% silica combined with oxides of magnesium, iron and other
metals.   There are six varieties of commercial asbestos and they
belong to two major families of mineral silicates, based on their
crystal structure [61].   One is the group of serpentine minerals
and the other the group of amphiboles. The major recognized
mineral species considered asbestos are chrysotile (a sheet
silicate) of the serpentine group and anthophyllite, tremolite
crocidolite, amosite and actinolite of the amphibole group (these
are chain silicates).   Each of these mineral species also occurs
in a non-asbestos form.   For example, antigorite is the non-
asbestos (or massive) variety of chrysotile, and riebeckite is
the non-asbestos variety of crocidolite.   Tremolite may occur as
a contaminant of commercial talc (a platy, non-fibrous form of
sheet silicate).   Because of their dissimilar crystal structure
and chemical composition, the two groups of asbestos minerals
differ in their physicochemical properties.   A complete description
of these chain and sheet silicates can be found in the literature
[62].   Methods for the determination of concentrations of asbestos
in the environment, such as in ambient air and in drinking water,
include x-ray diffraction, differential thermal analysis, infrared
spectrometry and other methods which do not provide fiber size
information.   Optical microscopy and electron microscopy techniques
have been used for fiber identification and fiber sizing.   Several
excellent reviews describing the application of these various
analytical techniques for identifying and distinguishing between
the various varieties of asbestos have appeared in the recent
literature [63,64].   Morphological, diffraction and elemental
information can be obtained from single particles of size well
below 1 μm by using the analytical electron microscope equipped
with energy-dispersive x-ray analyzer.   However, many ambiguities
can arise in the determination of the exact chemical (elemental)
composition.   Because of these problems, transmission electron
microscopy (TEM) coupled with selected area electron diffraction
(SAED) has come into use as an important method for identifying
and counting asbestos fibers [63,64].   The method is reported to

be specific with respect to chrysotile or amphibole fibers [65]. The non-asbestos mineral species show identical electron diffraction patterns as their asbestos analogs.

2.   Raman Characterization of Sheet and Chain Silicates

We set out in recent investigations to examine the utility of the Raman microprobe for the identification of individual microscopic crystals (or platelets) and fibers of both technologically and environmentally significant minerals. Analysis of individual microcrystallites is common in mineralogical studies [8]. Morphology, structure, optical properties and composition are determined by a variety of microanalytical techniques. It is important to note that many important mineral classes consist of a large number of species having closely related structural forms which are relatively similar optically, stoichiometrically and morphologically. For characterization and analysis of classes such as these, it is important to identify parameters which are uniquely associated with a particular mineral. In this regard, infrared and Raman molecular spectroscopy have proven to be quite important. These techniques provide information as to the identity of the basic molecular building blocks and their structural coordination in the mineral. To this point, they have been applicable only to macroscopic crystals or to multiparticle samples and thereby may be limited in scope by sample availability or may provide only average properties of the minerals.

A great many infrared spectroscopic studies have been made on bulk samples, whereas Raman spectra are available for relatively few common mineral species. Raman spectra provide an important complement to the infrared data on a given material. In certain cases involving centrosymmetric materials and microscopic particles, Raman spectra are generally more simple and thus more easily interpreted than infrared spectra. Obtaining the infrared spectrum of an individual microcrystal may not be possible. However, the capability offered by the Raman microprobe for extending molecular spectroscopic analysis to individual microscopic crystals should result in useful information for the mineralogist and holds promise for improved chemical analysis procedures.

A large part of our effort in micro-mineralogy by Raman spectroscopy has centered on sheet and chain silicates [62]. Examples of the latter are the amphibole minerals tremolite ($Ca_2Mg_5[Si_8O_{22}](OH)_2$) and anthophyllite (($Mg,Fe^{+2})_7[Si_8O_{22}](OH)_2$). The amphiboles are based on chemical and structural units common to the technologically important sheet silicates, talc ($Mg_3[Si_2O_5]_2(OH)_2$) and chrysotile ($Mg_3[Si_2O_5](OH)_4$). The sheet

and chain silicates can be interconverted in geochemical processes
and are often found as contiguous bands in mineral deposits.

The following will summarize the results of our spectroscopic
studies in this area of application.

Raman spectra have been obtained from individual microcrystals
and microfibers of sheet (talc and antigorite) and chain (anthophyllite, tremolite and actinolite) silicates.[2] These are the
first Raman spectra to be measured for many of these species
[66]. In the case of talc, a comparison can be made with the
spectra obtained from a macroscopic crystal [67]. As has been
the experience in other comparisons of this type [20], the spectra
of the macro- and micro-crystals are essentially identical. The
spectra of individual microcrystals of talc generally display
better resolved bands and are somewhat less ambiguous in the very
low frequency region than those of the macroscopic crystal.
Effects attributable to stoichiometric disorder can be identified
in the spectra of many of these minerals. Comparison of the
bands characteristic of each mineral species allows reasonably
unambiguous mineral identification. Spectra have been obtained,
for the first time, from individual microfibers of the chain
silicates. These spectra are essentially identical to those
obtained for the non-fibrous polymorphs. It is noted that attempts
to obtain Raman spectra by use of conventional techniques applied
to multiparticle, bulk samples of the high surface area polymorphs
(sheet talc and asbestos-form amphiboles) were unsuccessful. In
contrast to this result, it was found that in the vast majority
of the microprobe measurements on individual particles of all
forms of the minerals, a spectrum allowing definite mineral
identification was obtained. Over one hundred particles derived
from seventeen different bulk samples of the various minerals
were probed.[3] The mineral identity of the bulk, polycrystalline
samples was verified by x-ray powder diffraction.

---

[2] In this work, a particle is referred to as a fiber or microfiber
if it is greater than 5 μm in length and has a length/maximum
width ratio greater than 3.

[3] Documentation of mineral sources and identifying characteristics
for these minerals will be included in an extended version of
this report.

3. Discussion of Representative Spectra

A complete exposition of the results of this study and a detailed analysis of the Raman spectra of these sheet and chain silicates will be the subject of a separate publication [68]. Illustrative examples of these results and highlights of the analysis of the spectra are presented in the following discussion.

Spectra characteristic of tremolite (A), a variety of anthophyllite (B), and talc (C) are presented in Figure 3. In each case, the spectrum was obtained from a single, non-fibrous microcrystal approximately 10 μm in linear dimension. Measurements have been extended to single microcrystals as small as 2 μm. Spectra essentially identical to those in the figure were observed from many (10-30) individual microparticles of each mineral class. Each spectrum has from 10-20 major features which will most generally appear in any probe measurement. Comparison of the spectra shows that these minerals can be uniquely identified by their characteristic peaks.

Raman microprobe measurements have been successful for single micro-fibers of tremolite and anthophyllite. Figure 4 presents the spectra obtained from the two polymorphs of anthophyllite. In this and similar spectral comparisons of fibrous and non-fibrous polymorphs of the amphiboles no spectral differences have been found which can be associated with the crystal morphology.

The major bands in each of these spectra are referred to as "characteristic" in the sense that they are attributable to certain molecular (ionic) species in particular structural coordinations in the crystal lattice. It is also apparent from analysis of these data that the minerals have similar basic structural units [69]. They are all based on the cross-linking of pyroxene chains to form continuous double-chain silicate bands in the amphiboles and infinite sheets in talc. Two of these bands (sheets) are linked by metal and hydroxyl ions which form a brucite-like strip (layer) between the bands (sheets) yielding a 2:1 trioctahedral sandwich. Spectral features in the region from $\sim$450 cm^{-1} to 1200 cm^{-1} can be identified with stretching and deformation modes of the silicate network. For example, the most prominent feature at $\sim$680 cm^{-1} in each spectrum is characteristic of the Si-O-Si symmetric stretch of the pyroxene chain. Vibrations associated with the octahedrally coordinated metal ions in the brucite-like layer have frequencies from $\sim$450 to 150 cm^{-1}. In talc all the metal ions are in octahedral sites, whereas the amphiboles have metal ions in inter-band sites which give rise to additional modes in this region. The only major features in these Raman spectra which have been identified with the hydroxyl ions are OH^- stretching modes in the 3600-3700 cm^{-1} region (not shown).

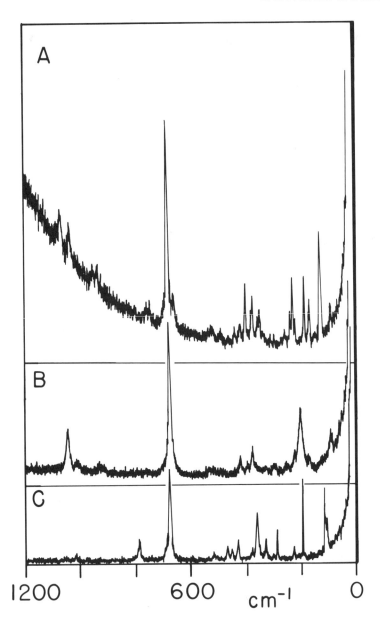

Figure 3. Raman spectra of single microparticles of tremolite (A),
           anthophyllite (B), and talc (C). Particle size ∿10 μm.
           Measurement parameters: excitation 514.5 nm; power
           ∿100 mW; 7 μm diam. beam spot; 3 cm⁻¹ spectral slit
           width; scan rate 50 cm⁻¹/min. Maxima at ∿680 cm⁻¹ are
           of the order of 500–800 counts per second.

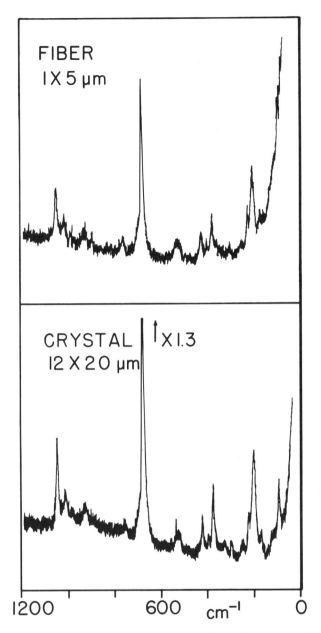

Figure 4. Raman spectra of a microcrystal and a microfiber of anthophyllite.
Measurement parameters: excitation 514.5 nm; power 120 mW (upper), 95 mW (lower); 7 μm diam. beam spot for both; time constant 2.0s (upper), 0.5s (lower); intensity 10^3 cts full scale for both.

The low frequency (below ∿250 cm^{-1}) modes have not previously
been analyzed and assigned for most sheet and chain silicates.
Assignments can be made on the basis of an intercomparison of
these Raman spectra and use of approximate frequencies calculated
in the literature [70]. These results can be summarized as
follows: modes in this region are ascribed to intra- and inter-
layer lattice modes. Features near 200 cm^{-1} are characteristic
of the "brucite" layer (sharp band in talc) and are sensitive to
disorder (broad-band in anthophyllite) and distortions of the
"talc-like" strip (split, shifted bands near 200 cm^{-1} in tremolite).
The strong features below 200 cm^{-1} are associated with inter-
layer lattice modes of the silicate strips or sheets. These are
less sensitive to the cation distribution in the "brucite" layer,
but are strongly affected by the distortion of the "talc-like"
strip in tremolite. All these spectral features appear to be
useful for identifying the basic structural assets and stoichiometric variations characteristic of the sheet and chain silicates.

The investigation of the amphibole minerals is far from
complete. Because of deleterious heating effects resulting from
absorption of the exciting laser radiation, measurements on the
iron-rich amphiboles have not yet been possible. Effects associated
with sample heating have been observed for the sheet silicates
antigorite and chrysotile. For these minerals, the absorption of
the laser most probably arises from impurities associated with
the exposed hydroxyl layers in these 1(silicate):1(brucite)
trioctahedral minerals. Only relatively large 10-30 μm crystal-
lites have been successfully probed to this point.

From this preliminary research it does appear that significant
information can be obtained by application of the Raman microprobe
to the study of microcrystalline minerals. This development can
provide a useful complement to other measurement techniques. It
is important to extend measurements of the type reported here to
a wide variety of well-characterized microspecimens of each
mineral class. The intention of such studies is to determine
those spectral features which uniquely establish mineral identity
and those which most clearly indicate the nature of the stoichiometric or structural variations characteristic of a mineral
subgroup or class. Continued efforts in this area should be very
useful for mineralogical and analytical applications.

C. Characterization of Particulate Emissions from
Oil-Fired Power Plants

1. Known Properties of Power Plant Particulate Emissions

To assess the effect of pollutant particles from power
plants on man and the environment, more information is needed on

the composition of fly ash. It is now well established that the high-temperature combustions involved emit particles containing toxic elements such as As, Cd, Pb, Sb, Se and Zn into the atmosphere. Most that is known about fly ash is known from the study of the particulate emissions from coal-fired power plants [12,71-74]. In comparison, much less has been learned about the emissions from oil-fired power plants. This may in part be due to the fact that the ash content of coal is approximately two orders of magnitude higher than that of fuel oil, so that coal combustion is responsible for the emission of much larger amounts of particulates in the air. Of the very few studies that have become available dealing with the characterization of particulates from fuel oil combustion, the emphasis has been on relating the physical/ chemical properties of bulk samples of ash to fuel (elemental) composition and the conditions of the combustion [75-77]. The crude oils burned in power generation exhibit wide variations in sulfur content (typically 0.5-2.5 wt % S) and in the inorganic trace element composition. Vanadium is usually the major heavy metal (typically 5-400 ppm V) component. Bulk elemental analyses of soot and oil ash are usually performed by x-ray fluorescence analysis [77]. The particulates produced in the combustion of heavy fuel oil have been characterized as to three basic types, i.e., smoke, cenospheres and ash residue [75].

Few studies have appeared on the characterization of single particles emitted by oil-fired power plants [11,78]. One study reports data on the elemental composition of particles with characteristic oil-soot morphology from ion microprobe analysis [11]. It was shown that the average oil-soot particle is characterized by high levels of oxygen, vanadium, carbon, sodium, calcium and potassium.

2. Investigation of Stack Particulates from an Oil-Fired Power Plant

We have applied the technique in preliminary experiments to the characterization of particulate emissions from coal-fired and oil-fired power plants. In the following we are reporting on some results from the study of stack-suspended particulates of an oil-fired power plant.

The source of these particulates is one of several power generating stations whose emissions are monitored by the Environmental Protection Agency (EPA). The samples analyzed in this study were collected by field sampling personnel of the EPA. Sample collection involved the use of the same type of five-stage cascade impactor as has been employed by us in the sampling of ambient air particulates [16]. Each impaction stage carried a sapphire substrate for the collection of the stack aerosol.

Particulate sampling of the stack aerosol was conducted isokineti-
cally out-of-stack. The sampling point was at a stack port where
the stack temperature was 164 °C. The sampler was positioned at
the end of a 8 ft. (2.4 m) long sampling train, with the probe
and sampler heated to 94 °C to prevent condensation of moisture
and acid mist. Short sampling times were employed to furnish low
yield particle deposits. The power plant monitored is one
operated without emission controls. During the test, the unit
was operated at an excess boiler oxygen level of about 0.2 percent,
using a fuel of sulfur content 2.5 percent with concentrations of
vanadium about 400 ppm. Chemical analysis of the fuel also showed
trace concentrations of Ni (16 ppm), Fe (6 ppm) and Mg (5 ppm).
The particulate emissions from this and other such generating
stations have been characterized by EPA investigations [76,77].
These studies have included the determination of particulate
mass, particle size distribution and trace element composition
from bulk collections obtained by in-stack and out-stack sampling
methods. Elemental analysis by x-ray fluorescence shows compounds
of sulfur, vanadium and nickel to be major components as expected
from the fuel analysis. The molecular form of these compounds
(e.g., oxides, sulfates, etc.) cannot be inferred from these data
on the elemental composition of bulk samples. The carbon content
of these samples is found to be typically 60-80 percent, depending
on the amount of excess boiler oxygen used in combustion.

Upon receipt from the power plant sampling site we have
characterized by light microscopy a set of five particulate
samples obtained with the cascade impactor. The object was to
take note of the optical properties and morphology of the particles
collected on each stage and to record their location on the
substrate for subsequent micro-Raman analysis. Representative of
the multi-particle deposit on the substrate of the fourth impaction
stage are the two areas of the sample shown in the optical micrographs of Figure 5. The low magnification (50X) micrograph shows
the central area of the substrate. This field of view contains
various types of solid particulates, some of which appear to have
formed or grown on the collection surface. Large differences are
noted in the size, shape and overall morphological appearance of
the particles. Observation of the sample in the light microscope
furthermore shows wide variations in optical properties, with
microparticles of dendritic morphology showing colors from brown
to green and particles of globular shape exhibiting similar
colors in a generally colorless, transparent matrix interspersed
with bits of black, opaque material. Most outstanding among the
dendrite-type particulates collected on this stage is the large
(>200 μm), four-leaf crystal seen in the 50X micrograph, as well
as the many groupings of microdendrites on this stage, of which a
typical one is shown in the 312X micrograph.

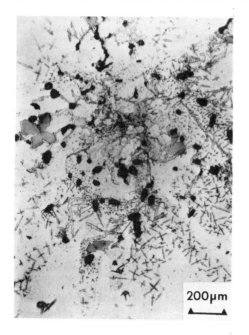 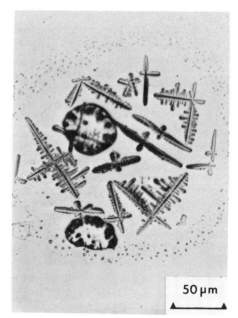

Figure 5.   Optical micrographs of oil-fired power plant stack
             particulates collected on a sapphire substrate.
             Emissions found on the fourth stage of a five-stage
             impaction sampler. Magnification (a) 50X, (b) 312X.

        The vast majority of the microparticles and microcrystals on
this and the other stages of the sampler are too large to be
accounted for by the size cut-off characteristics of each stage.
This implies that much of the particulate material found in these
samples has formed — or grew — on the collection surface through
the interaction of solid and/or liquid microparticles with other
components of the gas/vapor phase of the sampled stack aerosol.
Our observations of the particulate material collected on the
other stages similarly confirm the reactive (and corrosive)
nature of the stack emissions.

        Basically three representative types of particulates have
been probed in this sample. One type consists of brown microcrystals of dendrite morphology, another of globular microparticles.
Both of these two distinct types are shown in the 312X micrograph.
The globules do not appear to be porous, but rather seem to
consist of a fused mass of crystalline material — perhaps refractory
cenospheres — with various degrees of transparency. Spectroscopic
analysis indicates that these particles are inhomogeneous in

composition. A third type, or formation, of material analyzed appears in this sample as colorless, thin, transparent strands — sometimes sheets — connecting the two types of particulates described above.

3. Raman Spectra of Stack-Suspended Particles

In the following we describe the results of spectroscopic measurements which have led to the identification of one major molecular species present in particles of the sample referred to above. The kinds of particles which we have investigated most extensively — up to this point — have been those with dendrite morphology. In this sample of stage 4, the size of these microcrystals varies from 200 µm to below 5 µm. The true coloration (in transmitted light) may be various shades of light and dark brown, sometimes with a tinge of green.

Because of its outstanding appearance, the large, four-leaf crystal (shown at 50X magnification in Fig. 5) was one of the first to be analyzed. Several areas or segments of this crystal were probed, using varying measurement parameters (e.g., size of beam spot, irradiance, etc.), and a spectrum was obtained in each of these measurements. Representative of these spectra is that shown in Figure 6, obtained from probing a small area of one of the crystal leaves. The predominant features in this and the other spectra observed from this crystal are bands with Raman shifts around 145, 280 and 1000 cm^{-1}. Taken together these are characteristic of vanadium pentoxide (V_2O_5). To verify the existence of V_2O_5 in particles of this type, we have studied pure, crystalline V_2O_5 and obtained its spectrum from a particle

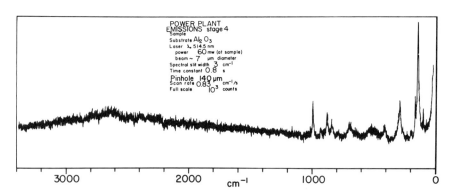

Figure 6. Raman spectrum recorded from probing a small region of the large, four-leaf microcrystal shown in the 50X micrograph of Figure 5.

approximately 5 μm in size. This result is shown in Figure 7.
Qualitative comparison of the spectrum of the unknown crystal
with that of the known particle shows good agreement, except for
several medium-intensity bands in the region 820 to 970 cm^{-1}
which are absent in pure V_2O_5. The vibrational Raman spectrum of
single-crystal V_2O_5 has been discussed in the literature [79,80].
The bands which are well resolved in the reference spectrum
obtained from the microparticle have been assigned frequency
shifts 104, 144, 285, 406, 701 and 995 cm^{-1}. These same bands
are observed in the spectrum of the crystal where they have
equally good definition and have also been found in the spectra
of many other microcrystals of this type which we have investigated.
The additional bands seen in the spectrum of the crystal and
which fall into the region 820-970 cm^{-1} are due to the presence
of a second component, so far unidentified. We have not detected
these features in the spectra of other types of particles in this
sample. These are, however, absent in the spectra of numerous
other brown, dendritic microcrystals (of sizes down to 5 μm)

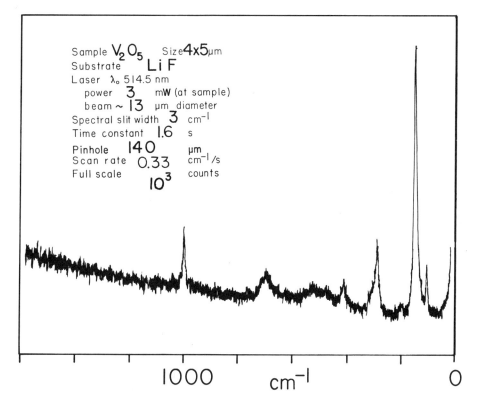

Figure 7. Raman spectrum of a microparticle of crystalline
vanadium pentoxide.

whose spectra — in all respects — are identical to that shown for
V_2O_5. In the environment, vanadium exists largely in the +5
oxidation state [81]. It is known that in the oxidation at high
temperatures of vanadium to V_2O_5 also one or more lower oxides
may be formed. These possible oxides are V_2O_3 and V_2O_4 (or VO_2).
Since no spectral data appear to exist for these species it is not
clear whether the bands in the 820-970 cm^{-1} region are due to
these lower oxides.

     In parallel with our study of V_2O_5, we have also examined
the spectra of other compounds containing oxygenated vanadium.
An example is the spectrum, shown in Figure 8, of a microparticle
of reagent grade, crystalline vanadyl sulfate ($VOSO_4 \cdot 2H_2O$), a salt
of bright blue color. Infrared data for this solid have been
reported [82] but Raman data do not appear to exist. The most
intense band in the spectrum we attribute to the symmetric S-O
stretching mode of the SO_4^{2-} ion with Raman shift $\sim$1010 cm^{-1}. Two
other sharp bands of lesser intensity appear at higher frequencies

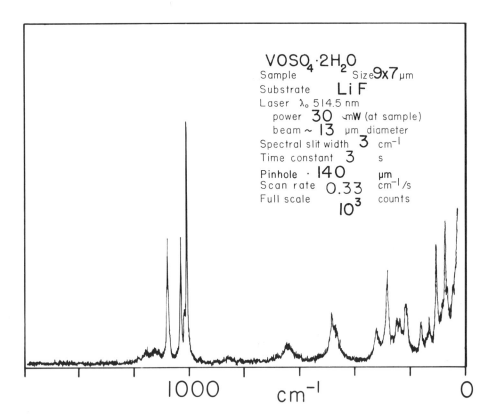

Figure 8. Raman spectrum of a microparticle of hydrated
           vanadyl sulfate.

of the symmetric sulfate stretch, and these may be due to sulfate asymmetric stretching modes. The bands in the region 420-700 cm^{-1} most probably arise from the symmetric and asymmetric bending modes of the SO_4^{2-} group in the crystal. The well-resolved sharp bands seen in the low frequency (<400 cm^{-1}) region may be assigned to various V-O vibrations and external modes of the crystalline lattice. None of the bands characteristic of $VOSO_4$ are present in the spectra of any of the stack microparticles we have analyzed thus far. We therefore feel that the vanadium present in these particles does not exist in the form of the vanadyl ion, VO^{2+}, which is one of the most stable species formed by tetravalent vanadium under many conditions.

We have also considered the vibrational Raman spectrum of the (ortho-) vanadate ion, VO_4^{3-}, — bearing pentavalent vanadium —, and are inclined to conclude that the existence of this species is not indicated in the spectra we have recorded from these particles. The vanadate anion is tetrahedral, as are SO_4^{2-} and PO_4^{3-}. Therefore, four fundamental modes are expected to be active in the Raman. For the "free" vanadate ion, these internal vibrational modes are observed [40] at Raman shifts 827, 780, and 341 cm^{-1}. Of these the strongest band is that due to the symmetric V-O stretch at 827 cm^{-1}, but this line is not observed in the spectra of the stack microparticles.

Because of the scarcity of Raman data in the literature for crystalline vanadates (including the meta-vanadate ion, VO_3^{-}), our studies in this area of application presently involve the spectro-scopic characterization of a number of these vanadium containing solids (e.g., Na_3VO_4, KVO_3).

Other types of microparticles contained in the stage 4 emissions sample have yielded spectra characteristic of other compounds. The particles of globular shape (cf. Figure 5) have furnished spectra many of which indicate the presence of crystal-line sulfate, — other than $VOSO_4$. Work continues on the identifi-cation of these solids with regard to the associated cation species. These spectra typically show four well resolved bands, with a strong, intense band around 990 cm^{-1}, and somewhat broader bands of lesser intensity centered around 625, 460 and 150 cm^{-1}. These spectral features are indicative of crystalline double sulfates, and in particular hydrated metal ammonium sulfates. These results require further study. Possible sources of sample contamination by NH_4^+ have also come under consideration.

Many of the spectra obtained from microparticles in this sample also show the kinds of carbon bands in the 1200-1700 cm^{-1} region which we have observed in the spectra of urban dust particles (cf. Figures 1 and 2). In these cases they probably derive

either from combustion–produced carbon (e.g., soot) or from
residual hydrocarbon matter (such as polycyclic organic matter)
associated with the particles. In many of these measurements,
sample heating effects have been observed.

Raman microprobe analysis of stack particulates sampled by
impaction methods is continuing. Of particular interest to us is
the spectroscopic study of liquid microparticles collected on the
fifth stage of the sample collection investigated to this point.
On this substrate, much of the impacted aerosol material exists
as a liquid under ambient atmospheric conditions. Based on
preliminary results obtained in the micro-Raman study of sulfate
(including sulfuric acid) aerosol microdroplets [16], we expect
to be able to acquire spectra of the liquid material found in
these samples.

Our finding of V_2O_5 as a component of these particulate
emissions is consistent with recent results reported from a study
of oil ash particles by Auger microprobe analysis [78]. From the
peak height ratios of V and O observed in the Auger spectrum, it
was concluded that these elements existed in these particles as
V_2O_5.

Vanadium (as V_2O_5) is not considered a highly toxic metal
and present levels of vanadium released by human activity do not
seem to represent a significant hazard to human health [81].
However, there is interest in the possible role of certain metal
oxides, — V_2O_5 included — as a constituent of soot particles, and
their potential catalytic role in the oxidation of sulfur dioxide
to sulfuric acid [48].

D.   Study of Microparticulate Organic Pollutants

1.   Characterization of Organic Compounds in the Environment

Many classes of organic compounds (e.g., polycyclic aromatic
hydrocarbons, chlorinated hydrocarbons) are suspected to be
carcinogens. For this reason, there is considerable interest in
determining their concentrations in the environment and their
chemical identity [49,83]. A large fraction of the organic
material in ambient air is believed to be associated with airborne
particulate material, and predominantly associated with atmospheric
particles in the respirable size range [84,85].

Existing microanalytical techniques are not capable of
chemically characterizing organic particles of micrometer size.
Identification of organic microparticles is possible in some
cases by making use of the (polarizing) light microscope, but

only when they are considerably larger than 10 µm. Many naturally occurring organics have characteristic physical features that are recognized by an experienced microscopist [8]. Sometimes micro-chemical methods can be employed to lead to successful identifica-tion of larger particles. The methods in use are infrared spectroscopy and luminescence (fluorescence and phosphorescence) spectroscopy. While infrared techniques have been extended to the analysis of microsamples, as a tool to identify small particles, however, the technique — with modern instrumentation — cannot be extended to particles less than 30-40 µm in size. Thus it appears that the application of Raman spectroscopy to the analysis and characterization of microparticulate organics and polymers can be useful [86,87].

2. Raman Probe Studies of Organic and Polymer Particles

Because of the importance of organic constituents in particu-late samples from environmental sources, we have in exploratory studies applied the Raman microprobe to the characterization of microparticulate organic, polymeric and biological materials. The object of these investigations was to establish practical limits of detection and specificity for microparticles obtained from pure compounds, and to assess from these measurements the potential for successful microanalysis of organic particulate matter in the environment.

In earlier work we have shown Raman spectra from microparticles of several classes of organic compounds and polymers [15,16]. These studies have included the characterization of several organic acids (e.g., benzoic acid), salts of organic acids (e.g., sodium oxalate), amides (e.g., urea), complex steroids (e.g., cholesterol) and halogenated hydrocarbons as solid, crystalline particles of size down to 2 µm. Among the polymers studied as microparticles are various Teflon-like polymers, polystyrene latex and poly(vinyl chloride). In recent measurements, we have extended these studies to other materials, including synthetic polymer fibers and selected biopolymers in microparticulate form.

We have shown in these measurements that analytical-quality spectra with good signal-to-noise ratio can be obtained from many of these materials, and that these spectra are in good agreement with the corresponding spectra of bulk samples reported in the literature. In some of these probe measurements it was necessary to employ low irradiance levels to prevent modification or slow destruction of the organic microparticle. In several other cases, where appreciable absorption of the exciting laser radiation was experienced, measurements were unsuccessful.

Various experimental difficulties must be anticipated in the micro-Raman analysis of organic or biological samples (whether they be fibers, films or microcrystals) which can be expected to be more pronounced than in the probing of most inorganic particles. The primary difficulty is frequently interference due to the fluorescent or luminescent background that accompanies intense laser irradiation of many of these materials. Fluorescence emission may be caused by the sample itself or by impurities. Often impurities alone absorb enough of the incident radiation, cause sample heating and produce a strong background that can easily overwhelm the Raman signal. Sometimes the background decays with time until a suitable Raman signal-to-noise level is reached. A potentially important factor in the high-intensity irradiation of organics is the possibility of photo-oxidation of the sample which is known to lead to all kinds of products which may give rise to fluorescence which would normally not derive from the parent compound.

Experimental difficulties have been encountered in probe measurements on microcrystals of benzo[a]pyrene, for example, a compound representative of a class of polynuclear aromatic hydro-carbons (PAH's) that have been shown to exist with air particulate matter [83]. Measurements performed with the lowest practical irradiance levels (typically 2 mW at sample in 20 μm beam spot) have caused enormous background signal levels with excitation at 514.5 nm. The fluorescence excitation and emission properties of this hydrocarbon have been discussed in the literature [88]. Thus, in the microprobing of organics, polymers and biological materials we have generally been successful when irradiances have been kept at minimum levels with a defocused beam, and in the absence of significant fluorescence by the sample and negligible absorption by colored samples. An example is the spectroscopic study of microparticles of highly purified, crystalline bovine serum albumin (BSA), a globular protein. Using lowest laser power densities, spectra were obtained that are in good agreement with that reported for BSA elsewhere [89].

3. Spectra of Pesticide Microparticles

Preliminary micro-Raman measurements have been conducted to explore the potential of the technique for the identification of microparticulate pesticides in the environment. We have therefore in recent studies examined the Raman spectra of microparticles of several chlorinated hydrocarbon pesticides.

Certain organochlorine insecticides have been recognized as potentially harmful, persistent environmental chemicals [90]. It is of interest to study their transport in the air and their transformation in the environment. For example, it is believed that in the atmosphere pesticides may be converted photochemically, and that certain types are strongly adsorbed onto air pollution particles.

Pesticides are usually analyzed as constituents of environmental samples by the combined techniques of chromatography and mass spectrometry. Limits of detection by this method are typically nanogram quantities of pesticides in solvent extracts of soils and other bulk samples [91]. Very few studies describe the application of spectrochemical methods, such as analysis by infrared and Raman spectroscopy [92-96], to the characterization of these chemicals. The available data on the infrared and Raman analysis of pesticides are limited to the investigation of bulk environmental samples where trace quantities of these compounds are determined. In one study [94], Fourier-transform infrared spectra of several pesticides adsorbed on thin layer chromatographic plates could be obtained from microgram quantities of sample, but Raman spectroscopy was reported to be less sensitive, requiring at least 200 μg of the pure pesticide sample.

For our studies we have chosen to investigate a number of chlorinated hydrocarbon pesticides. Because they are weak infrared absorbers, they could be expected to show strong Raman activity. They are colorless, crystalline solids and therefore will not heat and possibly give rise to characteristic lattice modes. Furthermore, fluorescence emission would not be expected from these compounds.

These pesticide materials were obtained as analytical reference standards (of minimum purity 99+%), from the Reference Standards Repository of the EPA Environmental Toxicology Division [97].

Among the compounds studied are the two closely related chlorocarbon insecticides kepone and mirex. Both compounds have received increasing concern in the recent literature as toxic environmental pollutants [91,98]. They are similar to other chlorinated organic compounds, such as the pesticides chlordane, aldrin, dieldrin and endrin. Both have cage-like molecules (cf. Figure 9) and the two chemicals have many similar properties. For example, they have very low vapor pressures and degrade in nature very slowly, persisting for many years in the environment. Their application, chemical action and transformation in the environment has been described [91,98].

We have obtained good spectra from microparticles of kepone and mirex of size down to 2 μm. An example of these results is shown in Figure 9. Included with the spectrum of each compound is the molecular structure of the solid. The close structural similarity is noted. In kepone, two carbon-chlorine bands are

replaced by doubly-bonded oxygen. The two particles are of about
equal size (∿5.5 μm) and are supported by a LiF substrate.   The
spectrum of the kepone particle was obtained at somewhat higher
irradiance.   Other differences in measurement parameters are
indicated in Figure 9.

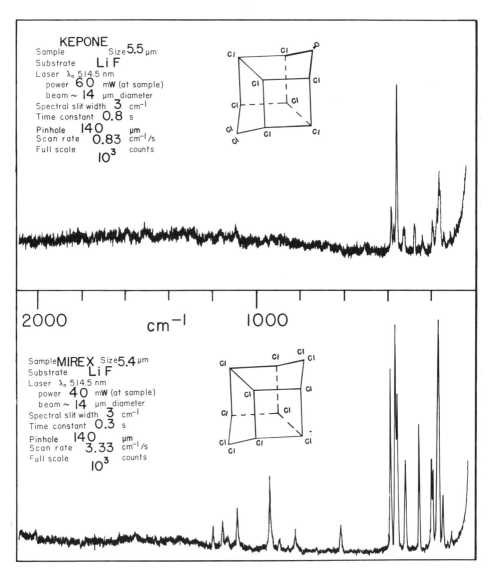

Figure 9.   Raman spectra recorded from microparticles of the
            chlorocarbon insecticides kepone and mirex.

A number of well-resolved, sharp bands of good intensity are seen in the spectrum of each microparticle. The background signal level is higher for the kepone particle. Many of the strong bands in both spectra at frequency shifts lower than 600 cm^{-1} can probably be ascribed to normal vibrational modes of the C-Cl linkages in these compounds. Laser Raman data do not seem to exist for either compound. Some infrared and (non-laser) Raman work has been reported for mirex in solution as part of the structural elucidation of the compound [99]. Although they are structurally very similar, the two compounds may readily be distinguished on the basis of the Raman spectrum. While in fact it may be comforting to know the origin of the Raman bands, it is not essential for qualitative identification. Our experiments and those of other investigators [96,100] show that the spectra of the common pesticides can serve as a fingerprint for a given compound. Thus, Raman spectroscopy for these types of compounds can be very specific. For example, the two geometrical isomers dieldrin and endrin can be distinguished on the basis of their spectra in the low frequency region [100]. This shows that even though closely related compounds have similar spectra, spectral regions can always be found which allow positive discrimination between the various possibilities.

Work on microparticulate pesticides is continuing in our laboratory. The Raman spectra of other types of pesticidal compounds are being investigated to serve as reference spectra in the analysis of environmental samples. Air particulate samples have been obtained for micro-Raman study which indicate (from bulk analyses) the presence of high concentrations of specific chlori-nated hydrocarbon pesticides. The objective of these studies is to detect and identify these compounds in microparticulate form or as adsorbates on airborne dust particles. Detection of these compounds as surface films, for example, may be more difficult. This is likely to require future work addressed to the spectro-scopic study of these materials as adsorbates on various types of mineral adsorbents (e.g., silica, alumina). It would also require examining the effect of adsorption on the Raman bands of these molecules.

The above examples of application of the probe to the spectro-scopic characterization of microparticulate organic compounds, demonstrate the potential for identification of environmental organic pollutants. A separate area of future investigation is the application of the technique to biological specimens (e.g., tissue samples). Raman spectroscopy has evolved as an effective method for studying the structure and conformation of biological molecules and has become a conformational probe in biochemistry for the study of bulk samples (solids and liquids) and of intact biological specimens [101,102]. The hope is, therefore, that

Raman microprobing can be employed to obtain molecular information
at the cellular level. With the development of the technique and
future advances in micro-Raman instrumentation there appear to be
opening up great possibilities of application to pathology and
tissue research.

IV. CONCLUSIONS

In applying the Raman microprobe to four areas of particle
microanalysis, we have given an overview of the kinds of problems
which can be successfully dealt with by micro-Raman spectroscopy.
The results obtained in each of these fields of study indicate
both the unique capabilities of the technique as well as its
inherent limitations as an analytical tool. We perceive as one
of the strengths of the technique the fact that it is emerging as
a valuable complement to other microprobe techniques, such as
electron probe and ion probe microanalysis, which furnish elemental
data on microsamples.

At this point in the development of the micro-Raman method
entirely new information can be obtained from particles or microscopic regions of a specimen no smaller than 1 μm. Molecular
microprobing in the submicrometer domain is not feasible on a
routine basis at this time, either with the NBS developed instrument or with commercially available instrumentation introduced
recently [25].

Various factors determine the outcome of a Raman microprobe
measurement and the amount of useful analytical information
obtained from the sample. The important variables are the
molecular nature of the species itself, its concentration or
abundance in the particle as well as the compositional complexity
of the matrix in which the species resides. These factors and
their relationship to single particle analysis are only qualitatively understood at best, yet they are fundamental to the phenomena
observed in the Raman scattering from microparticles. We cannot,
from the examples discussed, draw firm conclusions regarding the
ultimate limits of detection imposed by either particle size,
particle heating, the identity of the species, its concentration
or distribution, or the compositional complexity of the microparticle. These questions have not been addressed in this work
but will be examined in future studies.

We hope to have shown that Raman spectroscopy permits the
molecular analysis of a broad variety of environmentally significant
materials existing in microparticulate form. In any program of
environmental research, this kind of microanalytical capability
can be expected to yield new answers to problems that could only
be partially addressed before.

V.   ACKNOWLEDGMENTS

We are indebted to the Environmental Protection Agency, Research Triangle Park, N.C., for financial support of our research on particulate emissions from stationary sources, provided under Interagency Agreement EPA-IAG-D6-F012. We thank K.T. Knapp and R.L. Bennett of the EPA/RTP for collecting and making available to us the power plant stack particulates examined in this study. Finally, we are pleased to acknowledge numerous helpful discussions with K.F.J. Heinrich of the Microanalysis Section at the NBS.

VI.   REFERENCES

[1] CADLE, R.D.:   The Measurement of Airborne Particles.   John Wiley & Sons, Inc., New York (1975).

[2] STERN, A.C., Ed.:   Air Pollution, 3rd Edit., Vol. 3.   Academic Press, Inc., New York (1976).

[3] MALISSA, H.:   Angew. Chem. Int. Edit. Engl. 15 (1976) 141.

[4] DAMS, R., ROBBINS, J.A., RAHN, K.A. and WINCHESTER, J.W.: Anal. Chem. 42 (1970) 861.

[5] DZUBAY, T.G. and STEVENS, R.K.:   Environ. Sci. Technol. 9 (1975) 663.

[6] COLLINS, G.C.S. and NICHOLAS, D.:   Analyst (London) 101 (1976) 901.

[7] BIGG, E.K., ONO, A. and WILLIAMS, J.A.:   Atmos. Environ. 8 (1974). 1.

[8] McCRONE, W.C. and DELLY, J.G.:   The Particle Atlas, Edition Two.   Ann Arbor Science Publishers, Inc., Ann Arbor, Michigan (1973).

[9] DINGLE, A.N. and JOSHI, B.M.:   Atmos. Environ. 8 (1974) 1119.

[10] ARMSTRONG, J.T. and BUSECK, P.R.:   Anal. Chem. 47 (1975) 2178.

[11] McHUGH, J.A. and STEVENS, J.F.:   Anal. Chem. 44 (1972) 2187.

[12] LINTON, R.W., LOH, A., NATUSCH, D.F.S., EVANS, C.A. and WILLIAMS, P.:   Science 191 (1976) 852.

[13] GAVRILOVIC, J. and MAJEWSKI, E.:   Amer. Lab. (April 1977) 19.

[14] McDONALD, N.C., RIACH, G.E. and GERLACH, R.L.: Res. & Devel. (Aug. 1976), 42.

[15] ROSASCO, G.J. and ETZ, E.S.: Res. & Devel. 28 (June 1977) 20.

[16] ETZ, E.S., ROSASCO, G.J. and CUNNINGHAM, W.C.: The Chemical Identification of Airborne Particles by Laser Raman Spectroscopy, in Environmental Analysis (Ewing, G.W. Ed.), Academic Press, New York (1977), in press.

[17] ANDERSEN, A., Ed.: The Raman Effect, Vols. 1 and 2. Marcel Dekker, Inc., New York (1971).

[18] COLTHUP, N.B., DALY, L.H. and WIBERLEY, S.E.: Introduction to Infrared and Raman Spectroscopy, 2nd Edit., Academic Press, Inc., New York (1975).

[19] HENDRA, P.J.: Laser Raman Spectroscopy, Ch. 2, Vibrational Spectra and Structure, Vol. 2 (Durig, J.R. Ed.). Marcel Dekker, Inc., New York (1975).

[20] ROSASCO, G.J., ETZ, E.S. and CASSATT, W.A.: Appl. Spectrosc. 29 (1975) 396.

[21] KERKER, M.: J. Colloid Interface Sci. 58 (1977) 100.

[22] BENNET, H.S. and ROSASCO, G.J.: Resonances in the Efficiency Factors for Absorption, Mie Scattering Theory. Paper submitted to Appl. Opt.

[23] DELHAYE, M. and DHAMELINCOURT, P.: J. Raman Spectrosc. 3 (1975) 33.

[24] BRIDOUX, M. and DELHAYE, M.: Time-Resolved and Space-Resolved Raman Spectroscopy, Ch. 4, Advances in Infrared and Raman Spectroscopy (Clark, R.J.H. and Hester, R.E. Eds.). Heyden & Son Ltd., London (1976).

[25] DELHAYE, M., DaSILVA, E. and HAYAT, G.S.: Amer. Lab. (April 1977) 83.

[26] LIU, B.Y.H., Ed.: Fine Particles: Aerosol Generation, Measurement, Sampling and Analysis, Academic Press, New York (1976).

[27] MARPLE, V.A. and WILLEKE, W.: Atmos. Environ. 10 (1976) 891.

[28] MITCHELL, R.I. and PILCHER, J.M.: Ind. Engng. Chem. 51 (1959) 1039.

[29] TANNER, R.L. and NEWMAN, L.: J. Air Poll. Control. Assoc. 26 (1976) 737.

[30] FUCHS, N.A.: Atmos. Environ. 9 (1975) 697.

[31] KING, R.B., FORDYCE, J.S., ANTOINE, A.C., LEIBECKI, H.F., NEUSTADTER, H.E. and SIDIK, S.M.: J. Air Poll. Control Assoc. 26 (1976) 1073.

[32] DHAMARAJAN, V., THOMAS, R.L., MADDALONE, R.F. and WEST, P.W.: Sci. Total Environ. 4 (1975) 279.

[33] CHARLSON, R.J., VANDERPOL, A.H., COVERT, D.S., WAGGONER, A.P. and AHLQUIST, N.C.: Atmos. Environ. 8 (1974) 1257.

[34] CUNNINGHAM, P.T. and JOHNSON, S.A.: Science 191 (1976) 77.

[35] ROBERTS, P.T. and FRIEDLANDER, S.K.: Atmos. Environ. 10 (1976) 403.

[36] PARK, K.: Phys. Lett. 22 (1966) 39.

[37] DONOGHUE, M., HEPBURN, P.H. and ROSS, S.D.: Spectrochim. Acta 27A (1971) 1065.

[38] PORTO, S.P.S., GIORDMAINE, J.A. and DAMEN, T.C.: Phys. Rev. 147 (1966) 608.

[39] ROSS, S.D.: Inorganic Infrared and Raman Spectra. McGraw-Hill Book Co. (UK) Limited, Maidenhead, Berkshire, England (1972).

[40] GRIFFITH, W.P.: Raman Spectroscopy of Terrestrial Minerals, Ch. 12, Infrared and Raman Spectroscopy of Lunar and Terrestrial Minerals (Karr, C. Jr. Ed.), Academic Press, New York (1975).

[41] KRISHNAN, R.S. and KUMARI, C.S.: Proc. Indian Acad. Sci. (Sect. A.) 32 (1951) 105.

[42] ANANTHANARAYANAN, V.: Indian J. Pure Appl. Phys. 1 (1963) 58.

[43] KRISHNAMURTHY, N. and SOOTS, V.: Can. J. Phys. 49 (1971) 885.

[44] ROSEN, H. and NOVAKOV, T.: Nature 266 (1977) 708.

[45] ROSEN, H. and NOVAKOV, T.: Chemical Characterization of Atmospheric Aerosol Particles using Raman Spectroscopy, submitted to Atmos. Environ. (May 1977).

[46] TUINSTRA, F. and KOENIG, J.L.: J. Chem. Phys. 53 (1970) 1126.

[47] SOLIN, S.A. and KOBLISKA, R.J.: Raman Scattering from Carbon
     Microcrystallites and Amorphous Carbon, pp. 1251-1258,
     Amorphous and Liquid Semiconductors, Vol. 2 (Stuke, J. and
     Brenig, W. Eds.) Taylor & Francis Ltd., London (1974).

[48] BARBARAY, B., CONTOUR, J.P. and MOUVIER, G.: Atmos. Environ.
     11 (1977) 351.

[49] SUESS, M.J.: Sci. Total Environ. 6 (1976) 239.

[50] PORTO, S.P.S. and KRISHNAN, R.S.: J. Chem. Phys. 47 (1967)
     1009.

[51] GRIFFITH, W.P.: J. Chem. Soc. (A) (1970) 286.

[52] MROZ, E., CUNNINGHAM, W.C., ETZ, E.S. and ZOLLER, W.H., paper
     in preparation for publication in J. Geophys. Res.

[53] HANSEN, L.D., WHITING, L., EATOUGH, D.J., JENSEN, T.E. and
     IZATT, R.M.: Anal. Chem. 48 (1976) 634.

[54] SAWYER, R.N.: Environ. Res. 13 (1977) 146.

[55] ROHL, A.N., LANGER, A.M. and SELIKOFF, I.J.: Science 196
     (1977) 1319.

[56] BRUCKMAN, L., RUBINO, R.A. and CHRISTINE, B.: J. Air Poll.
     Control Assoc. 27 (1977) 121.

[57] MEURMAN, L.O., KIVILUOTO, R. and HAKURA, M.: Br. J. Ind.
     Med. (1974) 105.

[58] LIGHT, W.G. and WEI, E.T.: Nature 265 (1977) 537.

[59] FLICKINGER, J. and STANDRIDGE, J.: Environ. Sci.
     Technol. 10 (1976) 1028.

[60] BROWN, A.L., TAYLOR, W.F. and CARTER, R.E.: Environ. Res.
     12 (1976) 150.

[61] HARINGTON, J.S., ALLISON, A.C. and BADAMI, D.V.: Adv.
     Pharmac. Chemother. 12 (1975) 291.

[62] DEER, W.A., HOWIE, R.A. and ZUSSMAN, J.: Rock Forming
     Minerals, Vol. 2 Chain Silicates and Vol. 3 Sheet Silicates.
     Longmans, Green and Co. Ltd., London (1963).

[63] RUUD, C.O., BARRETT, C.S., RUSSELL, P.A. and CLARK, R.L.:
Micron 7 (1976) 115.

[64] CHAMPNESS, P.E., CLIFF, G. and LORIMER, G.W.: J. Microscopy
108 (1976) 231.

[65] BEAMAN, D.R. and FILE, D.M.: Anal. Chem. 48 (1976) 101.

[66] WHITE, W.B.: Structural Interpretation of Lunar and
Terrestrial Minerals by Raman Spectroscopy, Ch. 13, Infrared
and Raman Spectroscopy of Lunar and Terrestrial Minerals
(Karr, C. Jr. Ed.), Academic Press, New York (1975). The
spectrum of a bulk, polycrystalline powder of actinolite has
been reported, cf. p. 325.

[67] LOH, E.: J. Phys. (C) 6 (1973) 1091.

[68] ROSASCO, G.J. and BLAHA, J.J., to be published.

[69] The molecular (primarily infrared) spectra of sheet and chain
silicates are discussed in the following two reference texts:
LAZAREV, A.N.: Vibrational Spectra and Structure in Silicates,
Consultants Bureau, New York (1972); and FARMER, V.C., Ed.:
The Infrared Spectra of Minerals, Mineralogical Society
Monograph 4. Mineralogical Society, London (1974).

[70] ISHII, M., SHIMANOUCHI, T. and NAKAHIRA, M.: Inorg. Chim.
Acta 1 (1967) 387.

[71] H.J. WHITE: J. Air Poll. Control Assoc. 27 (1977) 114.

[72] BLOCK, C. and DAMS, R.: Environ. Sci. Technol. 10 (1976)
1011.

[73] GLADNEY, E.S., SMALL, J.A., GORDON, G.E. and ZOLLER, W.H.:
Atmos. Environ. 10 (1976) 1071.

[74] FISHER, G.L., CHANG, D.P.Y. and BRUMMER, M.: Science 192
(1976) 553.

[75] GOLDSTEIN, H.L. and SIEGMUND, C.W.: Environ. Sci. Technol.
10 (1976) 1109.

[76] KNAPP, K.T., CONNER, W.D. and BENNETT, R.L.: Physical
Characterization of Particulate Emissions from Oil-Fired
Power Plants. In: Proc. of the 4th Nat'l Conf. on Energy
and the Environment, Cincinnati, Ohio, Oct. 1976.

[77] BENNET, R.L. and KNAPP, K.T.: Chemical Characterization of
     Particulate Emissions from Oil-Fired Power Plants. In:
     Proc. of the 4th Nat'l Conf. on Energy and the Environment,
     Cincinnati, Ohio, Oct. 1976.

[78] CHENG, R.J., MOHNEN, V.A., SHEN, T.T., CURRENT, M. and
     HUDSON, J.B.: J. Air Poll. Control Assoc. 26 (1976) 787.

[79] GILSON, T.R., BIZRI, O.F. and CHEETHAM, N.: J. Chem. Soc.
     (Dalton) (1973) 291.

[80] BEATTIE, I.R. and GILSON, T.R.: J. Chem. Soc. (A) (1969)
     2322.

[81] HOPKINS, L.L.: Vanadium, Ch. 9, Geochemistry and the
     Environment, Vol. 2, Report of a Workshop held at Capon
     Springs, W. Va., May 1973; National Academy of Sciences,
     Washington, D.C. (1977).

[82] LADWIG, G.: Z. anorg. allg. Chemie 364 (1969) 225.

[83] TOMINGAS, R., VOLTMER, G. and BEDNARIK, R.: Sci. Total
     Environ. 7 (1977) 261.

[84] KETSERIDIS, G., HAHN, J., JAENICKE, R. and JUNGE, C.:
     Atmos. Environ. 10 (1976) 603.

[85] DONG, M.W., LOCKE, D.C. and HOFFMANN, D.: Environ. Sci.
     Technol. 11 (1977) 612.

[86] DOLLISH, F.R., FATELEY, W.G. and BENTLEY, F.F.: Characteristic
     Raman Frequencies of Organic Compounds. John Wiley & Sons,
     Inc., New York (1974).

[87] GALL, M.J., HENDRA, P.J., WATSON, D.S. and PEACOCK, C.J.:
     Appl. Spectrosc. 25 (1971) 423.

[88] SAWICKI, E.: Talanta 16 (1969) 1231.

[89] FRUSHOUR, B.G. and KOENIG, J.L.: Raman Spectroscopy of
     Proteins, Ch. 2, Advances in Infrared and Raman Spectroscopy,
     Vol. 1 (Clark, R.J.H. and Hester, R.E. Eds.) Heyden & Son
     Ltd., London (1975).

[90] KLEIN, W.: Environmental Pollution by Insecticides, pp. 65-
     95, The Future for Insecticides, Needs and Prospects.
     Advances in Environmental Science and Technology, Vol. 6
     (Metcalf, R.L. and McKelvey, J.J. Jr. Eds.) Wiley-Interscience,
     New York (1976).

[91] CARLSON, D.A., KONYHA, K.D., WHEELER, W.B., MARSHALL, G.P. and ZAYLSKIE, R.G.: Science 194 (1976) 939.

[92] ALLEY, E.G., LAYTON, B.R. and MINYARD, J.P. Jr.: J. Agr. Food Chem. 22 (1974) 727.

[93] CHEN, J.-Y.T. and DORITY, R.W.: J. Assoc. Offic. Anal. Chem. 55 (1972) 15.

[94] GOMEZ-TAYLOR, M.M., KUEHL, D. and GRIFFITHS, P.R.: Appl. Spectrosc. 30 (1976) 447.

[95] VICKERS, R.S., CHAN, P.W. and JOHNSEN, R.E.: Spectrosc. Letters 6 (1973) 131.

[96] NICHOLAS, M.L., POWELL, D.L., WILLIAMS, T.R., THOMPSON, R.Q. and OLIVER, N.H.: J. Assoc. Offic. Anal. Chem. 59 (1976) 1266.

[97] Analytical Reference Standards and Supplemental Data for Pesticides and Other Organic Compounds. Publication No. EPA-600/9-76-012, prepared by Analytical Chemistry Branch, Environmental Toxicology Division, Environmental Protection Agency, Research Triangle Park, N.C. (issued May 1976).

[98] STERRETT, F.S. and BOSS, C.A.: Environment 19 (1977) 30.

[99] ZIJP, D.H. and GERDING, H.: Rec. Trav. Chim. Pays-Bas 77 (1958) 682.

[100] NICHOLAS, M.L., POWELL, D.L., WILLIAMS, T.R. and BROMUND, R.A.: J. Assoc. Offic. Anal. Chem. 59 (1976) 197.

[101] YU, N.-T.: CRC Crit. Revs. Biochem. 4 (1977) 229.

[102] LORD, R.C.: Appl. Spectrosc. 31 (1977) 187.

Discussion

ROBILLARD - When you do your analysis of particles, are you seeing
the composition of the entire particle or are you limited to the
surface and subsurface regions?

ETZ - As you might expect, it is difficult to predict from theoretical considerations the relative contributions to the Raman
intensity of the surface and the bulk of a small, irregular-shaped
particle. From microprobe measurements on spherical particles,
we conclude that the Raman signal is approximately proportional
to the volume of the particle. In most work on irregular particles we see no evidence that the Raman microprobe is especially
sensitive to surface layers.

BEAMAN - Can you map the composition distribution in a sample of
many particles? Do you have any problems with thermal damage of
organic materials?

ETZ - In response to your first question, I can set the wavelength
of the monochromator to a particular Raman frequency of interest.
If I wanted to look for ammonium sulfate in the particle sample, I
would select to 976 wavenumber line and then scan the multiparticle sample - as we are doing under computer control - to identify
those particles that give rise to a signal at this frequency. I
should also mention that recently, in parallel with the instrument
development at NBS, a similar Raman microscope/microprobe has been
developed in France which is now commerically available. These
instruments will, no doubt, allow the micro-Raman technique to be
more widely applied. With regard to your second question, in the
case of the measurements on organic microparticles that I have
discussed, there were no problems due to absorption. Also, no
significant heating occurred. In other cases where we have looked
at some polynuclear aromatics, we observed strong fluorescence
emission and, in the extreme case, destruction of the particle.
The only recourse in these cases is to examine the usefulness of
other excitation frequencies, but we simply have not done this to
this point.

X-RAY ANALYSIS OF ENVIRONMENTAL POLLUTANTS

Thomas A. Cahill

Department of Physics and Crocker Nuclear Laboratory

University of California, Davis, California 95616

ABSTRACT

Excitation of characteristic x-rays by electrons, photons, or fast ion beams provides a method of elemental analysis characterized as quantitative, multielement, and nondestructive. Methods of excitation, selection of target configuration, and methods of x-ray detection all lead to diverse optimizations of analytical performance, allowing these methods to be used in a large variety of environmental problems.

INTRODUCTION

X-ray based methods of elemental analysis depend on excitation of atoms through x-rays, electrons, or heavy ions, and subsequent on-line detection of characteristic x-rays emitted during prompt de-excitation. It is a very old field, dating back to the beginning of the 20th century, that has been greatly expanded through improvements in x-ray excitation methods and introduction of energy-dispersive x-ray detectors. In the past three years, it has come to dominate research in the area of atmospheric particulates. This paper will discuss strengths and limitations of the technique in order to understand the reasons for past successes and failures and to predict its impact on areas other than air pollution.

METHODS OF EXCITATION AND DETECTION

Excitation of the atom through removal of K, L, or M shell electrons is usually accomplished by either photons (x-rays or gamma rays), electrons, or heavy ions (protons, alpha particles, etc.).[1] Energies must be adequate to efficiently create vacancies in the required shells, which leads to photon and electron energies in the 2 to 100 keV range and ion energies of a few MeV/amu. The method of excitation is closely coupled to the information desired from the analysis, the nature of the target to be irradiated, and the method of detection of the characteristic x-rays. The major question regarding the desired result of the analyses concerns which elements are required and in what amounts are they present. The target is generally either thin, intercepting a small fraction of the excitation, or thick, in which the excitation is extinguished. Intermediate cases cause calculational complexities that are avoided, if possible. The elemental nature of the target is likewise critical, especially for energy dispersive detection of x-rays. The method of detection is generally either wavelength dispersive, depending on Bragg scattering, or energy dispersive, depending on ionization of a gaseous or solid medium. The former is characterized by excellent energy resolution (to a few electron volts) and poor geometrical solid angle, while the latter has intermediate resolution (around 150 eV), and excellent geometrical solid angle.

Historically, the combination of rather weak photon excitation sources and inefficient wavelength dispersive detection encouraged use of thick targets. This, in turn, yielded difficulties in making analyses quantitative, due to problems of x-ray attenuation and refluorescence which depended upon target composition and morphology. The method, while definitive in elemental assignment, was often only semi-quantitative.

The 1960's and early 1970's saw development of solid state detectors, based on Si(Li), Ge(Li), or intrinsic Ge crystals, that had adequate energy resolution to separate K x-rays of adjacent light elements, about 300 eV; this allowed one to use thin targets in which only a small fraction of the excitation flux was used to excite the sample, since the efficiency of the detectors was close to 100% and solid angles could approach 2π steradians. In addition, excitation methods were advanced through x-ray tubes matched to detector geometry, use of quasi-monochromatic x-ray sources, use of focussed electron beams, and introduction of highly ionizing heavy ions from accelerators. The use of thin targets has resulted in x-ray analysis becoming one of the most accurate absolute methods of analysis in the analyst's arsenal, with absolute values of ±5% easily achieved in routine operations.

Much of the recent excitement in x-ray analysis is based on the possibilities of quantitative nondestructive multielement analyses, of minute amounts of target material, available at the cost of a destructive analysis for a single element by more traditional methods. The key question, in my opinion, is whether multielement analyses are useful in the analyst's problem. The answer, in the case of atmospheric particulates, is "Yes!", and this is probably the most important single reason for the growth of the technique. In the discussion that follows, each of the desirable characteristics of x-ray analysis will be discussed, in order to identify both advantages and limitations of the method.

Multielement Analyses

If the energy of a photon exceeds about 40 keV, it can excite all K x-rays from H to LA, and all L, M, and higher shell x-rays. An energy dispersive detector placed next to a target so excited can, in principle, detect any and all elements simultaneously, using K x-rays up to the rare earths and L x-rays for heavier elements. The same effect can be achieved by a "white" x-ray spectrum up to the same energy, an electron beam of 40 keV or more, or a heavy ion beam of a few MeV/amu, such as a 1 to 6 MeV proton.

In practice, while a monochromatic photon source provides optimum excitation near the excitation energy, for characteristic x-rays between about $E_{EXCIT} \gtrsim E_x \gtrsim \frac{1}{2}E_{EXCIT}$, several such sources (3) may be needed to efficiently cover all elements with x-rays above 1 keV (Na). For an analysis of a single element, such a multiplication of sources is not required. Thus, the range of elements desired becomes important in choice of excitations. Generally speaking, "white" x-ray sources, electrons, or ions provide more uniform excitation when a wide range of elements are of interest. Atmospheric particulates fall into this category, as any element (except noble gases) may be present in particulates, all contribute to total mass loadings, soiling, and light scattering, and many are toxic. The following table illustrates the prevalence of elements in particulates in an urban area:

Elements		% of Total Mass
Very light	H → F	65%
Light	Na → Cl	18%
Medium	K → Ba	15%
Rare earths	La → Lu	<0.1%
Heavy	Hf → U	2%

It is the importance of the elements Na → Cl to particulates
that has encouraged use of ions from accelerators in work on
atmospheric particulates.

Sensitivity

Detection of x-rays is a statistical process, and standard
criterion are used to tell when a peak is statistically significant at the 99.7% confidence level, as compared to the continuum
of x-rays that form a background and/or nearby or interfering
peaks. Thus, the intense background present from electron
bremsstrahlung due to incident electron beams reduces sensitivity
from this excitation method relative to photon and heavy ion
sources. Major use of electron beams involves focussed beams of
micron dimensions, where poor trace capabilities (~0.1%) are
offset by small excitation volumes to produce high mass sensitivity (nanogram to picogram).

The best ion beam systems and the best x-ray fluorescence
systems are roughly comparable in sensitivity, with an edge going
to ion beams for thin targets and light elements, Na → Cl, and
to x-ray excited systems for heavier elements K → U, and thick
samples. The importance of sensitivity for light elements is
emphasized for atmospheric particulates by two considerations:

1. Elements such as Na have such soft x-rays that they
    are quickly absorbed in matrix material unless total
    areal densities are held to no more than a few hun-
    dred micrograms/cm^2, and

2. Most devices that deliver vital information on
    particulate size only work quantitatively if no
    more than a few monolayers of particles are de-
    posited on a filter or impaction surface. A
    monolayer of 1 μ particles of density 2 gm/cm^3
    is about 200 μg/cm^2.

Thus, little mass is available, but little mass can be tolerated!

An important corollary to sensitivity to areal density,
rather than to mass, is that particle collection devices can be
made very small, leading to increased availability and utilization.

While sensitivities calculated for standards can be very
good, to picograms/cm^2, fractional mass sensitivities are normally limited to about 1 ppm of the target present due to

interfering lines and matrix background. Thus, x-ray methods
have only modest trace element capabilities.

Nondestructive

It is always difficult to predict the most important element
in any pollution sample, as interests vary. The ability to
measure all elements (except very light elements) and to retain
the sample for future analyses for elemental or chemical information is a valuable option possessed by all x-ray based methods.
Care must be used in electron or ion-beam analyses, as vacuums
are usually used and loss of volatile components may occur.

Elemental

Lack of sensitivity to chemical species may be a detriment
to some analytical programs. However, the obverse of this fact
is that x-ray analysis is insensitive to the chemical state of
the element, which can introduce serious errors into some other
methods of analysis. Accurate elemental analyses are achieved
in part because of chemical insensitivity.

Quantitative

X-ray analyses can easily be made quantitative when targets
are thin, as the ratio of observed x-ray counts in the elements
of interest can be normalized to beam flux and calibrated through
thin gravimetric elemental standards. The latter are accurate
to about ±5%, absolute. The major problems with the excitation
method involve selective attenuation of incident radiation in the
target and use of spatially small electron and ion beams. The
former problem can be handled, but the latter can be very difficult to handle, as almost any target has a high degree of
spatial variability on a micron scale. For larger excitation
areas that effectively integrate over target spatial inhomo-
geneities, one requires either a uniform excitation area entirely
covering the sample, or a uniform sample extending beyond the
limits of the area of excitation. Either will give an effective
areal density, gm/cm^2, over the area of integration in the first
case or of the uniform sample in the second case. Nonuniform
beams and nonuniform samples have the same problems as finely
focussed beams, and quantitative analyses are difficult.

Target thickness effects induce major corrections due to x-
ray attenuation and refluorescence, unless samples are kept very
thin. This is very important for light elements.

Interference effects in x-ray detection are a major source of nonquantitative analyses for energy dispersive x-ray detectors. Several transitions of different elements may occur together, and even if the elements involved are correctly identified, the amounts present are often only approximate. This effect sets limits of around 1 ppm for the trace capability of x-ray analysis in real targets.

Cost

The cost of such analyses typically range from about $10.00 to $25.00, independent of whether ion beams or x-rays are used to excite the sample. If one is interested in a single element, this cost is similar to that of many other methods. If one is interested in many elements, the cost per element drops sharply, and the method becomes very inexpensive on this basis, at less than $1.00/element for most systems.

SUMMARY

The systems used in work at the University of California, Davis,[2] include x-ray fluorescence with two anodes, ion beam excitation with protons and alpha particles, and electron beams on a scanning electron microscope and an electron microprobe. Each system excels in its own area, and the incremental cost of using more than one excitation source is small compared to information gained.

Almost all the air pollution work is done with ion beams, especially since very light elements can be also detected using ion scattering analysis. Work on plant nutrition is also done with the accelerator's ion beams, as elements Na → Ca are vital to plant growth, along with elements up to about Mo. Most trace metal work in geological samples and water are done with x-ray fluorescence, as abundant sample is available and heavier elements are generally of the greatest interest. Electron beams are used for single particle analyses of selected particulate samples, such as fly ash.

Use of these methods has yielded a great simplification in particulate collection devices, wherein they have become both more quantitative and less cumbersome and expensive.

Generation of enormous numbers of x-ray data has encouraged (almost forced!) us to use statistical techniques to interpret the results. However, this is also the area that is yielding the greatest scientific rewards in the study of the sources, transport, transformations, and sinks of atmospheric particulates.

LITERATURE

REFERENCES

1   JOHANSSON, S.A.E and JOHANSSON, T.: Nucl. Instr. and
    Methods 137 (1976) 173.

2   CAHILL, T.A.: Ion-Excited X-Ray Analyses of Environmental
    Samples, Ch. 1, New Uses of Ion Accelerators (Ziegler, J.,
    Ed.). Plenum. New York (1975).

Discussion

<u>FELDMAN</u> - Did I understand you correctly that the man-made aerosol
is much more dangerous because of its small particle size than the
natural aerosol of the same composition?

<u>CAHILL</u> - That's what I'm told from people who claim to know.

<u>FELDMAN</u> - The reason I ask is because one frequently hears the
idea that a given man-made pollutant isn't significantly dangerous
because it occurs naturally at higher concentrations. You're
suggesting that idea is not always valid.

<u>CAHILL</u> - Good point. We have a situation, for instance, where
there's a lot of sulfur in California at the shoreline and if you
stand in the ocean spray you are exposed to enormous amounts of
natural sulfur coming off the ocean. But the very large particles
settle quickly and are not inspired. Another sulfate aerosol
occurs inland, which appears to be anthropogenic. The aerosol has
a very different particle size and, we assume, a very different
chemistry. So sulfate isn't just sulfate. Its effect depends on
size and associated metals as well as other factors. I have a
feeling that when the jury is finally in on the epidemiological
studies of the effect of sulfate aerosols, the critical factor will
be the fine associated metals with sulfates and not the sulfate
itself. I just don't think the measurements made until now can
really separate those two possibilities. I think we have a lot to
learn.

<u>NEWBURY</u> - What sort of peak overlap corrections do you make when
your energy resolution is poor and you have a complicated spectrum?

<u>CAHILL</u> - We've put a lot of effort into these corrections. Cer-
tainly the most important effort in our whole program for the last
three years has been the automatic computer reduction of the measured spectra. However, there is always a sample that cannot be
analyzed. Our computer is smart enough to know when that happens
and punt. But most of the samples are pretty simple because the
major overlaps are lines of light elements overlapping with L lines
of heavy elements. Since every element has multiple lines, our
philosophy is not to hide these overlaps in a computer program but
to actually print out all the lines seen by the computer. Every
line is an independent measurement of some element present. For
instance, lead gives between three and five lines all identifying
lead, and if the amounts of lead calculated from all these five
lines agree, the computer is very happy and says, "Lead". But for
a mixture of barium, titanium, and chromium we do the best we can
in terms of peak location and amplitudes of secondary lines, both
of which are of course known to very high precision. But every so
often for light loadings close to threshhold the computer will

print out probability of barium 0.65, versus some mixture of titanium, vanadium, and chromium. So there's always a few cases in which confusion persists. I don't think the computer throws out more than one sample in a thousand in routine analyses of aerosols.

GOVE - Do you know the status of the PIXE technique for trace element measurements in human organs?

CAHILL - The groups at Duke, Newport News, and Gainsville have done work with the PIXE technique and organ samples. People are using this technique to do analyses and having success. The only problem is that in the cases I am aware of they really don't have very large programs. They do not seem to be getting the support the aerosol people have been getting. Does it work? The answer is yes and it works well, but I don't know the field well enough to know whether the PIXE technique is competitive with the competition for measuring nickel in organs.

BEAMAN - What kind of count rate do you run with your energy dispersive detector?

CAHILL - We like to run at about 6,000 counts per second. To control count rate we turn off our accelerator for 40 microseconds every time an x-ray hits the detector. This method allows us to run a pretty good count rate and still not have problems with dead time and gain shifts.

MODERN TECHNIQUES OF ACTIVATION ANALYSIS FOR THE MEASUREMENT OF

ENVIRONMENTAL TRACE METALS

R. H. Filby and K. R. Shah

Washington State University
Department of Chemistry and Nuclear Radiation Center
Pullman, Washington 99164

ABSTRACT

The principles of neutron activation, nuclear track, photon activation, and charged particle activation methods are discussed. The application of these techniques to the determination of trace elements in environmental materials is reviewed and recent developments in nuclear methods of environmental trace analysis are illustrated by several examples. The uses of nuclear activation methods in air pollution studies, water pollution studies, and energy production emissions are taken as specific examples of recent work. Reference is made to the accuracy and precision of nuclear activation methods and the need for environmental trace element standards.

Introduction

The presence of such toxic metals as Hg, As and Cd in the environment has caused concern about the distribution of these and other trace elements in biogeochemical systems and has increased the need for more sensitive and accurate methods of trace element analysis. Examples of serious environmental contamination by elements such as Hg and Cd have been well documented and defined, but more research is still needed on natural or background levels of these and other trace metals in order to understand their biogeochemical behavior. It is now realized that, in addition to determining total elemental contents of environmental materials, we need to be able to determine chemical forms, or species, of the elements. The distribution and the differences in toxicity of the inorganic and alkylated forms of Hg illustrate the importance of chemical speciation and the need for analytical methods.

In addition to present sources of environmental contamination, the increasing reliance on coal burning and coal conversion to solve our future energy needs will greatly add to the flux of trace metals in the environment.

The purpose of this paper is to outline the principles of activation analysis as applied to modern environmental analysis and to review some applications to the determination of trace elements in environmental materials. In this review we have included elements that are not regarded as toxic in their forms normally found in the environment (e.g. Mo, P) but which are environmentally important.

The determination of trace elements in environmental materials is often complicated by the low concentrations of the elements, the unknown nature of the chemical species present, and the wide range of chemically complex materials encountered. The determination of chemical species in natural materials, e.g. waters, is an even more difficult task. Water, for example, may contain an element in neutral, cationic, anionic or organic and inorganic complex species as well as colloidal or particulate forms. The nature of the analytical problem thus places some severe restrictions on methods of trace element analysis of environmental materials. A good analytical method for environmental trace analysis should satisfy the following criteria

a) high sensitivity for the element such that concentrations at the sub-ng/g region can be measured;
b) high selectivity in the presence of other elements encountered in the sample
c) good precision and accuracy, preferably better than 5-10% for precision (relative standard deviation);
d) freedom from interferences such as matrix effects, contamination problems, interelement effects;
e) be applicable to a wide range of materials, e.g. waters, biota, sediments, air particulates.

In addition to these necessary criteria, other desirable properties are:
a) the method is a instrumental multi-element technique,
b) identifies the chemical form of the element,
c) the method is rapid and requires a minimum of complex equipment.

No analytical method currently in use for environmental analysis satisfies all of the above. Methods such as flame and flameless atomic absorption spectroscopy (AAS), atomic fluorescence (AAF), optical emission spectroscopy (OES), X-ray fluorescence (XRF), aniodic stripping voltammetry (ASV), and spark source mass spectroscopy (SSMS) have certain advantages and disadvantages and there is no universally applicable trace analysis method.

Activation analysis is a nuclear method of analysis which has developed rapidly during the last twenty years and is now one of the most useful methods of trace analysis. The technique meets all of the analytical criteria for many elements of the periodic table and sensitivities attained are often much higher than for other methods.

TABLE I: Activation Analysis Methods for Environmental Materials

Neutron Activation Methods

    a)   Thermal Neutron Activation
    b)   Fast Neutron Activation
    c)   Epithermal Neutron Activation
    d)   Nuclear Track Methods
    e)   Capture Gamma-Ray Analysis

Photon Activation Methods

Charged Particle Methods

The development of Ge(Li) gamma-ray spectrometry has made multi-element analysis possible in a wide variety of materials. Table I lists the most important methods in current use. Most of the early applications of activation analysis were in materials science and in geochemical, cosmochemical or biological research, but in the past decade activation analysis has been applied to a variety of environmental problems. Activation analysis is now a routine method in many laboratories and recent advances in the field have been refinements of the technique or new applications. A recent review of applications to environmental science up to 1971 has been published by the authors (9). The advantages and disadvantages of activation analysis in environmental research will be described later, but it should be pointed out that the technique does not compare unfavorably in cost with many other methods. A complete multielement analysis system, excluding irradiation facilities, costs less than $30,000. For many applications it is not necessary to possess a nuclear reactor as more than sixty research and production reactors with irradiation facilities available to outside users are to be found in the United States and Canada alone. Many facilities perform irradiations at nominal costs, or at not cost under Reactor Sharing Programs. Portable neutron generators ($10,000-$20,000) are available in many institutions and the increasing availability of intense ^{252}Cf neutron sources will increase the application of activation analysis in many laboratories not having access to a nuclear reactor. Photon activation analysis still requires access to a LINAC or cyclotron.

Principles of Activation Analysis

The principle of activation analysis is the bombardment of an element with neutrons, charged particles, or photons thus causing a nuclear reaction to take place. The radioactive products of the reaction may be identified by gamma-ray energies and intensities,

TABLE II: Nuclear Reactions Used in Activation Analysis

Thermal Neutron   Fast Neutron

(n,γ), (n,f) (n,p), (n,α), $(n,2n)$, $(n,n'\gamma)$

Photon Reactions

(γ,n), (γ,γ'), (γ,n)

Charged Particle Reactions

(p,n), (p,γ), (α,n), (α,p)

(d,n), (d,γ), $(^3He,n)$, $(^3He,p)$

beta-ray energies, or half lives.   The most important activation
reactions are shown in Table II.   The analytical procedure falls
into two parts: irradiation of the sample and the measurement of
the induced radionuclides.   In radiochemical activation analysis,
the nuclide of interest (or group of nuclides) is separated from
other activities and measured, and in instrumental activation analysis the nuclide (or nuclides) is measured without separation.
Most activation analysis methods for trace element determination
employ neutrons because cross sections for thermal neutron induced
reactions are generally much higher than those of charged particle
or photonuclear (γ) reactions, although for some elements (e.g. Pb)
this is not the case.   Also, high neutron fluxes are available in
nuclear reactors and convenient portable neutron generators and
^{252}Cf sources can be used for specific applications.
The reactions commonly used in neutron activation analysis
(NAA) are the (n,γ), (n,α), $(n,2n)$ and $(n,n'\gamma)$, and these are illustrated with reference to the nuclear reactions of arsenic shown
in Table III.

TABLE III: Neutron Reactions of Arsenic

$$^{75}As + {}'n \rightarrow ({}^{76}As)$$

$^{75}As + {}'n$	(n,n)
$^{75m}As + {}'n'$	$(n,n'\gamma)$
$^{76}As + \gamma$	(n,γ)
$^{75}Ge + {}'H$	(n,p)
$^{72}Ga + {}^4He$	(n,α)
$^{74}As + 2{}'n$	$(n,2n)$

The reaction probability depends on the nature of the target nuclide
and the neutron energy and the most useful reaction in NAA is the
(n,γ) reaction with thermal neutrons (0.025 eV most probable energy).
All known stable nuclides (except ^{4}He) undergo the (n,γ) reaction
and cross sections (i.e. reaction probabilities) for thermal neutrons
are often very large (up to 240,000 barns). This fact, and the high
reactor thermal neutron fluxes makes the (n,γ) reaction the most generally applicable to activation analysis. Particle emission reactions (n,p), (n,α), $(n,2n)$, and $(n,n'\gamma)$, have appreciable cross sections only at high neutron energies (>1 MeV) because most reactions
have positive threshold energies. In most research reactors fast
neutron (>1 MeV) fluxes are at least an order of magnitude lower
than thermal neutron fluxes and 14 MeV neutron generators generally
have available fluxes orders of magnitude lower than reactor thermal
neutron fluxes.

For a nuclear reaction of the type $^AX(n,\gamma)^{A+1}X$, the induced
activity, A_t, in disintegrations/sec, at the end of irradiation of
time, t, is given by:

$$A_t = N\sigma\phi(1-e^{-\lambda t}) = N\sigma\phi(1-e^{-0.693t/T}) \tag{1}$$

where σ = thermal neutron capture cross section in cm^2. The
unit of cross section is the barn (1 barn = 10^{-24} cm^2).

ϕ = thermal neutron flux in neutrons cm^{-2}sec^{-1}.

λ = decay constant of the product nuclide.

T = half-life of the product nuclide = $\dfrac{0.693}{\lambda}$

N = number of atoms of A_X in the target.

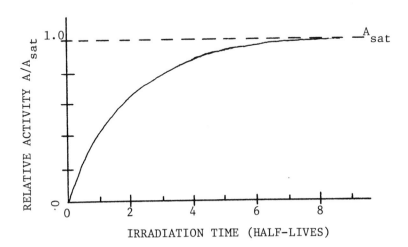

FIGURE 1:   Induced Activity as a Function of Irradiation Time

Figure 1 shows the variations of A_t as a function of irradiation time, t. The expression for the activity, A_t, of the product nuclide at any time, t', after the end of irradiation is given by:

$$A_{t'} = N\sigma\phi(1-e^{-\lambda t}).e^{-\lambda t'} \qquad (2)$$

Similar equations apply to photon and charged particle activation analysis.

Consideration of equation (2) indicates that the following criteria should be met for a practical, sensitive neutron activation method.

1. The cross section, σ, (and the isotopic fraction, f) of the parent stable nuclide should be as large as possible.
2. The half-life of the product nuclide should be neither very long nor very short compared to the irradiation time.
3. The induced radionuclide should be readily detectible and measurable.

The disintegration rate of the product nuclide, which is proportional to the weight of the element in the sample, may be determined by measurement of beta-ray or gamma-ray intensities. For instrumental multielement analysis beta rays are not used because the beta rays from a given nuclide are not monoenergetic. Most radionuclides emit gamma rays of characteristic energies and gamma-ray spectrometry is the preferred detection method for both radiochemical and instrumental activation analysis.

Equation (2) indicates that, in principle, neutron activation is an absolute method of analysis. If t, t', σ, ϕ, λ, and A_t are known, or can be measured, then N, the number of atoms of the target nuclide (and hence the weight of the element present) can be calculated. In practice, most neutron activation methods use the comparator technique. This involves irradiating a standard, which contains a known amount of the element of interest, with the sample. Thus, for sample and standard:

$A_{sa} = N_{sa}\sigma\phi(1-e^{-\lambda t})$ A_{sa} = activity of the nuclide in the sample.

$A_{st} = N_{st}\sigma\phi(1-e^{-\lambda t})$ A_{st} = activity of the nuclide in the standard.

Hence $\dfrac{A_{sa}}{A_{st}} = \dfrac{N_{sa}}{N_{st}} = \dfrac{W_{sa}}{W_{st}} = \dfrac{C_{sa}}{C_{st}}$

where W_{sa} = weight of the element in the sample in µg.

W_{st} = weight of the element in the standard in µg.

C_{sa} = measured activity of nuclide in the sample.

C_{st} = measured activity of nuclide in the standard.

TABLE IV: Neutron and Photon Reactions of Environmental Interest

Element	Reaction	Half-Life	Gamma Energies (keV)
	a) Thermal Neutron (9)		
As	$^{75}As(n,\gamma)^{76}As$	23.6h	559
Se	$^{74}Se(n,\gamma)^{75}Se$	120d	264,279
Se	$^{76}Se(n,\gamma)^{77m}Se$	17s	161
Sb	$^{121}Sb(n,\gamma)^{122}Sb$	67h	564
Sb	$^{123}Sb(n,\gamma)^{124}Sb$	60d	603,1692
Cd	$^{110}Cd(n,\gamma)^{111m}Cd$	49m	150,247
Cd	$^{114}Cd(n,\gamma)^{115}Cd$	54h	530
Hg	$^{196}Hg(n,\gamma)^{197}Hg$	65h	77
Hg	$^{202}Hg(n,\gamma)^{203}Hg$	47d	280
U	$^{238}U(n,\gamma)^{239}U$	24m	75
U	$^{235}U(n,f)^{140}La$	40h	1596
Pu	$^{239}Pu(n,f)FP$	–	–
	b) Photon Reactions (1)		
Ni	$^{58}Ni(\gamma,n)^{57}Ni$	36h	1378,1919
As	$^{75}As(\gamma,n)^{74}As$	17.9d	596,635
Sb	$^{123}Sb(\gamma,n)^{122}Sb$	67h	564
Zr	$^{90}Zr(\gamma,n)^{89}Zr$	78.4h	909
I	$^{127}I(\gamma,n)^{126}I$	12.8d	389,666
Pb	$^{204}Pb(\gamma,\gamma')^{204m}Pb$	67m	899
Pb	$^{204}Pb(\gamma,n)^{203}Pb$	52h	279

Therefore   concentration of element $(\mu g/g) = \dfrac{C_{sa.}W_{st}}{C_{st}.W}$

where W = weight of sample in g.

Table IV lists thermal neutron and photon reactions of some important elements used in activation analysis of envirnmental materials. Table V lists some charged particle (proton) reactions.

The nuclear track technique is being increasingly used in environmental trace analysis for U, Pu, and is also suitable for B, Li, N, and O. In this method the sample is placed in contact with a track detector (mica, quartz, lexan, cellulose nitrate) and irradiated with neutrons. Uranium and plutonium undergo fission and the fission fragments entering the detector produce chemically active ionization paths that can be etched with a suitable reagent (e.g. NaOH) to form visible tracks. The track density is proportional to the concentration of the element in the sample and quantitative measurements can be made either absolutely or more commonly by reference to a known standard. The reactions used for environmental materials are shown in Table VI.

Instrumentation

Irradiation Sources

(i) Nuclear Reactors. Nuclear reactors are the most intense neutron sources available for activation analysis and most applications of activation analysis to environmental problems have utilized research reactors where high thermal neutron fluxes are available. Nearly all research reactors use ^{235}U enriched uranium as the fissionable material for fuel and have the core immersed in water (or less commonly D_2O) which acts as a coolant, and a neutron moderator, and provides flexibility of experimental facilities. Most reactors have a variety of experimental irradiation positions and also pneumatic tube access to the core for rapid transit of a sample from the core to an experimental facility.

TABLE V: Proton Activation Reactions of Environmental Interest(6,23)

Element	Reaction	Half-lifes	Gamma Energies (keV)
As	$^{75}As(p,n)^{75}Se$	120d	264,279
Se	$^{76}Se(p,n)^{76}Br$	4.4h	559,657
Cd	$^{111}Cd(p,n)^{111}In$	2.8d	245
Sb	$^{121}Sb(p,n)^{121}Te$	17d	573,508
Tl	$^{203}Tl(p,n)^{203}Pb$	52h	279
Pb	$^{204}Pb(p,p')^{204m}Pb$	67m	899

TABLE VI: Nuclear Reactions Used in the Track Method[a] (3)

Element	Reaction	Neutron Energy
U	$^{235}U(n,f)FP$	thermal
U	$^{238}U(n,f)FP$	fast
Pu	$^{239}Pu(n,f)FP$	fast
Th	$^{232}Th(n,f)FP$	fast
Li	$^{6}Li(n,\alpha)^{3}H$	thermal
B	$^{10}B(n,\alpha)^{7}Li$	thermal

[a]FP = fission product

   The neutrons emitted in the ^{235}U fission process range in
energy from 0 to 25 MeV and the average fission neutron energy is
2.0 MeV. As thermal neutrons (most probable energy 0.025 eV) are
used for most NAA, irradiation positions are generally placed in
graphite reflectors or in the water region where significant neutron moderation has occurred. Typical values for thermal neutron
fluxes in research reactors range from 10^{11}-10^{14} n.cm^{-2}sec^{-1} depending on the reactor and irradiation position.
   In thermal NAA using (n,γ) reactions, the fast neutron component of the neutron spectrum can give rise to interfering (n,p) or
(n,α) reactions. On the other hand, the fast neutron component can
be used for fast reactions where thermal reactions are not suitable.
An example is the determination of Pb by the $^{204}Pb(n,n'\gamma)^{204m}Pb$ fast
neutron reaction (12). Where fast neutron activation is carried out
with reaction neutrons, the thermal neutron component must be eliminated or reduced to prevent competing (n,γ) reactions. Irradiations
are generally performed in cadmium containers (α_{th} = 2460 barns) or
in ^{10}B enriched boron containers (α_{th} = 3.9 x 10^3 barns) which act
as thermal neutron filters but which are essentially transparent to
fast neutrons.
   Of considerable interest is the development of a small low cost
reactor "SLOWPOKE" by Atomic Energy of Canada. This relatively simple facility should make NAA more available for environmental research (5).
(ii) Accelerators as Neutron Sources. Accelerators, which are
sources of charged particles, may be used to produce monoenergetic
neutrons through (p,n), (d,n), (^{3}He,n) and (γ,n) reactions on light
elements. The most useful of these reactions is the $^3H(d,n)^4$He
reaction which is highly exothermic and produces 14-Mev neutrons.
This reaction forms the basis of most commercial neutron generators
and by bombarding targets of ^{3}H diffused into a metal (usually Pd)
with deuterons, neutron outputs of up to 10^{12} n.sec^{-1} can be obtained, with useful 14-MeV neutron fluxes of up to 5 x 10^{10} n.cm^{-2}
sec^{-1}.

Generally cross sections for fast neutron reactions are much lower
than for thermal (n,γ) reactions. The low cross sections and low
fast neutron fluxes ($<10^{11}$ n.cm^{-2}sec^{-1}) available from generators
limits their application to problems where: a) no suitable (n,γ)
product exists; b) the lower (n,γ) contribution gives much lower
background; and c) light elements which have low (n,γ) cross sections, e.g. oxygen, silicon, nitrogen, and carbon. However, high
output tubes (up to 5×10^{13} n.cm^{-2}sec^{-1}) are now under development
and show promise for environmental applications (27).

(iii) Cf-252 Sources. Californium-252 (half-life 2.55 years) decays by spontaneous fission (3%) and by α-particle emission (97%).
It is thus a neutron emitter with a total neutron output of $2.34 \times$
10^{12} n.sec^{-1}g^{-1} ^{252}Cf and with subcritical (U or Pu) multiplication,
neutron fluxes of the order of 10^{12} n.cm^{-2}sec^{-1} per g ^{252}Cf are possible. At present ^{252}Cf sources of approximately 50 mg are available and effective thermal neutron fluxes are of the order of $5 \times$
10^8 n.cm^{-2}sec^{-1} which is much lower than most reactor thermal neutron fluxes. The ^{252}Cf source does have the advantage of porta-
bility and a number of specific environmental activation methods
have been published using ^{252}Cf. As source strengths increase,
mobile activation analysis facilities and remote activation analy-
sis--including environmental sensors, will be possible.

(iv) Charged Particle Sources. For charged particle activation,
sources of high energy protons, deuterons, or alpha particles are
necessary. For trace analysis relatively high beam currents may
be necessary and most charged particle work must be done with a
cyclotron or linear accelerator.

(v) Photon Sources. For photon activation analysis, use is made
of the bremsstrahlung produced in the slowing down of high energy
(up to 40 MeV) electrons from a LINAC in a stopping medium. The
gamma rays produced must in general have energies of 8 MeV or
greater since neutron emission from (γ,n) reactions requires approximately 8 MeV to overcome the neutron binding energy.

Measurement of Radionuclides

Several radiation detectors can be used to measure the radionuclides produced by irradiation of the element of interest. Proportional counters, Geiger counters, and liquid scintillation techniques, although highly efficient for β-particles have low efficiencies for γ-rays. The β-particles emitted by a radionuclide
show a continuous energy distribution from 0 to E_{max}, thus proportional and Geiger counters can only be used when the radionuclide
of interest has been separated in a radiochemically pure state.
Fortunately most radionuclides produced by (n,γ), (γ,n), (p,n),
(p,γ) and other reactions emit γ-rays of characteristic energies
and γ-ray spectrometry forms the basis of most modern activation
analysis techniques used in environmental research. Large volume
NaI(Tl) detectors have high efficiencies for γ-ray measurement but
resolutions [$(\Delta E/E)100$] are poor (5-10 per cent), and multielement

analysis is best used for materials giving relatively few γ-rays. More complex spectra, such as those commonly obtained with environmental samples, result in many overlapping peaks and spectrum analysis techniques using computers must be employed to resolve individual radionuclide contributions.

The excellent resolution of Ge(Li) detectors (FWHM, 1.5-3 at 1333 keV) allows even complex spectra to be resolved with very few overlapping peaks. At present the size and efficiency of Ge(Li) detectors is governed by Li-drifting limitations and volumes of high resolution (<2 keV) detectors are limited to less than 100 cm^3. Intrinsic Ge detectors which do not need Li compensation are now available in efficiencies approaching the largest Li-drifted detectors. The high resolution of Ge(Li) detectors not only allows more radionuclides to be measured but also permits much simpler computer programs than are possible with NaI(Tl) detectors.

Typical spectra of environmental materials obtained by INAA using Ge(Li) are shown in Figures 2 and 3. Figure 2 shows the spectrum of an irradiated coal and Figure 3 shows irradiated coal miner's lung. The similarities are clearly indicated by the ^{46}Sc, ^{182}Ta, ^{181}Hf and ^{124}Sb, elements not normally seen in non-miner lung tissue.

To achieve the high resolution capabilities of Ge(Li) detectors, it is essential to employ stable low noise electronics. Cooper (7) has reviewed the performance of gamma-ray spectrometers in NAA. The most important contributor to electronic noise is the preamplifier and improvements in preamplifier design permit close to theoretical detector resolution in certain cases. The multichannel analyzer for collecting the spectral data commonly has at least 4096 channels for high resolution γ-ray spectrometry and possesses a range of output devices. The most satisfactory output devices are magnetic tape systems which allow direct input to a computer for data reduction. Recent developments in multichannel analyzers have resulted in minicomputer or microprocessor based systems that permit simultaneous data collection and data reduction.

Analytical Procedures

Typical activation analysis procedures for environmental samples can be divided into:

a) Preparation of samples and elemental standards of known concentrations.

b) Irradiation of samples and standards followed by a decay period to allow unwanted short-lived radionuclides to decay.

c) Post-irradiation treatment of the sample. For INAA this step is usually confined to transfer of the irradiated material to a given counting geometry but chemical separation of the radionuclide of interest may be necessary to improve sensitivity.

d) Measurement of radionuclide activity and computation of analytical results.

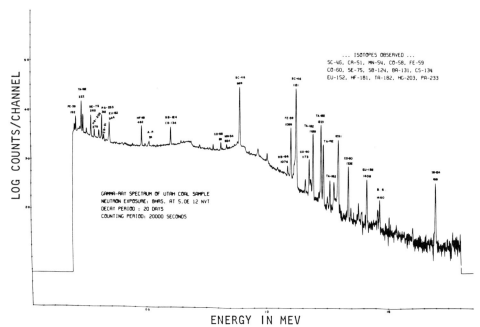

FIGURE 2:   Gamma Ray Spectrum of Utah Coal Sample

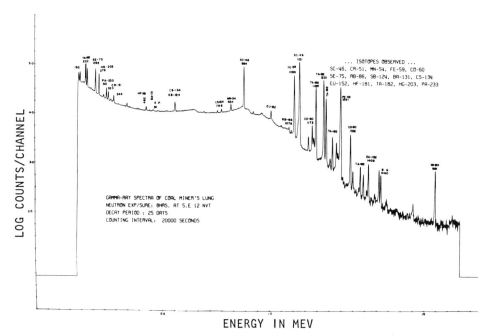

FIGURE 3:   Gamma Ray Spectrum of Coal Miner's Lung

Sample Preparation

One of the most important advantages of activation analysis in environmental research is that a minimum of sample treatment is necessary prior to irradiation. Thus, contamination of the sample is minimized. In the preparation of environmental samples for irradiation, however, certain sources of error must be eliminated or reduced. In many INAA or IPAA methods, the samples must be packaged in a container and the sample plus the container is irradiated and then counted. It is important that the container material contain very low concentrations of the elements of interest and high purity polyethylene or quartz vials are generally used. Sealed quartz vials are essential if the sample contains Hg or other volatile elements that may be lost during irradiation.

The analysis of environmental and biological materials for trace elements presents some special sample preparation problems. The analysis of water is perhaps the most difficult and for elements present at very low concentrations, adsorption on suspended material or on the walls of the container may be significant. Robertson (26) has shown that serious losses of In, Sc, Fe, Ag, U and Co occur by adsorption from seawater onto polyethylene or glass containers over periods up to 90 days. For fresh waters, adsorption of trace elements may be more serious than for seawater because of the lower total electrolyte concentration of fresh waters and HNO_3 is often added to prevent trace element precipitation and adsorption on container surfaces. The addition of acids to aqueous samples will, however, change the chemical nature of the medium. For example, certain trace elements may be present as suspended matter (e.g. silicates), finely divided precipitates of colloids (Fe compounds, Zn hydroxides, etc.), or bound to organic ligands such as humic acids and addition of acid will change equilibria in the solution. Thus the preirradiation treatment selected for water samples will depend on whether the total content of the element is required, or the distribution of the element among different phases in the sample.

Most water samples are filtered through 0.45μ Millipore (Millipore Corp.) or 0.4μ Nucleopore (General Electric Co.) filters and acidified to pH 2 with HNO_3 prior to analysis. Several authors (13,19) have proposed immediate freezing of samples to prevent adsorption on the container and some reactors have facilities for irradiation of frozen samples. A very satisfactory method for water samples is to freeze samples in a thin polyethylene bag and then freeze-dry the sample. The bag and its contents are then irradiated. The advantage of this technique is that a large volume can be reduced to a small convenient sample for irradiation. This technique was proposed by Harrison, LaFleur and Zoller (19) and Jackson and Filby (20). These authors also investigated losses of trace elements during freeze-drying and in both studies losses were found to be negligible except for Hg and I. Jackson and Filby (20), however found no losses for Hg on freeze-drying waters. This method is now

in routine use in our laboratory for environmental water analyses.
    Air particulates or aerosols are generally obtained impreg-
nated on a filter or impactor foil and the accuracy and sensitivity
of the analysis will be affected by the composition of the filters
and foils used. Fiberglass filters are commonly used but contain
high impurity levels and for low trace element samples filter paper,
polystyrene, Millipore, or Nucleopore filters are preferred.
    Biological materials may be irradiated either in polyethylene
containers or in quartz ampoules and it is often desirable to
freeze-dry biological samples prior to activation to prevent high
pressure buildup of water radioanalysis products. Pillay et al.
(25) have presented evidence that Hg is lost from biological samples during freeze-drying, perhaps as $Hg(CH_3)_2$ formed in the tissues.
Experiments on [203]Hg-tagged tissues in this laboratory (8) and by
LaFleur (22), however, indicated no losses of Hg during freeze-
drying. Little information on the behavior of other elements during
freeze-drying is available.
    Standards of accurately known composition are essential for
precise analysis. Aqueous solutions of known elemental composition
are commonly used as flux monitors in INAA for analysis of a wide
variety of materials. Thermal neutron attenuation due to the major
constituents (except Cl^- in seawater) of most environmental materials is similar to that in water and the nuclide specific activities
obtained do not depend on chemical composition. For Ge(Li) spectrometry it is convenient to prepare multiple element solution standards, provided no chemical interactions among elements occur. For
certain elements, care must be taken to ensure that the chemical
form used is stable under the irradiation conditions used and the
case of mercury is particularly interesting. Several authors (2,
21) have noted that irradiation of Hg(II) solutions in polyethylene
containers results in considerable [203]Hg on the container walls.
The mechanism of this loss is probably:

    Hg(II) → Hg(I)   reduction

    2Hg(I) → Hg(II) + Hg(0)   disproportionation

The Hg(0) then diffuses out of the containers or adsorbs on the
wall. An alternate explanation is that recoil [203]Hg atoms produced
by [202]Hg neutron capture react with alkyl groups in the polyethylene
to give volatile dialkyl mercury compounds. The problem is avoided
in NAA methods by irradiation of samples and standards in quartz
vials or by use of natural standards. Other chemical (or radiolytic)
changes that occur during irradiation of aqueous standards include
Se(IV) → Se(0), Ag(I) → Ag(0), Au(III) → Au(0), and Mn(VII) → MnO_2.
Many of the problems associated with aqueous standards can be solved
by use of natural standards of well characterized composition. The
National Bureau of Standards has taken the lead in preparing a large
number of well calibrated trace element standards of environmental
materials. These are listed in Table VII, together with other
materials which, although not certified, have sufficient trace element data available for use as standards.

TABLE VII: Trace Element Environmental Standards

Material	Organization	Elements Certified
SRM 1571 Orchard Leaves	NBS	23 elements plus 15 information values
SRM 1577 Bovine Liver	NBS	12 elements plus 10 information
SRM 1630 Coal	NBS	Hg
SRM 1632 Coal	NBS	14 elements plus 7 information
SRM 1633 Fly Ash	NBS	12 elements plus 7 information
SRM 1570 Spinach	NBS	16 elements plus 11 information
SRM 1573 Tomato Leaves	NBS	14 elements plus 13 information
SRM 1575 Pine Needles	NBS	15 elements plus 11 information
SRM 1634 Fuel Oil	NBS	6 elements plus 11 information
Blood, dried	I.A.E.A.	no elements certified
Wheat Flour	I.A.E.A.	no elements certified
Kale	H. J. M. Bowen	no elements certified but many data from different methods

Other NBS standards, including sediment, flour, yeast, etc., are in preparation

Irradiation

When samples and standards are irradiated it is important that all receive the same exposure. In large graphite moderated reactors, flux gradients in irradiation positions are generally small but in most research reactors there are large thermal neutron flux gradients. In many reactors, groups of samples and standards are irradiated in a given position and are generally in tiers, each tier containing a number of samples and standards. In the Washington

State University reactor graphite reflector the thermal neutron
flux changes by a factor of 2.5 over a horizontal distance of 3.0
inches. To eliminate horizontal flux differences among samples on
a given tier the samples are often rotated about a vertical axis
during irradiation. Vertical differences among tiers are corrected
for by irradiating standards on each tier, or by irradiating a flux
monitor (Cu-wire, Au-foil, etc.) on each tier to obtain relative
fluxes. This latter practice should be used with care because relative magnitudes of thermal and epithermal (n,γ) activation will be
different for different nuclides.

Accurate flux calibration and standardization is more difficult
using 14-MeV neutron generators, ^{252}Cf sources, and photon sources
(LINAC) where flux gradients are large. The problem of calibration
is even more difficult for charged particle irradiations because of
the small ranges of charged particles in the samples and standards.

Post Irradiation Treatment

After irradiation and a suitable decay period, the radionuclide
activities are determined using Geiger, proportional, NaI(Tl), or
Ge(Li) detectors. For INAA or IPAA using gamma-ray spectrometry no
chemical treatment is performed on the sample. Samples and standards may be counted in the irradiation containers provided constant
counting geometry is maintained.

If chemical separation of the radionuclides is necessary because of insufficient sensitivity by INAA or IPAA then conventional
chemical separations can be used. The sample is dissolved and known
amounts of non-radioactive carriers of the elements of interest are
added; care must be taken to ensure that the chemical form of the
radionuclide and the carrier are identical. The yield of separation
is measured by the amount of carrier recovered – thus quantitative
separations are not required. This fact often allows highly specific, but not quantitative, separations to be made (e.g. many solvent
extraction methods).

The high resolution of Ge(Li) detectors allows complex mixtures
of radionuclides to be measured, but often one or more nuclides present at high activities will mask most others. In environmental,
biological and geochemical materials, the ^{24}Na activities of irradiated samples are often high enough to mask other nuclides of similar
or shorter half-lives, e.g. ^{72}Ga, ^{42}K, ^{64}Cu. Removal of the interfering nuclide by a simple separation technique allows other nuclides present to be measured by Ge(Li) spectrometry. Girardi and
Sabbioni (28) have shown that hydrated antimony pentoxide (HAP) is
highly selective for removing ^{24}Na from 8M H_2SO_4 and other highly
specific ion adsorbers have been developed, e.g. Al_2O_3, SnO_2, MnO_2,
etc., (29). Several recent papers (10,14,30) have discussed the use
of group chemical separation procedures for activation analysis of
environmental materials and these often can be automated. An
example of a group separation scheme for trace elements in foods
(14) is discussed in the Applications Section.

Measurement of Radionuclide Activities

The irradiated material (or separated fractions) are generally counted for γ-ray emitting nuclides by Ge(Li) gamma-ray spectrometry. The parameter used as a function of nuclide activity is generally the area of a peak corresponding to a given gamma ray. As such peaks are superimposed on a continuous Compton background, peak areas must be obtained after background subtraction. For Ge(Li) spectra, the peaks occupy a relatively small portion of the total energy range and except where peaks overlap or fall on a Compton edge, the background can be assume to vary linearly from one side of the peak to the other. A number of authors have discussed techniques for the measurement of Ge(Li) spectra peak areas (7,9). For Ge(Li) spectra computer programs are generally used for spectrum analysis and the computational steps generally involve: a) spectrum smoothing, b) peak identification, c) background subtraction, d) peak area and multiplet analysis, e) decay corrections, and f) calculation of nuclide activities or elemental concentrations. The complexity of the program will depend on the nature of the spectra and the computer available and many computer programs are in routine use. Considerable flexibility is now available in the new microprocessor based spectrometers (e.g. Nuclear Data ND 6600) in which data reduction can be carried out independently of data collection.

Errors and Interferences

Due to the nuclear nature of the method, the NAA technique is subject to few systematic sources of error compared to many chemical or physicochemical methods of analysis. There are, however, several sources of error in NAA which may be divided into errors (or interferences) associated with irradiation and those associated with the measurement of induced radionuclide activities. The errors associated with irradiation involve: a) neutron flux perturbation in samples or standards and b) interfering nuclear reactions on other elements which produce the nuclide of interest. These sources of error are generally small for environmental materials and have been adequately discussed elsewhere (9).

Precision and Accuracy

In activation analysis the precision is often reported as the standard deviation of a single measurement derived from counting statistics. This error term is often the largest contribution to the overall precision of the analysis. Relative standard deviations of 5-10% can be obtained in careful trace analysis, provided the concentration of the element is not close to the detection limit. Accuracy is more difficult to determine and can only be adequately assessed by analysis of calibrated natural materials. In this

respect the National Bureau of Standards Standard Reference Materials (SRM's) are invaluable. An indication of the accuracy and
precision of activation analysis of environmental samples relative
to atomic absorption (AAS) and emission spectroscopy (OES) is provided by the results of the NBS-EPA round robin analyses of NBS
Coal (SRM 1632), Fly Ash (SRM 1633) and Fuel Oil (SRM 1634) in
which more than 40 laboratories participated prior to certification
of the standards. Ondov et al. (24) have compared the INAA and
IPAA results obtained by four laboratories (Lawrence Livermore
Laboratory, Battelle Northwest Laboratories, the University of Mary-
land and Washington State University) with those obtained by AAS
and OES. The results clearly show the good precision and accuracy
obtainable by INAA and IPAA on real environmental samples, and for
most elements, the results are superior in accuracy and precision
to those obtained by other techniques. Figure 4 shows data for Ni
and V in Coal and Fly Ash compared to NBS values and the ranges
obtained by AAS and OES. The data for fuel oil obtained by Filby
and Shah (11) are also included. The results for Ni and V in fuel
oil are particularly surprising because AAS and OES are standard
methods for these elements in the petroleum industry.

Applications

        It is not the purpose of this article to review in detail the
applications of activation analysis to environmental research.
Filby and Shah (9) have reviewed applications of NAA up to 1971
and several recent conferences have dealt with applications of
nuclear methods to environmental research. Several recent articles
have considered photon activation methods applied to environmental
problems (1) and the uses of charged particle activation techniques
have also received attention (6,23). The nuclear track technique,
although limited to U, Pu, Li, B, N, and O, has important applications and several novel procedures have been developed (3,4,17).
        To illustrate some recent uses of NAA applied to environmental
problems, the authors have chosen to present work performed in this
laboratory.
(i) Trace Metal Forms in Water. Activation analysis and atomic
absorption have been used to determine metal concentrations in
water, sediments and biota in the Coeur d'Alene River-Lake system
in Northern Idaho, a system that is polluted by mining wastes from
the large Kellogg (Ag-Zn-Pb) mining area (13,18).
        In addition to determining the concentrations of Hg, Sb, As,
Fe, Zn, Cr, Co, Cs, Rb and other elements in the waters (18) an
attempt was made to determine the chemical forms of Fe, Zn, and Sb
(related to the pollution source) and Sc (natural weathering product) in the polluted waters (South Fork Coeur d'Alene River) and
similar unpolluted source (North Fork). This was done by using a
combination of filtration (0.45μ Millipore), anion (Dowex 1) and
cation (Dowex 50) exchange, and freeze drying of water samples (13).

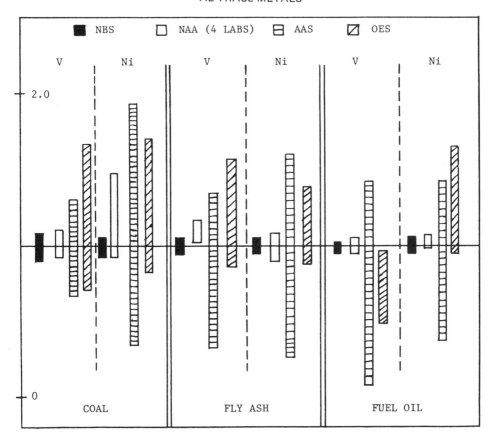

FIGURE 4:   Comparison of Results of NBS-EPA Round-Robin on Three
            Standards

Table VIII shows some of the results obtained. The distribution of
species pertains only to the metals in solution (or in colloidal
form).

     These data indicate the importance played by colloidal forms,
e.g. $Fe(OH)_3$, and/or complexes because, except for Zn, the forms
found are not the predicted cationic forms ($Fe(OH)^{2+}$, Zn^{2+}, SbO^+,
Sc^{3+}).   The combination of activation analysis and separations
chemistry can thus provide information on chemical species of trace
metals although the NAA method itself provides no direct information
on speciation.

TABLE VIII: Distribution of Fe, Zn, Sb, and Sc in South and North
              Forks, Coeur d'Alene River (13).

Concentration (μg/g)

Location	Element	Total	Solution	%Anionic	%Cationic	%Neutral
South	Fe	93.4	16.4	100	–	–
North		41.2	11.6	100	–	–
South	Zn	2740	2720	1	97	2
North		18.4	16.7	5	50	44
South	Sb	2.3	1.1	100	–	–
North		0.1	0.05	–	–	–
South	Sc	79	61.4	–	–	100
North		112	102	–	–	100

(ii) Trace Elements in Foods. The dietary intake of toxic and
nutritionally important elements is monitored by the Food and Drug
Administration. In this laboratory we have developed a group radiochemical separation and INAA scheme for the determination of 22 elements in FDA food composites (14). In this procedure Ge(Li) spectrometry is used and the radionuclides are separated into groups
using a combination of inorganic and organic ion-exchanges. In this
procedure the irradiated sample was decomposed by O_2 ashing in an
oxygen bomb followed by dissolution of the residues in 1 M HCl. The
solution was passed through a 5 cm SnO_2 column to remove As, Mo, Se,
and Sb. The solution was evaporated, adjusted to 12 M HCl and Na
removed by passing through a 5 cm hydrated antimony pentoxide (HAP)
column. The elements Co, Cu, Zn, Fe, Hg and Cd were then removed
by a Dowex 1 column in 9 M HCl and the eluate contained K, Rb, Cs,
Cr, REE. The elements Fe, Co, Cu and Zn were eluted from the column
with 0.0 1 M HCl and the column was counted for Cd and Hg. Thus the
sample was separated into 5 groups of elements. Because all fractions from the separation were counted, it was not necessary to
determine the chemical yields of each element.

    The concentrations of the toxic metals As, Sb, Cd, and Hg in 11
market basket composites are shown in Table IX. The data thus provide information on the sources of toxic metals in the "average"
U.S. diet.

(iii) Trace Elements in Coal Conversion Process. The development
of coal liquefaction and gasification processes requires that the
environmental impact of such processes be established. One important concern is the fate of trace elements during coal conversion,
during combustion of synfuels, and another concern is the formation
of toxic organometallics during coal conversion. In an initial

TABLE IX: Mean Concentrations of As, Sb, Hg, and Cd in Market Basket Composites (14).

Food Composite	Elemental Concentration (ng/g)			
	As	Sb	Cd	Hg
Dairy Products	40	47	<20	<5
Meat, Fish, Poultry	1140	128	49	23.3
Grains and Cereals	49	30	<20	28.1
Potatoes	580	42	163	8.4
Leafy Vegetables	120	47.8	658	50.9
Legume Vegetables	<10	13.3	<20	7.0
Root Vegetables	91.3	341	115	16.6
Garden Fruits	85.3	31.5	242	16.6
Fruits	36	37	<20	14.8
Oils and Fats	<10	9.3	23	1
Sugar Adjunts	<10	88	<20	5

study of the Solvent Refined Coal (SRC) Process (Pittsburg and Midway Coal Co.), the material balance of 22 trace elements in the 50 ton/day pilot plant (Tacoma, WA) was determined by INAA (15,16). In this study, input coal, solvent, SRC, insoluble residues (FC), by-product solvents, sulfur, and process waters were analyzed. The concentrations of some environmentally important elements together with data on other fuels are shown in Table X.

Present research involves the combination of AAS, INAA and separation procedures such as gel permeation chromatography and high pressure liquid chromatography to determine specific organometallic species.

Conclusions

Activation Analysis (INAA, IPAA, charged particle, nuclear track) is a sensitive and accurate method of determining trace elements in a wide variety of environmental materials. Although relatively sophisticated in terms of instrumentation, the technique is now a routine method of trace analysis and new applications are appearing rapidly.

TABLE X: Concentrations (μg/g) of Some Elements in Solvent Refined
          Coal Process Fractions and Fuels (16).

| | | Process Fraction or Fuel | | |
Element	Coal	SRC	FC[c]	No. 6 Fuel Oil[a]	Crude[b]
As	13.6	1.39	33.3	0.056	0.284
Sb	0.50	0.074	1.3	0.010	0.303
Se	1.53	0.148	8.1	0.17	0.369
Hg	0.436	0.025	0.447	0.01	0.027
Br	3.51	3.95	6.22	0.039	–
Ni	20.0	2.7	5.15	37.4	117
Co	3.7	0.31	12.8	0.301	0.178
Cr	14.0	2.68	56.4	0.093	0.430
Fe,%	1.73	0.068	5.70	0.0023	0.004

[a]Washington State University
[b]Reference (16)
[c]FC-Filter Cake

REFERENCES

1   ARAS, N.K., ZOLLER, W.H., GORDON, G.E. and LUTZ, G.J.:   Anal.
    Chem. 45 (1973) 1481.

2   BATE, L.C.:   Radiochem. Radioanal. Letters 6 (1971) 139.

3   CARPENTER, B.S. and REIMER, G.M.:   N.B.S. Special Publication
    422 (1976) 457.

4   CARPENTER, B.S., SAMUEL, D., WASSERMAN, I. and YUWILER, A.:
    Proc. 1976 Int. Conf. Modern Trends Activation Analysis 1
    (1976) 338.

5   CHATTOPADHYAY, A.:   Proc. 1976 Int. Conf. Modern Trends
    Activation Analysis 1 (1976) 493.

6   CHAUDHRI, M.A., LEE, M., ROUSE, J.L. and SPICER, B.M.:   Proc.
    1976 Int. Conf. Modern Trends Activation Analysis 1 (1976)
    566.

7   COOPER, J.A.:   U.S.A.B.C. Report BNWL-SA-3603 (1971).

8   FILBY, R.H., and SCHMIDT, J.O., unpublished data.

9   FILBY, R.H. and SHAH, K.R.:   Toxicol. Environ. Chem. Reviews
    2 (1974).

10 FILBY, R.H. and SHAH, K.R.:   HEW Publ. (NIOSH) 75-187 (1975).

11 FILBY, R.H. and SHAH, K.R.:   Neutron activation methods for
    trace elements in crude oils, Ch. 5, Role Trace Metals in
    Petroleum (Yen, T.F. Ed.) Ann Arbor Press, Ann Arbor (1975).

12 FILBY, R.H., SHAH, K.R. and DAVIS, A.J.:   Radiochem. Radioanal.
    Letters 5 (1970) 9.

13 FILBY, R.H., SHAH, K.R. and FUNK, W.H.:   Proc. Second Int.
    Conf. Nuclear Methods in Environmental Research, CONF-740701
    (1975) 10.

14 FILBY, R.H., SHAH, K.R. and PALMER, C.A.:   (in press).

15 FILBY, R.H., SHAH, K.R. and SAUTTER, C.A.:   Proc. 1976 Int.
    Conf. Modern Trends Activation Analysis 1 (1976) 664.

16 FILBY, R.H., SHAH, K.R. and SAUTTER, C.A.:   J. Radioanal.
    Chem. 37 (1977) 693.

17 FLEISCHER, R.L., PRICE, P.B. and WALKER, R.M.: Nuclear Tracks
    in Solids: Principles and Applications (1975) Univ.
    California Press, Berkeley, CA.

18 FUNK, W.H., RABE, F.W., FILBY, R.H. and others: Office
    Water Research Technology (Title II Project C-4145) Final
    Report (1975), Washington State Univ., Pullman, WA.

19 HARRISON, S.H., LAFLEUR, P.D. and ZOLLER, W.H.: Anal. Chem.
    47 (1975) 1685.

20 JACKSON, K.E. and FILBY, R.H.: Abstract NW Regional ACS
    Meeting 1974.

21 JERVIS, R.E. and TIEFEMBACH, B.: Proc. Amer. Nucl. Soc.
    Topical Meeting Nuclear Methods in Environmental Research
    (1971) 188.

22 LAFLEUR, P.D.: Anal. Chem. 45 (1973) 1534.

23 McGINLEY, J.R. and SCHWEIKERT, E.A.: Anal. Chem. 48 (1976)
    429.

24 ONDOV, J.L., ZOLLER, W.H., OLMEZ, I, ARAS, N.K., GORDON, G.E.,
    RANCITELLI, L.A., ABEL, K.H., FILBY, R.H., SHAH, K.R. and
    RAGAINI, R.C.: Anal. Chem. 47 (1975) 1102.

25 PILLAY, K.K.S., THOMAS, C.C., SONDEL, J.A. and HYCHE, C.M.:
    Anal. Chem. 43 (1971) 1419.

26 ROBERTSON, D.E.: Anal. Chim. Acta 42 (1968) 533.

27 WAINERDI, R.E., ZEISLER, R. and SCHWEIKERT, E.A.: Proc. 1976
    Int. Conf. Modern Trends Activation Analysis 1 (1976) 198.

28 GIRARDI, F., and SABBIONI, E.: J. Radioanal. Chem. 1 (1968)
    168.

29 GIRARDI, F., PIETRA, R., and SABBIONI, E.: EURATOM Report
    EUR-4287e (1969).

30 SCHUMACHER, J., MAIER-BORST, W., and HAUSER, H.: Proc. 1976
    Int. Conf. Modern Trends Activation Analysis 1 (1976) 322.

SESSION CHAIRMEN

SESSION I: SPECIFICATION OF ANALYTICAL PROBLEMS

                James P. Lodge, Ph.D.
                Consultant in Atmospheric Chemistry
                Boulder, Colorado 80303

SESSION II: MORE FAMILIAR PRINCIPLES

                Isaac Feldman, Ph.D.
                The University of Rochester
                Rochester, New York 14642

SESSION III: METHODS SUITABLE FOR FIELD USE

                Taft Y. Toribara, Ph.D.
                The University of Rochester
                Rochester, New York 14642

SESSION IV: HIGH SPATIAL RESOLUTION MICROPROBE METHODS

                Kurt Heinrich, Ph.D.
                National Bureau of Standards
                Washington, D.C. 20234

SESSION V: PHYSICAL ANALYTICAL METHODS

                Barton E. Dahneke, Ph.D.
                The University of Rochester
                Rochester, New York 14642

SPEAKERS

Marti Bancroft
Celdat Design Associates
Rochester, New York

Donald Beaman, Ph.D.
Dow Chemical Company
Midland, Michigan 48640

Chris W. Brown, Ph.D,
University of Rhode Island
Kingston, Rhode Island 02881

Thomas Cahill, Ph.D.
University of California
Davis, California 95616

Lucian Chaney
University of Michigan
Ann Arbor, Michigan 48105

Robert Collin, Ph.D.
Dept. Environmental Conservation
Albany, New York 12233

Barton E. Dahneke, Ph.D.
The University of Rochester
Rochester, New York 14642

William Davis, Ph.D.
General Electric Company
Schenectady, New York 12301

Edgar S. Etz, Ph.D.
National Bureau of Standards
Washington, D.C. 20234

Roy H. Filby, Ph.D.
Washington State University
Pullman Washington 99164

Friedrich Geiss, Ph.D.
European Communities Joint
  Research Center
I-21020 Ispra, Italy

Kurt F. J. Heinrich, Ph.D.
National Bureau of Standards
Washington, D.C. 20234

Harvey B. Herman, Ph.D.
University of North Carolina
Greensboro, North Carolina 27412

E. David Hinkley, Ph.D.
Laser Analytics
Lexington, Massachusetts 02172

James J. Huntzicker, Ph.D.
Oregon Graduate Center
Beaverton, Oregon 97005

Michael Isaacson, Ph.D.
University of Chicago
Chicago, Illinois 60637

Paul A. Lindfors, Ph.D.
Physical Electronic Industries
Eden Prairie, Minnesota 55343

James P. Lodge, Ph.D.
385 Broadway
Boulder, Colorado 80303

Dale Newbury, Ph.D.
National Bureau of Standards
Washington, D.C. 20234

Eugene Sawicki, Ph.D.
National Environmental Research
Center
Research Triangle Pk.,
North Carolina 27711

Laurant Van Haverbeke, Ph.D.
University of Rhode Island
Kingston, Rhode Island 02881
                              and
Laboratory for Inorganic Chemistry
State University of Antwerp
Groenenborgerlaan, 171, B2020
Antwerp (Belgium)

Bernard Weiss, Ph.D.
The University of Rochester
Rochester, New York 14642

Philip W. West, Ph.D.
Louisana State University
Baton Rouge, Louisana 70803

PARTICIPANTS

Patricia F. Lynch
University of Rhode Island
Kingston, Rhode Island 02881

Rudolph H. Stehl, Ph.D.
Dow Chemical Company
Midland, Michigan 48640

Kenneth Robillard, Ph.D.
Eastman Kodak Company
Rochester, New York 14650

Victorio Wee, Ph.D.
Procter and Gamble Company
Cincinnati, Ohio 45217

Stanley B. Smith, Jr.
Instrumentation Laboratory Inc.
Lexington, Massachusetts 02173

INDEX